differential equations with
applications in biology, physics,
and engineering

# PURE AND APPLIED MATHEMATICS

*A Program of Monographs, Textbooks, and Lecture Notes*

### EXECUTIVE EDITORS

Earl J. Taft
Rutgers University
New Brunswick, New Jersey

Zuhair Nashed
University of Delaware
Newark, Delaware

### CHAIRMEN OF THE EDITORIAL BOARD

S. Kobayashi
University of California, Berkeley
Berkeley, California

Edwin Hewitt
University of Washington
Seattle, Washington

### EDITORIAL BOARD

M. S. Baouendi
University of California, San Diego

Donald Passman
University of Wisconsin-Madison

Jack K. Hale
Georgia Institute of Technology

Fred S. Roberts
Rutgers University

Marvin Marcus
University of California, Santa Barbara

Gian-Carlo Rota
Massachusetts Institute of Technology

W. S. Massey
Yale University

David L. Russell
Virginia Polytechnic Institute
and State University

Leopoldo Nachbin
Centro Brasileiro de Pesquisas Físicas
and University of Rochester

Jane Cronin Scanlon
Rutgers University

Anil Nerode
Cornell University

Walter Schempp
Universität Siegen

Mark Teply
University of Wisconsin-Milwaukee

# LECTURE NOTES

# IN PURE AND APPLIED MATHEMATICS

1. *N. Jacobson*, Exceptional Lie Algebras
2. *L.-Å. Lindahl and F. Poulsen*, Thin Sets in Harmonic Analysis
3. *I. Satake*, Classification Theory of Semi-Simple Algebraic Groups
4. *F. Hirzebruch, W. D. Newmann, and S. S. Koh*, Differentiable Manifolds and Quadratic Forms (out of print)
5. *I. Chavel*, Riemannian Symmetric Spaces of Rank One (out of print)
6. *R. B. Burckel*, Characterization of C(X) Among Its Subalgebras
7. *B. R. McDonald, A. R. Magid, and K. C. Smith*, Ring Theory: Proceedings of the Oklahoma Conference
8. *Y.-T. Siu*, Techniques of Extension on Analytic Objects
9. *S. R. Caradus, W. E. Pfaffenberger, and B. Yood*, Calkin Algebras and Algebras of Operators on Banach Spaces
10. *E. O. Roxin, P.-T. Liu, and R. L. Sternberg*, Differential Games and Control Theory
11. *M. Orzech and C. Small*, The Brauer Group of Commutative Rings
12. *S. Thomeier*, Topology and Its Applications
13. *J. M. Lopez and K. A. Ross*, Sidon Sets
14. *W. W. Comfort and S. Negrepontis*, Continuous Pseudometrics
15. *K. McKennon and J. M. Robertson*, Locally Convex Spaces
16. *M. Carmeli and S. Malin*, Representations of the Rotation and Lorentz Groups: An Introduction
17. *G. B. Seligman*, Rational Methods in Lie Algebras
18. *D. G. de Figueiredo*, Functional Analysis: Proceedings of the Brazilian Mathematical Society Symposium
19. *L. Cesari, R. Kannan, and J. D. Schuur*, Nonlinear Functional Analysis and Differential Equations: Proceedings of the Michigan State University Conference
20. *J. J. Schäffer*, Geometry of Spheres in Normed Spaces
21. *K. Yano and M. Kon*, Anti-Invariant Submanifolds
22. *W. V. Vasconcelos*, The Rings of Dimension Two
23. *R. E. Chandler*, Hausdorff Compactifications
24. *S. P. Franklin and B. V. S. Thomas*, Topology: Proceedings of the Memphis State University Conference
25. *S. K. Jain*, Ring Theory: Proceedings of the Ohio University Conference
26. *B. R. McDonald and R. A. Morris*, Ring Theory II: Proceedings of the Second Oklahoma Conference
27. *R. B. Mura and A. Rhemtulla*, Orderable Groups
28. *J. R. Graef*, Stability of Dynamical Systems: Theory and Applications
29. *H.-C. Wang*, Homogeneous Branch Algebras
30. *E. O. Roxin, P.-T. Liu, and R. L. Sternberg*, Differential Games and Control Theory II
31. *R. D. Porter*, Introduction to Fibre Bundles
32. *M. Altman*, Contractors and Contractor Directions Theory and Applications
33. *J. S. Golan*, Decomposition and Dimension in Module Categories
34. *G. Fairweather*, Finite Element Galerkin Methods for Differential Equations
35. *J. D. Sally*, Numbers of Generators of Ideals in Local Rings
36. *S S. Miller*, Complex Analysis: Proceedings of the S.U.N.Y. Brockport Conference
37. *R. Gordon*, Representation Theory of Algebras: Proceedings of the Philadelphia Conference
38. *M. Goto and F. D. Grosshans*, Semisimple Lie Algebras
39. *A. I. Arruda, N. C. A. da Costa, and R. Chuaqui*, Mathematical Logic: Proceedings of the First Brazilian Conference

40. *F. Van Oystaeyen*, Ring Theory: Proceedings of the 1977 Antwerp Conference
41. *F. Van Oystaeyen and A. Verschoren*, Reflectors and Localization: Application to Sheaf Theory
42. *M. Satyanarayana*, Positively Ordered Semigroups
43. *D. L. Russell*, Mathematics of Finite-Dimensional Control Systems
44. *P.-T. Liu and E. Roxin*, Differential Games and Control Theory III: Proceedings of the Third Kingston Conference, Part A
45. *A. Geramita and J. Seberry*, Orthogonal Designs: Quadratic Forms and Hadamard Matrices
46. *J. Cigler, V. Losert, and P. Michor*, Banach Modules and Functors on Categories of Banach Spaces
47. *P.-T. Liu and J. G. Sutinen*, Control Theory in Mathematical Economics: Proceedings of the Third Kingston Conference, Part B
48. *C. Byrnes*, Partial Differential Equations and Geometry
49. *G. Klambauer*, Problems and Propositions in Analysis
50. *J. Knopfmacher*, Analytic Arithmetic of Algebraic Function Fields
51. *F. Van Oystaeyen*, Ring Theory: Proceedings of the 1978 Antwerp Conference
52. *B. Kedem*, Binary Time Series
53. *J. Barros-Neto and R. A. Artino*, Hypoelliptic Boundary-Value Problems
54. *R. L. Sternberg, A. J. Kalinowski, and J. S. Papadakis*, Nonlinear Partial Differential Equations in Engineering and Applied Science
55. *B. R. McDonald*, Ring Theory and Algebra III: Proceedings of the Third Oklahoma Conference
56. *J. S. Golan*, Structure Sheaves over a Noncommutative Ring
57. *T. V. Narayana, J. G. Williams, and R. M. Mathsen*, Combinatorics, Representation Theory and Statistical Methods in Groups: YOUNG DAY Proceedings
58. *T. A. Burton*, Modeling and Differential Equations in Biology
59. *K. H. Kim and F. W. Roush*, Introduction to Mathematical Consensus Theory
60. *J. Banas and K. Goebel*, Measures of Noncompactness in Banach Spaces
61. *O. A. Nielson*, Direct Integral Theory
62. *J. E. Smith, G. O. Kenny, and R. N. Ball*, Ordered Groups: Proceedings of the Boise State Conference
63. *J. Cronin*, Mathematics of Cell Electrophysiology
64. *J. W. Brewer*, Power Series Over Commutative Rings
65. *P. K. Kamthan and M. Gupta*, Sequence Spaces and Series
66. *T. G. McLaughlin*, Regressive Sets and the Theory of Isols
67. *T. L. Herdman, S. M. Rankin, III, and H. W. Stech*, Integral and Functional Differential Equations
68. *R. Draper*, Commutative Algebra: Analytic Methods
69. *W. G. McKay and J. Patera*, Tables of Dimensions, Indices, and Branching Rules for Representations of Simple Lie Algebras
70. *R. L. Devaney and Z. H. Nitecki*, Classical Mechanics and Dynamical Systems
71. *J. Van Geel*, Places and Valuations in Noncommutative Ring Theory
72. *C. Faith*, Injective Modules and Injective Quotient Rings
73. *A. Fiacco*, Mathematical Programming with Data Perturbations I
74. *P. Schultz, C. Praeger, and R. Sullivan*, Algebraic Structures and Applications Proceedings of the First Western Australian Conference on Algebra
75. *L. Bican, T. Kepka, and P. Nemec*, Rings, Modules, and Preradicals
76. *D. C. Kay and M. Breen*, Convexity and Related Combinatorial Geometry: Proceedings of the Second University of Oklahoma Conference
77. *P. Fletcher and W. F. Lindgren*, Quasi-Uniform Spaces
78. *C.-C. Yang*, Factorization Theory of Meromorphic Functions
79. *O. Taussky*, Ternary Quadratic Forms and Norms
80. *S. P. Singh and J. H. Burry*, Nonlinear Analysis and Applications
81. *K. B. Hannsgen, T. L. Herdman, H. W. Stech, and R. L. Wheeler*, Volterra and Functional Differential Equations

82. *N. L. Johnson, M. J. Kallaher, and C. T. Long,* Finite Geometries: Proceedings of a Conference in Honor of T. G. Ostrom
83. *G. I. Zapata,* Functional Analysis, Holomorphy, and Approximation Theory
84. *S. Greco and G. Valla,* Commutative Algebra: Proceedings of the Trento Conference
85. *A. V. Fiacco,* Mathematical Programming with Data Perturbations II
86. *J.-B. Hiriart-Urruty, W. Oettli, and J. Stoer,* Optimization: Theory and Algorithms
87. *A. Figa Talamanca and M. A. Picardello,* Harmonic Analysis on Free Groups
88. *M. Harada,* Factor Categories with Applications to Direct Decomposition of Modules
89. *V. I. Istrățescu,* Strict Convexity and Complex Strict Convexity: Theory and Applications
90. *V. Lakshmikantham,* Trends in Theory and Practice of Nonlinear Differential Equations
91. *H. L. Manocha and J. B. Srivastava,* Algebra and Its Applications
92. *D. V. Chudnovsky and G. V. Chudnovsky,* Classical and Quantum Models and Arithmetic Problems
93. *J. W. Longley,* Least Squares Computations Using Orthogonalization Methods
94. *L. P. de Alcantara,* Mathematical Logic and Formal Systems
95. *C. E. Aull,* Rings of Continuous Functions
96. *R. Chuaqui,* Analysis, Geometry, and Probability
97. *L. Fuchs and L. Salce,* Modules Over Valuation Domains
98. *P. Fischer and W. R. Smith,* Chaos, Fractals, and Dynamics
99. *W. B. Powell and C. Tsinakis,* Ordered Algebraic Structures
100. *G. M. Rassias and T. M. Rassias,* Differential Geometry, Calculus of Variations, and Their Applications
101. *R.-E. Hoffmann and K. H. Hofmann,* Continuous Lattices and Their Applications
102. *J. H. Lightbourne, III, and S. M. Rankin, III,* Physical Mathematics and Nonlinear Partial Differential Equations
103. *C. A. Baker and L. M. Batten,* Finite Geometries
104. *J. W. Brewer, J. W. Bunce, and F. S. Van Vleck,* Linear Systems Over Commutative Rings
105. *C. McCrory and T. Shifrin,* Geometry and Topology: Manifolds, Varieties, and Knots
106. *D. W. Kueker, E. G. K. Lopez-Escobar, and C. H. Smith,* Mathematical Logic and Theoretical Computer Science
107. *B.-L. Lin and S. Simons,* Nonlinear and Convex Analysis: Proceedings in Honor of Ky Fan
108. *S. J. Lee,* Operator Methods for Optimal Control Problems
109. *V. Lakshmikantham,* Nonlinear Analysis and Applications
110. *S. F. McCormick,* Multigrid Methods: Theory, Applications, and Supercomputing
111. *M. C. Tangora,* Computers in Algebra
112. *D. V. Chudnovsky and G. V. Chudnovsky,* Search Theory: Some Recent Developments
113. *D. V. Chudnovsky and R. D. Jenks,* Computer Algebra
114. *M. C. Tangora,* Computers in Geometry and Topology
115. *P. Nelson, V. Faber, T. A. Manteuffel, D. L. Seth, and A. B. White, Jr.* Transport Theory, Invariant Imbedding, and Integral Equations: Proceedings in Honor of G. M. Wing's 65th Birthday
116. *P. Clément, S. Invernizzi, E. Mitidieri, and I. I. Vrabie,* Semigroup Theory and Applications
117. *J. Vinuesa,* Orthogonal Polynomials and Their Applications: Proceedings of the International Congress
118. *C. M. Dafermos, G. Ladas, and G. Papanicolaou,* Differential Equations: Proceedings of the EQUADIFF Conference
119. *E. O. Roxin,* Modern Optimal Control: A Conference in Honor of Solomon Lefschetz and Joseph P. Lasalle
120. *J. C. Díaz,* Mathematics for Large Scale Computing
121. *P. S. Milojević,* Nonlinear Functional Analysis

122. *C. Sadosky*, Analysis and Partial Differential Equations: A Collection of Papers Dedicated to Mischa Cotlar
123. *R. M. Shortt*, General Topology and Applications: Proceedings of the 1988 Northeast Conference
124. *R. Wong*, Asymptotic and Computational Analysis: Conference in Honor of Frank W. J. Olver's 65th Birthday
125. *D. V. Chudnovsky and R. D. Jenks*, Computers in Mathematics
126. *W. D. Wallis, H. Shen, W. Wei, and L. Zhu*, Combinatorial Designs and Applications
127. *S. Elaydi*, Differential Equations: Stability and Control
128. *G. Chen, E. B. Lee, W. Littman, and L. Markus*, Distributed Parameter Control Systems: New Trends and Applications
129. *W. N. Everitt*, Inequalities: Fifty Years On from Hardy, Littlewood and Pólya
130. *H. G. Kaper and M. Garbey*, Asymptotic Analysis and the Numerical Solution of Partial Differential Equations
131. *O. Arino, D.E. Axelrod, and M. Kimmel*, Mathematical Population Dynamics: Proceedings of the Second International Conference
132. *S. Coen*, Geometry and Complex Variables
133. *J. A. Goldstein, F. Kappel, and W. Schappacher*, Differential Equations with Applications in Biology, Physics, and Engineering
134. *S. J. Andima, R. Kopperman, P. R. Misra, J. Z. Reichman, and A. R. Todd*, General Topology and Applications

*Other Volumes in Preparation*

# differential equations with applications in biology, physics, and engineering

Edited by
Jerome A. Goldstein
**Department of Mathematics**
**Tulane University**
**New Orleans, Louisiana**

Franz Kappel
Wilhelm Schappacher
**Institute for Mathematics**
**University of Graz**
**Graz, Austria**

**Marcel Dekker, Inc.**  New York • Basel • Hong Kong

ISBN 0-8247-8571-1

This book is printed on acid-free paper. It was typeset by $\mathcal{A}_{\mathcal{M}}\mathcal{S}$-T$_E$X, the T$_E$X macro system of the American Mathematical Society.

Copyright © 1991 by MARCEL DEKKER, INC. All Rights Reserved

Neither this book nor any part may be reproduced or transmitted in any form or by any means, electronic or mechanical, including photocopying, microfilming, and recording, or by any information storage and retrieval system, without permission in writing from the publisher.

MARCEL DEKKER, INC.
270 Madison Avenue, New York, New York 10016

Current printing (last digit):
10 9 8 7 6 5 4 3 2 1

PRINTED IN THE UNITED STATES OF AMERICA

# Preface

The International Conference on Differential Equations and Applications was held at the Volksbildungshaus Retzhof in Leibnitz, Austria. It was a pleasurable and exciting conference, with participants from all over Europe (Czechoslovakia, Finland, France, Germany, Italy, Netherlands, Yugoslavia, Austria) as well as the United States, Canada and People's Republic of China.

This volume provides an adequate record of this meeting. Since most of the papers are concerned with more than one aspect of differential equations, it is rather difficult to give a one-to-one correspondence between these papers and a catalogue of different areas in differential equations. In the following we attempt to overcome this difficulty partially:

**Population dynamics.** C. Castillo-Chavez, S. Busenberg and K. Gerow analyse pair formation, which is an important aspect in modelling sexually transmitted diseases, in particular AIDS. J. Metz and O. Diekmann address the question 'when do input-output maps corresponding to the infinite dimensional state linear system arising in modelling structured populations allow equivalent finite dimensional representations?' A. De Roos and J. Metz clarify the numerics of the escalator boxcar train, a method used for numerical approximations.

**Plate and beam equations.** D. Ang, K. Schmitt and L. Vy study the problem of the contact of two elastic plates by formulating this problem as a system of noncoercive variational inequalites. H.T. Banks and D.A. Rebnord analyse the inverse problem for analytic semigroups describing the dynamics of flexible structures. The stabilization problem for structurally damped wave and plate equations with Neumann boundary control is investigated by A. Lunardi. J. Lagnese and G. Leugering prove uniform energy decay for a class of nonlinear beams.

**General theory.** P. Charissiadis and R. Nagel investigate semigroups generated by operator matrices. In particular, they characterize positivity and stability of these semigroups. K. Engel applies the theory of polynomial operator matrices to a model for the transversal vibration of a string. Aspects of asymptotic convergence are also the main topic of the papers by W.W. Farr, W.E. Fitzgibbon, J.J. Morgan and S.J. Waggoner and R.H. Martin and H.L. Smith, respectively. A. Favini analyses a second order equation, G. Rieder establishes the existence of limit solutions for a degenerate diffusion problem. J. von Below derives a maximum principle for a semilinear parabolic network equation. In the paper by K. Ito and F. Kappel the semigroup approach is used to study a class of integrodifferential equations with a weakly singular kernel. Evolution equations with nondensely defined operators are treated in the contributions by G. da Prato and E. Sinestrari, who consider time dependent equations and R. Grimmer and H. Liu, who develop an approach to Volterra integrodifferential equations.

**Applications in physics.** Several aspects of problems associated with Schrödinger and/or Dirac equations are investigated in the papers by B. Najman, M. and T. Hoffmann-Ostenhof and B. Thaller. F. Gesztesy gives a comprehensive review of recent developments on the modified Korteweg-de Vries equation and H. Kaper and M. Kwong prove a result on the ground states of semilinear diffusion equations.

We gratefully acknowledge the financial support for the conference provided by the National Science Foundation of the United States, the Fonds zur Förderung der wissenschaftlichen Forschung of Austria, and the government of the state of Styria, Austria. The typing of this volume was done efficiently and expertly by Gerlinde Krois. Providing great help with the local arrangements were Wolfgang Desch, Michael Kroller, Gunther Peichl and Georg Propst. To these agencies and individuals we extend our deepest appreciation.

Jerome A. Goldstein<br>
Franz Kappel<br>
Wilhelm Schappacher

# Contents

| | |
|---|---|
| Preface | iii |
| List of Participants | ix |
| D.D. ANG, K. SCHMITT AND L.K. VY:<br>    Variational Inequalities and the Contact of Elastic Plates | 1 |
| H.T. BANKS AND D.A. REBNORD:<br>    Analytic Semigroups: Applications to Inverse Problems for Flexible Structures | 21 |
| JOACHIM VON BELOW:<br>    A Maximum Principle for Semilinear Parabolic Network Equations | 37 |
| CARLOS CASTILLO–CHAVEZ, STAVROS BUSENBERG AND KEN GEROW:<br>    Pair Formation in Structured Populations | 47 |
| PANOS CHARISSIADIS AND RAINER NAGEL:<br>    Positivity for Operator Matrices | 67 |

GIUSEPPE DA PRATO AND EUGENIO SINESTRARI:
    Time Dependent Differential Equations in Non Reflexive Banach Spaces    79

A.M. DE ROOS AND J.A.J. METZ:
    Towards a Numerical Analysis of the Escalator Boxcar Train    91

KLAUS-J. ENGEL:
    An Application of Polynomial Operator Matrices to a Second Order Cauchy Problem    115

W.W. FARR, W.E. FITZGIBBON, J.J. MORGAN AND S.J. WAGGONER:
    Asymptotic Convergence for a Class of Autocatalytic Chemical Systems    121

ANGELO FAVINI:
    Second Order Parabolic Equations in Banach Space    129

F. GESZTESY:
    On the Modified Korteweg-deVries Equation    139

RONALD GRIMMER AND HETAO LIU:
    Integrodifferential Equations with Nondensely Defined Operators    185

M. HOFFMANN-OSTENHOF AND T. HOFFMANN-OSTENHOF:
    On Nodes of Local Solutions to Schrödinger Equations    201

KAZUFUMI ITO AND FRANZ KAPPEL:
    On Integro-Differential Equations with Weakly Singular Kernels    209

HANS G. KAPER AND MAN KAM KWONG:
    Ground States of Semi–Linear Diffusion Equations    219

J. LAGNESE AND G. LEUGERING:
    Uniform Energy Decay of a Class of Cantilevered Nonlinear Beams with Nonlinear Dissipation at the Free End    227

ALESSANDRA LUNARDI:
    Neumann Boundary Stabilization of Structurally Damped Time Periodic Wave and Plate Equations    241

R.H. MARTIN AND H.L. SMITH:
  Convergence in Lotka-Volterra Systems with Diffusion and Delay ... 259

J.A.J. METZ AND O. DIEKMANN:
  Exact Finite Dimensional Representations of Models for Physiologically Structured Populations.I: The Abstract Foundations of Linear Chain Trickery ... 269

BRANKO NAJMAN:
  The Nonrelativistic Limit of Klein–Gordon and Dirac Equations ... 291

GISÈLE RUIZ RIEDER:
  Spatially Degenerate Diffusion with Periodic-Like Boundary Conditions ... 301

BERND THALLER:
  Scattering Theory of a Supersymmetric Dirac Operator ... 313

Index ... 327

# List of Participants

Includes those who attended the Conference but did not necessarily contribute to this volume. Individual contributor's addresses may be found at the conclusion of each chapter.

| | |
|---|---|
| W. ARENDT | Equipe de Mathématiques |
| | Université Besançon |
| | F-25030 Besançon-Cedex, France |
| H.T. BANKS | Center for Applied Mathematical Sciences |
| | University of Southern California |
| | Los Angeles, CA 90089-1113, USA |
| | E-mail:HTBANKS@USCVM.BITNET |
| J.V. BELOW | Lehrstuhl für Biomathematik |
| | Universität Tübingen |
| | Auf der Morgenstelle 10 |
| | D-7400 Tübingen 1, Germany |
| R. BÜRGER | Institut für Mathematik |
| | Universität Wien |
| | Strudlhofgasse 4, A-1090 Wien, Austria |
| | E-mail:A8131DAH@AWIUNI11 |
| C. CASTILLO–CHAVEZ | Biometric Unit, Cornell University |
| | 341 Warren Hall |
| | Ithaca, NY 14853–7801, USA |
| | E-mail:(BITNET):NEP@CORNELLC |

| | |
|---|---|
| G. Da Prato | Scuola Normale Superiore di Pisa<br>Piazza dei Cavalieri 7<br>I-56126 Pisa, Italy<br>E-mail:DAPRATO@IPISNSIB.BITNET |
| A. De Roos | Department of Pure and Applied Ecology<br>University of Amsterdam<br>Kruislaan 302<br>1098 SM Amsterdam, The Netherlands |
| W. Desch | Institut für Mathematik<br>Karl–Franzens–Universität Graz<br>Heinrichstraße 36, A-8010 Graz, Austria<br>E-mail:DESCH@EDVZ.UNI-GRAZ.ADA.AT |
| O. Diekmann | Institute of Theoretical Biology<br>University of Leiden<br>Kaiserstraat 63, NL-2311 GP Leiden<br>and<br>CWI, Kruislaan 413<br>NL-1098 SJ Amsterdam, The Netherlands |
| K.J. Engel | Mathematisches Institut<br>Universität Tübingen<br>Auf der Morgenstelle 10<br>D-7400 Tübingen, Germany<br>E-mail:MINA001@CONVEZ.ZDV.UNI-TUEBINGEN.DE |
| A. Favini | Dipartimento di Matematica<br>Università di Bologna<br>Piazza di Porta S. Donato 5<br>I-40126 Bologna, Italy<br>E-mail:POST@DM.UNIBO.IT |
| W.E. Fitzgibbon | Department of Mathematics<br>University of Houston<br>Houston, Texas 77204-3476, USA |
| F. Gesztesy | Department of Mathematics<br>University of Missouri<br>Columbia, MO 65211, USA<br>E-mail:MATHFG@UMCVMB.BITNET |
| J.A. Goldstein | Department of Mathematics<br>Tulane University<br>New Orleans, LA 70118, USA<br>E-mail:MT0NAMF@VM.TCS.TULANE.EDU |
| R. Grimmer | Department of Mathematics<br>Southern Illinois University<br>Carbondale, Illinois 62901, USA<br>E-mail:GA3582@SIUCVMB.BITNET |

*Participants*

| | |
|---|---|
| M. GYLLENBERG | Department of Mathematics<br>Helsinki University of Technology<br>Otakaari 1, SF-02150 Espoo, Finland<br>E-mail:MAT-MG@FINHUT.BITNET |
| J. HEJTMANEK | Institut für Mathematik<br>Universität Wien<br>Strudlhofgasse 4, A-1090 Wien, Austria |
| J. HOFBAUER | Institut für Mathematik<br>Universität Wien<br>Strudlhofgasse 4, A-1090 Wien, Austria<br>E-mail:A8131DAI@AWIUNI11 |
| M. HOFFMANN-OSTENHOF | Institut für Mathematik<br>Universität Wien<br>Strudlhofgasse 4, A-1090 Wien, Austria |
| T. HOFFMANN-OSTENHOF | Institut für Theoretische Chemie<br>Universität Wien<br>Währinger Straße 17, A-1090 Wien, Austria |
| M. IANNELLI | Dipartimento di Matematica<br>Università di Trento<br>I-38050 Povo (TN), Italy<br>E-mail:IANNELLI@ITNCISCA.BITNET |
| H. INABA | CWI, Kruislaan 413<br>NL-1098 SJ Amsterdam<br>and<br>Institute of Theoretical Biology<br>University of Leiden, Kaiserstraat 63<br>NL-2311 GP Leiden, The Netherlands |
| K. ITO | Center for Applied Mathematical Sciences<br>University of Southern California<br>Los Angeles, CA 90089-1113, USA<br>E-mail:KITO@USCVM.BITNET |
| H.G. KAPER | Mathematics and Computer Science Division<br>Argonne National Laboratory<br>Argonne, IL 60439-4844, USA<br>E-mail:KAPER@MCS.ANL.GOV |
| F. KAPPEL | Institut für Mathematik<br>Karl–Franzens–Universität Graz<br>Heinrichstraße 36, A-8010 Graz, Austria<br>E-mail:KAPPEL@EDVZ.UNI-GRAZ.ADA.AT |
| M. KRETZSCHMAR | CWI, Kruislaan 413<br>NL-1098 SJ Amsterdam, The Netherlands |

| | |
|---|---|
| F. KROLLER | Institut für Mathematik<br>Karl–Franzens–Universität Graz<br>Heinrichstraße 36, A-8010 Graz, Austria |
| J. LAGNESE | Department of Mathematics<br>Georgetown University<br>Washington, DC 20057, USA<br>E-mail:LAGNESE@NBS.BITNET |
| A. LUNARDI | Dipartimento di Matematica<br>Università di Cagliari<br>Via Ospedale 72<br>I-09124 Cagliari, Italy |
| R.H. MARTIN | Department of Mathematics<br>North Carolina State University<br>Raleigh, NC 27695-8205, USA |
| J.A.J. METZ | Institute of Theoretical Biology<br>University of Leiden, Kaiserstraat 63<br>NL-2311 GP Leiden, The Netherlands |
| J. MILOTA | Department of Mathematics<br>MFF UK, Sokolovska 83<br>CS-18600 Praha 8, Czechoslovakia |
| K. MURPHY | Department of Mathematics<br>University of North Carolina<br>at Chapel Hill, CB 3250<br>Chapel Hill, NC 27599-3250, USA<br>E-mail:UKAMSW@UNC.BITNET |
| R. NAGEL | Mathematisches Institut<br>Universität Tübingen<br>Auf der Morgenstelle 10<br>D-7400 Tübingen, Germany<br>E-mail:MINA001@CONVEX.ZDV.UNI-TUEBINGEN.DE |
| B. NAJMAN | Department of Mathematics<br>University of Zagreb<br>P.O.Box 187<br>YU-41000 Zagreb, Yugoslavia |
| E. OBRECHT | Dipartimento di Matematica<br>Università di Bologna<br>Piazza di Porta S. Donato 5<br>I-40126 Bologna, Italy |

*Participants*

| | |
|---|---|
| G. PEICHL | Institut für Mathematik<br>Karl–Franzens–Universität Graz<br>Heinrichstraße 36, A-8010 Graz, Austria<br>E-mail:PEICHL@EDVZ.UNI-GRAZ.ADA.AT |
| G. PROPST | Universität Graz Institut für Mathematik<br>Karl–Franzens–Universität Graz<br>Heinrichstraße 36, A-8010 Graz, Austria<br>E-mail:PROPST@EDVZ.UNI-GRAZ.ADA.AT |
| J. PRÜSS | Fachbereich 17, GHS Paderborn<br>Warburgerstraße 100, D-4790 Paderborn, Germany |
| G. RIEDER | Department of Mathematics<br>Louisiana State University<br>Baton Rouge, Louisiana 70803, USA |
| W. RING | Institut für Mathematik<br>Technische Universität Graz<br>Steyrergasse 30, A-8010 Graz, Austria |
| W. SCHAPPACHER | Institut für Mathematik<br>Karl–Franzens–Universität Graz<br>Heinrichstraße 36, A-8010 Graz, Austria |
| K. SCHMITT | Department of Mathematics<br>University of Utah<br>Salt Lake City, UT 84112, USA<br>E-mail:MA.SCHMITT@SCIENCE.UTAH.EDU |
| J. WAI HUNG SO | Department of Mathematics<br>University of Alberta<br>Edmonton, Alberta, Canada T6G2G1<br>E-mail:JOSO@UALTMTS |
| B. THALLER | Institut für Mathematik<br>Karl–Franzens–Universität Graz<br>Heinrichstraße 36, A-8010 Graz, Austria<br>E-mail:THALLER@EDVZ.UNI-GRAZ.ADA.AT |
| R.L. WHEELER | Department of Mathematics<br>Virginia Polytechnic Inst. & State University<br>Blacksburg, VA 24061, USA<br>E-mail:WHEELER@VTVM1.BITNET |
| S. ZHANG | Department of Mathematics<br>Anhui University<br>Hofei, China |

# Variational Inequalities and the Contact of Elastic Plates

D.D. ANG
K. SCHMITT
L.K. VY

Department of Mathematics, Ho Chi Minh City University
Department of Mathematics, University of Utah
Department of Mathematics, Ho Chi Minh City University

## 1. Introduction

Khludnev [K] has considered the problem of contact of two thin elastic plates with clamped boundaries, placed one above the other, contacting each other. He reduces the problem to one involving *coercive* variational inequalities. He established existence and smoothness properties of the solution and furthermore established topological properties of the contact surface in the case that these plates have the same flexural rigidity.

It is the purpose of this paper to study the contact of two elastic thin plates subjected to various boundary conditions. The boundary conditions assumed are in a very general form to include unilateral boundary conditions, clamped boundary conditions, etc. .... In the case of clamped boundaries, our results give extensions of that of [K] in the sense that our plates need not have the same flexural rigidity.

The remainder of the paper consists of five sections. In sections 2, 3, 4, we shall formulate the problem as a variational inequality (which is not necessarily coercive) and investigate the problem of existence of solutions for the latter.

In section 5, we shall study smoothness properties of the solutions, and in the final section 6, we shall study partially the case of clamped boundaries, establish global smoothness properties and extend the result of [K] on the topological property of the contact surface.

## 2. Formulation of the problem

Consider two thin plates $P_1$, $P_2$ occupying a bounded horizontal domain $\Omega$

of $\mathbb{R}^2$ with smooth boundary and placed one above the other at a distance $\delta \geq 0$ apart. Let $D_1 > 0$, $D_2 > 0$ be the respective flexural rigidities of the two plates.

It is assumed that $P_1$, $P_2$ are subjected to the vertical forces $F, G$ respectively. We denote by $u$, $v$ be the vertical displacements of $P_1$, $P_2$ and assume the following boundary conditions:

$$F_3(u) = -g_1(u) \text{ on } \Gamma = \partial\Omega, \qquad (1)$$
$$M_\tau(u) = 0,$$

$$F_3(v) = -g_2(v) \text{ on } \Gamma, \qquad (2)$$
$$M_\tau(v) = 0.$$

where the $g_i$'s are increasing, continuous functions with $g_i(0) = 0 (i = 1, 2)$, and $F_3, M_\tau$ are the normal Kirchhoff shear force and twisting moment respectively (see §2.3, chapter 4, [DL]).

In what follows, we shall formulate, the problem as a variational inequality. Call $p$ the pressure exerted by $P_2$ on $P_1$. We have $p \geq 0$. Then $u$, $v$ satisfy the equations

$$D_1 \Delta^2 u = F + p \text{ on } \Omega, \qquad (3)$$
$$D_2 \Delta^2 v = G - p \text{ on } \Omega. \qquad (4)$$

Put

$$\psi_i(x) = \int_0^x g_i(\xi) d\xi, \quad x \in \mathbb{R}, \qquad (5)$$

$$J_i(v) = \int_\Gamma \psi_i(v) d\Gamma, \quad i = 1, 2. \qquad (6)$$

Since $g_i$ is increasing, $\psi_i$ is convex and

$$\psi_i(x) - \psi_i(y) \geq g_i(y)(x - y), \quad x, y \in \mathbb{R}. \qquad (7)$$

Likewise $v \mapsto J_i(v)$ is convex and

$$J_i(v) - J_i(u) \geq \int_\Gamma g_i(u)(v - u) d\Gamma. \qquad (8)$$

We adopt the following notation

$$(u, v) = \int_\Omega uv\, dx, \quad (u, v)_\Gamma = \int_\Gamma uv\, d\Gamma.$$

Let $u', v'$ be given functions on $\Omega$ such that

$$u' - v' \to -\delta \text{ on } \Omega, \tag{9}$$

and let $u, v$ be a solution of the contact problem (1) - (4), then

$$u - v \geq \delta \quad \text{on} \quad \Omega. \tag{10}$$

By Green's formula we obtain from (3) and (1)

$$D_1 a(u, u' - u) - (F_3(u), u' - u)_\Gamma + (M_\tau(u), \frac{\partial}{\partial m}(u' - u))_\Gamma$$
$$= (F + p, u' - u).$$

Hence

$$D_1 a(u, u' - u) + (g_1(u), u' - u)_\Gamma = (F + p, u' - u). \tag{11}$$

Here $a$ is the bilinear form (§4.1, chapter 4 [DL])

$$a(u, w) = \int_\Omega [\partial_{11} u \partial_{11} w + \partial_{22} u \partial_{22} w + \nu(\partial_{11} u \partial_{22} w + \partial_{22} u \partial_{11} w)$$
$$+ 2(1 - \nu)\partial_{12} u \partial_{12} w]\, dx, \ 0 < \nu < 1/2,$$

where $\partial_{ij} = \frac{\partial^2}{\partial x_i \partial x_j}$. From (8) and (11) we deduce

$$D_1 a(u, u' - u) + J_1(u') - J_1(u) \geq (F + p, u' - u). \tag{12}$$

Similarly for $v$

$$D_2 a(v, v' - v) + J_2(v') - J_2(v) \geq (G - p, v' - v). \tag{13}$$

Consider the term $(p, u' - u - v' + v)$. If $x \in \Omega$ is such that no contact occurs then $p(x) = 0$. If contact occurs at $x$, then $u'(x) - v'(x) \geq -\delta = u(x) - v(x)$. Since $p \geq 0$, one has $p(x)(u'(x) - u(x) - v'(x) + v(x)) \geq 0$. Hence $p(u' - u - v' + v) \geq 0$ on $\Omega$, which implies

$$(p, u' - u - v' + v) \geq 0. \tag{14}$$

Adding (12) to (13) gives in view of (14)

$$D_1 a(u, u' - u) + D_2 a(v, v' - v) + J_1(u') + J_2(v') - J_1(u) - J_2(v)$$
$$\geq (F, u' - u) + (G, v' - v), \tag{15}$$

for all $u', v'$ such that $u' - v' \geq -\delta$ on $\Omega$.

Conversely suppose $u, v$ satisfy (15) and $u - v \geq -\delta$ on $\Omega$. Let $\phi$ be an arbitrary function on $\Omega$ and $\theta \in (0,1)$. Then for $u' = u + \theta\phi$, $v' = v + \theta\phi$, one has $u' - v' \geq -\delta$ on $\Omega$. Substituting these expressions for $u', v'$ in (15) and dividing both sides by $\theta > 0$ we get:

$$D_1 u(u,\phi) + D_2 a(v,\phi) + \frac{1}{\theta}[J_1(u+\theta\phi) - J_1(u)]$$
$$+ \frac{1}{\theta}[J_2(v+\theta\phi) - J_2(v)] \geq (F,\phi) + (G,\phi). \quad (16)$$

Since

$$\frac{1}{\theta}[J_1(u+\theta\phi) - J_1(u)] = \int_\Gamma \frac{\psi_1(u+\theta\phi) - \psi_1(u)}{\theta} d\Gamma$$
$$\to \int_\Gamma g_1(u)\phi d\Gamma \quad \text{for} \quad \theta \to 0,$$

with a similar expression for $J_2$. As $\theta \to 0$ in (16), one obtains

$$D_1 a(u,\phi) + D_2 a(v,\phi) + \int_\Gamma g_1(u)\phi d\Gamma + \int_\Gamma g_2(v)\phi d\Gamma$$
$$\geq (F,\phi) + (G,\phi). \quad (17)$$

We replace $\phi$ by $-\phi$ in (17) to get

$$D_1 a(u,\phi) + D_2 a(v,\phi) + \int_\Gamma g_1(u)\phi d\Gamma + \int_\Gamma g_2(v)\phi d\Gamma$$
$$= (F,\phi) + (G,\phi), \quad \forall \phi. \quad (18)$$

Hence, $\forall \phi$,

$$D_1 a(u,\phi) + \int_\Gamma g_1(u)\phi d\Gamma - (F,\phi) = -\left[D_2 a(v,\phi) + \int_\Gamma g_2(v)\phi d\Gamma - (G,\phi)\right].$$

Equating each side of the preceding equality to $(p, \phi)$ we obtain

$$D_1 a(u,\phi) + \int_\Gamma g_1(u)\phi d\Gamma = (F+p, \phi) \quad (19)$$

$$D_2 a(uv,\phi) + \int_\Gamma g_2(v)\phi d\Gamma = (G-p, \phi). \quad (20)$$

For $\phi$ in $\mathcal{D}(\Omega)$, (19) gives

$$D_1 a(u,\phi) = (F+p, \phi)$$

and hence by Green's formula

$$D_1 \Delta^2 u = F + p. \quad (21)$$

One similarly has
$$D_2 \Delta^2 v = G - p.$$

From (21), one has by Green's formula for arbitrary $\phi$

$$D_1 a(u, \phi) = (D_1 \Delta^2 u, \phi) + (F_3(u), \phi)_\Gamma - \left(M_\tau(u), \frac{\partial \phi}{\partial n}\right). \tag{22}$$

Comparing (22) and (19), one has in view of (21):

$$(F_3(u), \phi)_\Gamma - \left(M_\tau(u), \frac{\partial \phi}{\partial n}\right)_\Gamma = (-g_i(u), \phi)_\Gamma, \quad \forall \phi. \tag{23}$$

Let $w$ be a function defined on $\Gamma$, chose $\phi$ such that $\phi = 0$ on $\Gamma$, $\frac{\partial \phi}{\partial n} = w$ on $\Gamma$ and $(M_\tau(u), w)_\Gamma = 0$. Hence

$$M_\tau(u) = 0 \quad \text{on} \quad \Gamma. \tag{24}$$

Substituting into (23) gives

$$(F_3(u), \phi) = (-g_1(u), \phi)_\Gamma, \quad \forall \phi,$$

and therefore

$$F_3(u) = -g_1(u) \quad \text{on} \quad \Gamma. \tag{25}$$

From (21), (24), and (25) and from the corresponding expressions for $v$, one sees that $u, v$ is a solution of the contact problem (1) - (4).

We have hence shown that the contact problem may be reformulated as the following variational inequality:

$$\begin{aligned} & D_1 a(u, u' - u) + D_2 a(v, v' - v) + J_1(u') + J_2(v') - J_1(u) - J_2(v) \\ & \geq (F, u' - u) + (G, v' - v), \\ & (u, v) \in K, \quad \forall (u', v') \in K, \end{aligned} \tag{26}$$

where
$$\begin{aligned} K &= \{(u', v') \in V : u' - v' \geq -\delta \quad \text{on} \quad \Omega\}, \\ V &= H^2(\Omega) \times H^2(\Omega), \\ F, G &\in L^2(\Omega), \end{aligned} \tag{27}$$

where $a$ is the bilinear form introduced earlier and

$$J_i(w) = \int_\Gamma \psi_i(w) d\Gamma, \quad i = 1, 2, \tag{28}$$

$\psi_i : \mathbb{R} \to \mathbb{R}$ is convex, continuous, $\psi_i(0) = 0$, $\psi_i \geq 0$ on $\mathbb{R}$, $i = 1, 2$. (29)

Now if for the contact problem (3) - (4), the boundary conditions (1) - (2) are replaced by the following boundary conditions.

$$\begin{aligned}\frac{\partial u}{\partial n} &= 0 \quad \text{on} \quad \Gamma, \\ F_3(u) &= -g_1(u),\end{aligned} \tag{30}$$

$$\begin{aligned}\frac{\partial v}{\partial n} &= 0 \quad \text{on} \quad \Gamma, \\ F_3(v) &= g_2(v),\end{aligned} \tag{31}$$

then one obtains the following variational inequality

$$\begin{aligned}&- D_1 a(u, u' - u) + D_2 a(v, v' - v) + J_1(u') + J_2(v') - J_1(u) - J_2(v) \\ &\geq (F, u' - u) + (G, v' - v), \\ &(u, v) \in K', \quad \forall \ (u', v') \in K'.\end{aligned} \tag{32}$$

Where $a, J_1, J_2, F, G$ have been defined at the beginning and where

$$K' = \left\{ \begin{aligned} &(u', v') \in V : \frac{\partial u'}{\partial n} = 0 \quad \frac{\partial v'}{\partial n} = 0 \quad \text{on} \quad \Gamma \\ &\text{and} \quad u' - v' \geq -\delta \quad \text{on} \quad \Omega \end{aligned} \right\}. \tag{33}$$

**Remark 1.** If the boundary conditions (1), (2) are replaced by conditions such as

$$\begin{cases} F_3(u) &= 0, \\ M_\tau(u) &= g_1(u) \end{cases} \quad \begin{cases} F_3(v) &= 0 \quad \text{on} \quad \Gamma, \\ M_\tau(v) &= g_2(v) \end{cases}$$

or

$$\begin{cases} u &= 0, \\ M_\tau(u) &= g_1(u) \end{cases} \quad \begin{cases} v &= 0 \quad \text{on} \quad \Gamma, \\ M_\tau(v) &= g_2(v) \end{cases}$$

then one obtains functional inequalities similar to (20) or (32).

## 3. Existence of solutions

We shall apply the results on noncoercive variational of [ASV] to the study of the variational inequalities derived earlier in this paper. We have the following result giving both necessary and sufficient conditions for the existence of solutions.

**Theorem 1.** *Let*

$$\mathcal{P} = \{p : \bar{\Omega} : p \text{ is a polynomial of degree one on } \mathbb{R}^2\}, \tag{34}$$

$$\mathcal{A} = \{(p,q) \in \mathcal{P} \times \mathcal{P} : p \geq q \text{ on } \Omega\}, \tag{35}$$

$$\psi_i^+ = \lim_{\lambda \to \infty} \frac{\psi_i(\lambda)}{\lambda} > 0 > \lim_{\lambda \to -\infty} \frac{\psi_i(\lambda)}{\lambda} = \psi_i^-, \ i = 1,2, \tag{36}$$

$$\chi_p^+ = \chi_{\{x \in \Gamma : p(x) \geq 0\}}, \ \chi_p^- = \chi_{\{x \in \Gamma : p(x) \leq 0\}} \tag{37}$$

($p$ is a function defined on $\Gamma$ and $\chi_B$ is the characteristic function of a set $B$). Then

a) *A necessary condition for* (26) *to have a solution is that*

$$(p, \psi_1^+ \chi_p^+ + \psi_1^- \chi_p^-)_\Gamma + (q, \psi_2^+ \chi_q^+ + \psi_2^- \chi_q^-)_\Gamma \geq (F, p) + (G, q)$$
$$\forall \ (p,q) \in \mathcal{A}. \tag{38}$$

b) *A sufficient condition for* (26) *to have a a solution is that*

$$(p, \psi_1^+ \chi_p^+ + \psi_1^- \chi_p^-)_\Gamma + (q, \psi_2^+ \chi_q^+ + \psi_q^- \chi_q^-)_\Gamma \geq (F, p) + (G, q)$$
$$\forall \ (p,q) \in \mathcal{A} \setminus \{(0,0)\}. \tag{39}$$

**Proof.** Put

$$A((u,v),(u',v')) = D_1 a(u,u') + D_2 a(v,v').$$

Then $A$ is a bilinear form on $V$ which is continuous, symmetric and positive, since $a$ is continuous, symmetric and positive on $H^2(\Omega)$. On the other hand, $a$ is $\mathcal{P}$-coercive on $H^2(\Omega)$ (cf. §2 and §5 of [ASV]). Hence $A$ is $\mathcal{P}$-coercive on $V$. We know that

$$\text{Ker} a \equiv \{x \in H^2(\Omega) : a(x,x) = 0\} = \mathcal{P}.$$

Hence
$$\text{Ker} A = \mathcal{P} \times \mathcal{P}. \tag{40}$$

Now clearly $K$ is a convex closed subset of $V$ containing $(0,0)$. On the other hand the recession cone of $K$ (cf. definitions 1, 1' of [ASV]) is

$$rcK = \{(u',v') \in V : u' \geq v' \text{ on } \Omega\}. \tag{41}$$

Define the functional

$$j(u',v') = J_1(u') + J_2(v').$$

By (28) and (29), $j$ is convex, and $j < \infty$ and is continuous on $V$. Let us calculate $j_\infty$ (cf. propositions 1, 2 of [ASV]),

$$j_\infty(p,q) = J_{1,\infty}(p) + J_{2,\infty}(q), \tag{42}$$

where

$$J_{1,\infty}(p) = \lim_{\lambda \to \infty} \frac{J_1(\lambda p)}{\lambda} = \lim_{\lambda \to \infty} \int_\Gamma \frac{\psi_1(\lambda p)}{\lambda} d\Gamma.$$

By proposition 1 of [ASV], $\left(\frac{\psi_1(\lambda p)}{\lambda}\right)_\lambda$ is an increasing family of functions (with respect to $\lambda$).

We observe

$$\begin{aligned} \frac{\psi_1(\lambda p(x))}{\lambda} &\to \psi_1^+ p(x) \quad \text{for} \quad \lambda \to \infty, \text{ where } p(x) > 0, \\ \frac{\psi_1(\lambda p(x))}{\lambda} &\to \psi_1^- p(x) \quad \text{for} \quad \lambda \to \infty, \text{ where } p(x) < 0. \end{aligned} \tag{43}$$

Hence (since $\psi_1(0) = 0$)

$$\int_\Gamma \frac{\psi_1(\lambda p)}{\lambda} d\Gamma = \int_{\{x \in \Gamma : p(x) \neq 0\}} \frac{\psi_1(\lambda p)}{\lambda} d\Gamma$$

$$= \int_{\{x \in \Gamma : p(x) > 0\}} \frac{\psi_1(\lambda p)}{\lambda} d\Gamma + \int_{\{x \in \Gamma : p(x) < 0\}} \frac{\psi_1(\lambda p)}{\lambda} d\Gamma$$

$$\to \int_{\{x \in \Gamma : p(x) > 0\}} p\psi_1^+ d\Gamma + \int_{\{x \in \Gamma : p(x) < 0\}} p\psi_1^- d\Gamma,$$

as follows from the monotone convergence theorem. We thus obtain

$$\int_\Gamma \psi_1^+ p\chi_p^+ d\Gamma + \int_\Gamma \psi_1^- p\chi_p^- d\Gamma = (p, \psi_1^+ \chi_p^+ + \psi_1^- \chi_p^-)_\Gamma. \tag{44}$$

Similarly we get

$$j_\infty(p,q) = (p, \psi_1^+ \chi_p^+ + \psi_1^- \chi_p^-)_\Gamma + (q, \psi_2^+ \chi_2^+ + \psi_2^- \chi_2^-). \tag{45}$$

To prove necessity, let (26) have a solution. Then by theorem 1 of [ASV],

$$j_\infty(p,q) \geq (F,p) + (G,q), \quad \forall (p,q) \in rcK \cap Ker A. \tag{46}$$

But by (40), (41) we have $rcK \cap Ker A = \mathcal{A}$. (38) follows then from (46) - (45).

**Proof of sufficiency.** Suppose (39) holds, then the resolvent set of (26) (cf. definition 2 of [ASV]) is

$$\mathcal{R}_{(F,G)} = \{(p,q) \in rcK \cap KerA : j_\infty(p,q) \leq (F,p) + (G,q)\} = \{(0,0)\}.$$

Since by (39),

$$j_\infty(p,q) > (F,p) + (G,q), \quad \forall\, (p,q) \in \mathcal{A}\setminus\{(0,0)\},$$

it follows from theorem 2 of [ASV] that (26) has a solution.

We next consider the variational inequality (32). We shall denote by $|\Gamma|$ the measure of $\Gamma$.

**Theorem 2.**

a) *A necessary condition for (32) to have a solution is that*

$$\begin{aligned}(\psi_1^- + \psi_2^-)|\Gamma| &\leq (F+G,1) \leq (\psi_1^+ + \psi_2^+)|\Gamma|,\\ \psi_1^+|\Gamma| &\geq (F,1),\\ \psi_2^-|\Gamma| &\leq (G,1).\end{aligned} \qquad (47)$$

b) *A sufficient condition for (32) to have a solution is that*

$$\begin{aligned}(\psi_1^- + \psi_2^-)|\Gamma| &< (F+G,1) < (\psi^+, +\psi_2^-)|\Gamma|,\\ \psi_1^+|\Gamma| &> (F,1),\\ \psi_2^-|\Gamma| &< (G,1).\end{aligned} \qquad (48)$$

**Proof.** For $K'$ given by (33), we have

$$rcK' = \left\{(u',v') \in V : \frac{\partial u'}{\partial n} + \frac{\partial v'}{\partial n} = 0 \text{ on } \Gamma \text{ and } u' \geq v' \text{ on } \Omega\right\}. \qquad (49)$$

Note that if $p \in \mathcal{P}$ and $\frac{\partial p}{\partial n} = 0$ on $\Gamma$ then $p \in \mathcal{R}$. Hence

$$rcK' \cap Ker A = rcK' \cap (\mathcal{P} \times \mathcal{P}) = \{(p,q) \in \mathbb{R}^2 : p \geq q\}. \qquad (50)$$

Let $p, q \in \mathbb{R}$, then

$$\begin{aligned}J_{1,\infty}(p) &= \lim_{\lambda \to \infty} \int_p \frac{\psi_1(\lambda p)}{\lambda} d\Gamma = \lim_{\lambda \to \infty} \frac{\psi_1(\lambda p)}{\lambda} |\Gamma|\\ &= \begin{cases}\psi_1^+ p|\Gamma| & \text{if } p \geq 0\\ \psi_1^- p|\Gamma| & \text{if } p < 0\end{cases}\\ &= (\psi_1^+ p^+ - \psi_1^- p^{-1})|\Gamma|,\end{aligned}$$

$(\psi_1^+, \psi_1^-)$ given by (36), and $p^+ = \max(p,0) p' = \max(-p,0)$. A similar calculation gives $J_{2,\infty}(q)$. We thus have

$$j_\infty(p,q) = |\Gamma|[(\psi_1^+ p^+ - \psi_1^- p^-) - (\psi_2^+ q^+ - \psi_2^- q^-)]. \tag{51}$$

Let (32) have a solution. Then by (50), (51), theorem 1 of [ASV] gives

$$|\Gamma|[(\psi_1^+ p^+ - \psi_1^- p^-) + (\psi_2^+ q^+ - \psi_2^- q^-)] \geq p(F,1) + q(G,1), \\ \forall\, p,q \in \mathbb{R}, p \geq q. \tag{52}$$

Now in (52), we take various values of $p,q$ and note the corresponding values for (52)

$$p = q = 1 : (\psi_1^+ + \psi_2^+)|\Gamma| \geq (F+G, 1),$$
$$p = 1, q = 0 : \psi_1^+ |\Gamma| \geq (F,1),$$
$$p = q = -1 : (\psi_1^- + \psi_2^-)|\Gamma| \leq (F+G, 1),$$
$$p = 0, q = -1 : \psi_2^- |\Gamma| \leq (G,1).$$

This latter set of expressions is just (47).

Now for the sufficiency. Suppose (48) holds. We claim that the resolvent set for (32) is given by

$$\mathcal{R}_{(F,G)} \equiv \{(p,q) \in \mathbb{R}^2 : p \geq q \text{ and } j_\infty(p,q) \leq (F,p) + (G,q)\} = \{(0,0)\}. \tag{53}$$

Indeed for $p, q, \in \mathbb{R}$ with $p \geq q$ and $(p,q) \neq (0,0)$ it can be shown that

$$|\Gamma|[(\psi^+, p^+ - \psi^-, p^-) + (\psi_2^+ q^+ - \psi_2^- q^-)] > p(F,1) + q(G,1). \tag{54}$$

(This is shown by examining the various possibilites that occur if $(p,q) \in \{\mathbb{R}^2 \setminus \{(0,0)\} : p \geq q\}$). Since (53) holds, it follows from theorem 2 of [ASV] that (32) has a solution.

## 4. Some special cases

If the plates were subjected to unilateral boundary conditions, then the functions $\psi_i$ become

$$\psi_i(x) = -g_i x^+ - g_i' x^-, \ g_i' < 0 < g_i \ \text{(constants)}, \quad i = 1,2, \tag{55}$$

(cf.(4.3), §4.1 chapter 4 [DL]). Then

$$\psi_i^+ = g_i, \quad \text{and} \quad \psi_i^- = g_i'. \tag{56}$$

Then theorems 1 and 2 assume the following special forms:

**Theorem 1'.** *Let $\psi_i$ be given by (55) $(i = 1, 2)$. Then*

a) *A necessary condition for (26) to have a solution is that*

$$(p, g_1\chi_p^+ + g_1'\chi_p^-)\Gamma + (q, g_2\chi_q^+ + g_2'\chi_q^-)\Gamma \geq (F, p) + (G, q)$$
$$\forall \ (p, q) \in \mathcal{A}.$$

b) *A sufficient condition for (26) to have a solution is that*

$$(p, g_1\chi_p^+ + g_1'\chi_p^-)\Gamma + (q, g_2\chi_q^+ + g_2'\chi_q^-)\Gamma > (F, p) + (G, q)$$
$$\forall \ (p, q) \in \mathcal{A}\setminus\{(0, 0)\}.$$

**Theorem 2'.** *Let $\psi_i$ be given by (55), $i = 1, 2$. Then*

a) *A necessary condition for (32) to have a solution is that*

$$(g_1' + g_2')|\Gamma| \leq (F + G, 1) \leq (g_1 + g_2(|\Gamma|),$$
$$g_1|\Gamma| \geq (F, 1),$$
$$g_2'|\Gamma| \leq (G, 1).$$

b) *A sufficient condition for (32) to have a solution is that*

$$(q_i' + g_2')|\Gamma| < (F + G, 1) < (g_1 + g_2)|\Gamma|,$$
$$g_1|\Gamma| > (F, 1),$$
$$g_2'|\Gamma| < (G, 1).$$

We next examine the cases $\psi_i = 0$, $i = 1, 2$. In this case $j = 0$ on $V$ and the following holds.

**Theorem 3.** *Let $\psi_i = 0 \ \ (i = 1, 2)$ then*

a) *A necessary condition for (26) to have a solution is that*

$$(G, \phi) \geq 0, \quad \forall \phi \in \mathcal{D}, \tag{57}$$
$$(F + G, 1) = (F + G, x_1) = (F + G, x_2) = 0.$$

b) *A sufficient condition (26) to have a solution is that*

$$(G, \phi) > 0, \quad \forall \phi \in \mathcal{D}\setminus\{0\}, \tag{58}$$
$$(F + G, 1) = (F + G, x_1) = (F + G, x_2) = 0.$$

Here

$$\mathcal{D} = \{\phi \in \mathcal{P} : \phi \geq 0 \ \text{ on } \ \Omega\}.$$

**Proof.** Note first that $rcK \cap KerA$ as defined in (35) can be rewritten

$$rcK \cap KerA = \{(p, p - \phi) : p \in \mathcal{P}, \phi \in \mathcal{D}\}. \tag{59}$$

A necessary condition for (26) to have a solution is

$$(F, p) + (G, q) \leq 0, \quad \forall \ (p, q) \in rcK \cap KerA.$$

From (59) follows

$$(F, p) + (G, p - \phi) \geq 0, \quad \forall \ p \in \mathcal{P}, \quad \forall \ \phi \in \mathcal{D}. \tag{60}$$

Let $\phi = 0$ and $p = \pm 1$, $p = \pm x_1$, $p = \pm x_2$ in (60) to get

$$(F + G, 1) = (F + G, x_1) = (F + G, x_2) = 0.$$

Next let $p = 0$, $\phi \in \mathcal{D}$. Then $(G, \phi) \geq 0$, $\forall \phi \in \mathcal{D}$, proving (57).

Next suppose (58) holds, then one also has

$$\mathcal{R}_{(F,G)} = \{(p, p) : p \in \mathcal{P}\}. \tag{61}$$

Indeed one has

$$\mathcal{R}_{(F,G)} = \left\{(p, q) \in rcK \cap KerA : (F, p) + (G, q) \geq 0\right\}.$$

(cf. proposition 2 of [ASV]).
In view of (59) and (58), one has

$$\{(p, p) : p \in \mathcal{P}\} \subset \mathcal{R}_{(F,G)}. \tag{62}$$

For if $(p, p - \phi) \in rcK \cap KerA$ and $\phi \neq 0$ one has $(F, p) + (G, p - \phi) = -(G, \phi) < 0$. Hence $(p, p - \phi) \notin \mathcal{R}_{(F,G)}$. Thus one has (61). Since $\mathcal{P}$ is a vector space, so is $\mathcal{R}_{(F,G)}$. By theorem 2 in [ASV], (26) has a solution.

**Theorem 4.** Suppose $\psi_i = 0$, $i = 1, 2$. Then a necessary and sufficient for (32) to have a solution is

$$\begin{aligned}(F + G, 1) &= 0 \\ (G, 1) &\geq 0.\end{aligned} \tag{63}$$

**Proof.** We already know that a necessary condition for (32) to have a solution is

$$(F, p) + (G, q) \leq 0, \quad \forall p, q \in \mathbb{R}, \quad p \leq q. \tag{63'}$$

Taking $p = q = 1$ and $p = 0, q = -1$ we immediately have (63). Conversely suppose we have (63). We distinguish several cases. If $(G, 1) > 0$ then

$$\mathcal{R}_{(F,G)} = \{(p, p) : p \in \mathbb{R}\}. \tag{64}$$

Indeed since $(F + G, 1) = 0$, then clearly

$$\{(p,p) : p \in \mathbb{R}\} \subset \mathcal{R}_{(F,G)}.$$

If
$$(F,p) + (G,q) = p(F + G, 1) + (q - p)(G, 1) = (q - p)(G, 1) < 0$$
$$\text{for} \quad p, q \in \mathcal{R} \quad \text{and} \quad p > q,$$

then we have (64). Since $\{(p,p) : p \in \mathcal{R}\}$ is a vector space, (32) has a solution. Next, if $(G, 1) = 1$ then $(F, 1) = 0$. Put

$$W = \left\{(u', v') \in H^2(\Omega) \times H^2(\Omega) : \frac{\partial w}{\partial n} + \frac{\partial v'}{\partial n} = 0 \text{ on } \Gamma\right\}. \quad (65)$$

Then $W$ is a Hilbert subspace of $H^2(\Omega) \times H^2(\Omega)$. Consider the variational inequality.

$$A((w,t), (u', v') - (w, t)) \geq (F, u' - w) + (G, v' - t),$$
$$(w, t) \in W, \quad \forall (u', v') \in W. \quad (66)$$

Let us consider the resolvent set of (66). Since $rcW = W$, one has

$$\mathcal{R} = \{(u', v') \in W \cap Ker A : (F, u') + (G, v') \geq 0\},$$
$$W \cap Ker A = W \cap (\mathcal{P} + \mathcal{P}) = \mathbb{R}^2.$$

By (65), we have at once $\mathcal{R} = \mathbb{R}^2$. Hence (66) has a solution. If $(w, t)$ is a solution of (66) then $(w + \lambda, t) \in W$ and is also a solution of (66). Since $w, t \in C(\bar{\Omega})$, then we have $(w + x) - t \geq -\delta$ on $\Omega$, if we pick $\lambda \geq \|w - t\|_{C(\bar{\Omega})} - \delta$. Hence $(w + \lambda, t) \in K'$. Since $K' \subset W$, $(w + \lambda, t)$ is a solution of (32) and we conclude that (32) always has a solution.

## 5. Smoothness properties of solutions of contact problems

**Theorem 5.** *Let $(u, v)$ be a solution of (26) (or of (32)). Then*

$$u, v \in H^3_{loc}(\Omega) \cap H^{2+\epsilon}(\Omega) \quad \forall \epsilon \in (0, 1).$$

**Proof.** Consider (26) for instance. To prove that $u, v \in H^3_{loc}(\Omega)$, we use the method of finite differences (cf. [K]). For each precompact subset $\Omega_2 \subset\subset \Omega$, we shall prove that the restrictions $u|_{\Omega_2}, v|_{\Omega_2} \in H^3(\Omega_2)$. Choose $\Omega_1$ so that $\Omega_2 \subset\subset \Omega_1 \subset\subset \Omega$ and $\phi \in C_0^\infty(\Omega), 0 \leq \phi \leq 1$ and $\phi \equiv 1$ on $\Omega_2$ and $\phi = 0$ outside $\Omega_1$. It is then sufficient to prove that

$$\phi u, \ \phi v \in H^3(\Omega). \quad (67)$$

For small $\tau \in \mathbb{R}$, put $h = \tau \ell_i$ ($i = 1, 2$   $\ell_i$ is a standard unit vector of $\mathbb{R}^2$). Put

$$d_h w(x) = \frac{1}{|h|}[w(x+h) - w(x)],$$
$$\Delta_h w(x) = \frac{1}{|h|^2}[w(x+h) + w(x-h) - 2w(x)]. \tag{68}$$

Let $q = d(\Omega_1, \partial\Omega)$ and let $|\tau| < q$. Choose $0 < x < \tau^2/2$ and put

$$u' = u + \lambda \phi^2 \Delta_h u,$$
$$v' = v + \lambda \phi^2 \Delta_h v. \tag{69}$$

Then clearly $(u', v') \in K$. Substituting into (26) yields

$$\begin{aligned}&D_1 a(u, \lambda \phi^2 \Delta_h u) + D_2 a(v, \lambda \phi^2 \Delta_h v) \\&+ J_1(u + \lambda \phi^2 \Delta_h u) - J_1(u) + J_2(v + \lambda \phi^2 \Delta_h v) - J_2(v) \\&\geq (F, \lambda \phi^2 \Delta_h u) + (G, \lambda \phi^2 \Delta_h v).\end{aligned} \tag{70}$$

Since $\phi = 0$ on $\Gamma$, one has

$$\begin{aligned}&J_1(u + \lambda \phi^2 \Delta_h u) + \int_\Gamma \psi_1(u + \lambda \phi^2 \Delta_h u) d\Gamma \\&= \int_\Gamma \psi_1(u) d\Gamma = J_1(u),\end{aligned} \tag{71}$$

and similarly $J_2(v + \lambda \phi^2 \Delta_h v) = J_2(v)$. On the other hand,

$$a(u, \phi^2 \Delta_h u) = (\Delta u, \Delta(\phi^2 \Delta_h u)). \tag{72}$$

Indeed, since $\phi^2 \Delta_h u \in H_0^2(\Omega)$, we have

$$\left.\begin{aligned}u_n &\to u \\ v_n &\to \phi^2 \Delta_h u\end{aligned}\right\} \quad \text{in} \quad H^2(\Omega),$$

since
$$u_n \in C^\infty(\bar\Omega), v_n \in C_0^\infty(\Omega).$$

By Green's formula ((4.2), §4.1, chapter 4 [DL])

$$a(u_n, v_n) = (\Delta^2 u_n, v_n) + (F_3(u_n), v_n)_\Gamma - (M_\tau(u_n), \frac{\partial v_n}{\partial n})_\Gamma = (\Delta^2 u_n, v_n). \tag{73}$$

But

$$\begin{aligned}(\Delta^2 u_n, v_n) &= (\Delta u_n, \Delta v_n) + \left(\frac{\partial}{\partial n}, u_n\right)_\Gamma - \left(\Delta u_n, \frac{\partial}{\partial n}(v_n)\right)_\Gamma \\&= (\Delta u_n, \Delta v_n).\end{aligned} \tag{74}$$

Comparing (73), (74) and letting $n \to \infty$ gives (72). From (70), (71), (72) and similar equalities involving $v'$, we deduce

$$D_1(\Delta u, \Delta(\phi^2 \Delta_h u)) + D_2(\Delta v, \Delta(\phi^2 \Delta_h v)) \geq (F, \phi^2 \Delta_h u) + (G, \phi^2 \Delta_h v). \quad (75)$$

The remaining part of this proof proceeds along similar lines as the proof of lemma 1 of [K]. For the sake of completeness, we give below a sketch of it.

By direct computation, it is verified that for the expressions to follow, each one either is equal to the preceding one or differs from it by a quantity

$$\leq c(\|u\|_{H^2(\Omega)}^2 + \|u\|_{H^2(\Omega)} \|d_h(\phi\phi)\|_{H^2(\Omega)})$$

(where $c$ is constant depending only on $\Omega$ and $\phi$)

$$(\Delta u, \Delta(\phi^2 \Delta_h u)) \to (\Delta(\phi u), \Delta_h \Delta(\phi u)) \to$$
$$\to -(\Delta(\phi u), d_{-h} d_h \Delta(\phi u)) \to$$
$$\to -(d_h(\Delta(\phi u)), d_h(\Delta(\phi u))).$$

Hence

$$\|\Delta(d_h(\phi u))\|_{L^2(\Omega)}^2 + (\Delta u, \Delta(\phi^2 \Delta_h u))$$
$$\leq c(\|u\|_{H^2(\Omega)}^2 + \|u\|_{H^2(\Omega)} \|d_h(\phi_h(\phi u))\|_{H^2(\Omega)}).$$

Since

$$\|d_h(\phi u)\|_{H^2(\Omega)}^2 \leq c\|\Delta(d_h(\phi u))\|_{L^2(\Omega)}^2, \ (d_h(\phi u) \in H_0^2(\Omega)),$$

we have

$$\|d_h(\phi u)\|_{H^2(\Omega)}^2 + D_1(\Delta u, \Delta(\phi^2 \Delta_h u))$$
$$\leq c(\|u\|_{H^2(\Omega)}^2 + \|u\|_{H^2(\Omega)} \|d_h(\phi u)\|_{H^2(\Omega)}), \quad (76)$$

and a similar inequality for $v$. Adding the two, we obtain using (75)

$$\|d_h(\phi u)\|_{H^2}^2 + \|d_h(\phi v)\|_{H^2}^2$$
$$\leq c(\|u\|_{H^2}^2 + \|v\|_{H^2}^2 + \|F\|_{L^2}^2 + \|G\|_{L^2}^2$$
$$+ \|u\|_{H^2} \|d_h(\phi u)\|_{H^2} + \|v\|_{H^2} \|d_h(\phi v)\|_{H^2}).$$

Applying Cauchy's inequality to $\|u\|_{H^2} \|d_h(\phi u)\|_{H^2}$ and to $\|v\|_{H^2} \|d_u(\phi v)\|_{H^2}$, we have

$$\|d_h(\phi u)\|_{H^2}^2 + \|d_h(\phi v)\|_{H^2}^2 \leq c,$$

where $c$ is a constant independent of $\tau$. Hence $\phi u, \phi v \in H^3(\Omega)$ as claimed. We now turn to the proof of $u, v \in H^{2+\epsilon}(\Omega)$ $(\forall \epsilon \in (0,1))$. Let $\phi \in \mathcal{D}(\Omega), \phi \geq 0$ on $\Omega$. Then

$$(u', v') = (u + \phi, v) \in K.$$

Inserting this into (26) gives
$$\mathcal{D}_1 a(u,\phi) - J_1(u) \geq (F,\phi).$$
It can be shown (as in the proof of (71) - (72) and (73) that
$$J_1(u+\phi) + J_1(u)\mathcal{D}_1 a(u,\phi) = \mathcal{D}_1(\Delta u, \Delta \phi) = \mathcal{D}_1(\Delta^2 u, \phi)_{\mathcal{D}'(\Omega), \mathcal{D}(\Omega)}.$$
Hence
$$(\mathcal{D}_1 \Delta^2 u - F, \phi)_{\mathcal{D}'(\Omega), \mathcal{D}(\Omega)} \geq 0$$
$$\forall \phi \in \mathcal{D}(\Omega), \ \phi \geq 0.$$

Now $\mathcal{D}_1 \Delta^2 u - F = p$ is a finite measure on $\Omega$, i.e. $p \in (C_0(\Omega))' \subset H^{-1-\sigma}(\Omega)$ $\forall \ \sigma > 0$. Let $\sigma = 1 - \epsilon > 0$. Then $p \in H^{-2+\epsilon}(\Omega)$. Hence $\mathcal{D}_1 \Delta^2 u = F + p \in H^{-2+\epsilon}(\Omega)$. By known smoothing properties for $\Delta^2$, we have $u \in H^{2+\epsilon}(\Omega)$. A similar conclusion holds for $v$.

## 6. Contact of two plates with clamped boundaries

Consider two plates as in section 2 but with clamped boundaries

$$u = 0 \ \frac{\partial u}{\partial n} = 0 \quad \text{on} \quad \Gamma$$
$$v = 0 \ \frac{\partial v}{\partial n} = 0 \quad \text{on} \quad \Gamma. \tag{77}$$

In [K], the problem was reduced to a variational inequality

$$D_1(\Delta u, \Delta u' - \Delta u) + D_2(\Delta v, \Delta v' - \Delta v) \geq (F, u' - u) + (G, v' - v)$$
$$(u,v) \in K_0, \quad \forall (u', v') \in K_0, \tag{78}$$

where
$$K_0 = \{(u', v') \in H_0^2 \times H_0^2 : u' - v' \geq -\delta \quad \text{on} \quad \Omega\}.$$
Since the nonlinear form
$$((u,v),(u',v')) \mapsto D_1(\Delta u, \Delta u') + D_2(\Delta v, \Delta v')$$
is continuous coercive on $H_0^2 \times H_0^2$, since $K_0$ is closed and convex, and since $F, G \in L^2$, (78) has a unique solution $(u,v)$.

In [K] a smoothness property of $(u,v)$ is proved (i.e. $u, v \in H^3_{loc}$) and furthermore a topological property of the constant surface is established in the case the plates have the same flexural rigidity $D_1 = D_2$, in which case the complement of the contact surface is connected. We shall show that the same property holds even if these plates have different flexural rigidities $D_2 \neq D_2$. This is achieved by reducing (78) to an obstacle problem in $H_0^2$ and a variational equation. Furthermore as is the case for solutions of elliptic equations, we shall prove $u, v \in H^3(\Omega)$.

**Theorem 6.** *Let $(u,v)$ be the solution of (78). Put $w = u - v$. Then $w$ is the unique solution of the obstacle problem.*

$$(\Delta w, \Delta t - \Delta w) \geq \left(\frac{F}{D_1} - \frac{G}{D_2}, t - w\right), \quad w \in M, \quad \forall t \in M, \tag{79}$$

where

$$M = \{t \in H_0^2 = t \geq -\delta \text{ on } \Omega\}.$$

**Proof.** Since $(u,v) \in K$, we have $w \in M$. For given $t$ in $M$, we shall look for $(u', v')$ such that

$$u' - v' = t \tag{80}$$

$$D_1(u' - u) + D_2(v' - v) = 0. \tag{81}$$

Since $u, v, t \in H_0^2, u', v' \in H_0^2$, solving (81) - (82) gives

$$u' = (D_1 + D_2)^{-1}(D_1 u + D_2 v + D_2 t)$$
$$v' = (D_1 + D_2)^{-1}(D_1 u + D_2 v - D_1 t).$$

By (81) and since $t \in M$, we have $(u', v') \in K$. Substituting for $(u', v')$ into (78) we have (note that $D_1(\Delta u' - \Delta u) = -D_2(\Delta v' - \Delta v)$)

$$D_1(\Delta u, \Delta u' - \Delta u) - D_1(\Delta v, \Delta u' - \Delta u) \geq (F, u' - u) - \frac{D_1}{D_2}(G, u' - u).$$

Simplifying yields

$$(\Delta w, \Delta u' - \Delta u) \geq \left(\frac{F}{D_1} - \frac{G}{D_2}, u' - u\right). \tag{82}$$

From (82) we obtain

$$t - w = (u' - u) - (u' - v) = \left(1 + \frac{D_1}{D_2}\right)(u' - u).$$

Substituting $u' - u = (D_2/(D_1 + D_2))(t - w)$ into (83) and multiplying both sides by $(D_1 + D_2)/D_2$, one gets (80). On the other hand, since $(w, t) \mapsto (\Delta w, \Delta t)$ is coercive on $H_0^2$, we conclude that (80) has a unique solution.

**Corollary 1.** *$(u,v)$ is a solution of (78) if and only if $w \equiv u - v$ satisfies (80) and*

$$(D_1 + D_2)(\Delta u, \Delta u') = (F + G, u') + D_2(\Delta w, \Delta u'), \quad \forall u' \in H_0^2. \tag{83}$$

**Proof.** Suppose $(u,v)$ is a solution of (1), then by theorem 6, $w \equiv u - v$ is a solution of (80). Note that $K_0$ can be expressed in the form

$$K_0 = \{(u', u' - t) : u' \in H_0^2, t \in M\}$$

and that $v = u - w$. For $t \in M$, $u' \in H_0^2$, one has from (78)

$$D_1(\Delta u, \Delta u' - \Delta u) + D_2(\Delta(u-w), \Delta(u'-t) - \Delta(u-w))$$
$$\geq (F, u' - u) + (G, (u' - t) - (u - w)).$$

For $t = w$, this becomes

$$D_1(\Delta u, \Delta u' - \Delta u) + D_w(\Delta(u-w), \Delta u' - \Delta u)$$
$$\geq (F, u' - u) + (G, u' - u), \forall u' \in H_0^2.$$

Hence

$$D_1(\Delta u, \Delta u') + D_2(\Delta(u-w), \Delta u') \geq (F + G, u'),$$
$$\forall u' \in H_0^2.$$

Substituting $\pm u'$ into the foregoing inequality, we get

$$(D_1 + D_2)(\Delta u, \Delta u') = (F + G, u') + D_2(\Delta w, \Delta u'), \forall u' \in H_0^2, \qquad (84)$$

which is (83). Conversely suppose (84) holds. By the coerciveness of the form $(u, u') \mapsto (\Delta u, \Delta u')$ on $H_0^2$, (83) has a unique solution. Let $(u_1, v_1)$ a solution of (78). By theorem 6, $u_1 - v_1$ is a solution of (79). By uniqueness $w = u_1 - v_1$. By the above proof, $u_1$ satisfies equation (83) corresponding to $w = u_1 - v_1$. By uniqueness $u = u_1$. Thus $v = v_1$ and $(u, v)$ is a solution of (78).

**Corollary 2.** *Let $(u, v)$ be a solution of (78). Then $D_1 u + D_2 v$ satisfies the variational equation*

$$(\Delta(D_1 u + D_2 v), \Delta z) = (F + G, z), \quad \forall z \in H_0^2, \qquad (85)$$

*and hence if $\partial \Omega \in C^2$ then*

$$D_1 u + D_2 v \in H_0^2 \cap H^2. \qquad (86)$$

**Proof.** From (85) and $w = u - v$, (79) follows. By smoothness properties of solutions of biharmonic equations, (86) follows from (85).

**Corollary 3.** *Let $\delta > 0$ and $F/D_1 - G/D_2 \leq 0$ on $\Omega$. Then the complement of the contact surface for (78) is connected.*

**Proof.** Consider the contact problem for 2 plates with the same flexural rigidity $D = 1$ and subjected to vertical forces $F_1 = F/D_1$, $G_1 = G/D_2$, i.e., consider the following variational inequality

$$(\Delta u_1, \Delta u' - \Delta u_1) + (\Delta v_1, \Delta v' - \Delta v_1) \geq (F_1, u' - u_1) + (G_1, v' - v_1)$$
$$(u_1, v_1) \in K_0, \quad \forall (u', v') \in K_0. \qquad (87)$$

Call $C_1$ the contact surface for (87).
$$C_1 = \{x \in \Omega : u_1(x) - v_1(x) = -\delta\}. \tag{88}$$
By results in [K], $\Omega \backslash C_1$ is connected.

Applying theorem 6 to (87), we see that $w_1 \equiv u_1 - v_1$ is a solution of the variational inequality
$$\begin{aligned}(\Delta w_1, \Delta t - \Delta w_1) &\geq (F_1 - G_1, t - w_1)\\ w_1 \in M, \quad &\forall t \in M.\end{aligned} \tag{89}$$

By the definition of $F_1, G_1$ (89) is precisely (79). By uniqueness of solutions of (80), we have $w_1 = w$, that is $u_1 - v_1 = u - v$ on $\Omega$.

Whence $C_1 = [u \in \Omega : u(x) - v(x) = -\delta]$ is precisely the contact surface for (78). Hence $\Omega \backslash C$, the complement of the contact surface for (78) is connected.

The next result provides some smoothness properties of solutions of (78).

**Theorem 7.** *Let $\delta > 0$ and $\partial\Omega \in C^3$. Then the solution $(u, v)$ of (78) satisfies $u, v \in H^3(\Omega)$.*

For the proof of this theorem, we rely on results in [A]. The computations and verifications are lengthy and therefore we shall not give them here.

## References

[A]   S. Agmon, *Lectures on Elliptic Boundary Value Problems*, Van Nostrand, New York, 1965.

[ASV] D. D. Ang, K. Schmitt, and L. K. Vy, *Noncoercive variational inequalities: Some applications*, Nonlinear Analysis, TMA (to appear).

[DL]  G. Duvaut and J. L. Lions, *Inequalities in Mechanics and Physics*, Springer Verlag, Berlin, 1976.

[K]   A. M. Khludnev, *The problem on the contact of two elastic plates*, PMM, USSR **47** (1984), 110–115.


D.D. Ang
Department of Mathematics
Ho Chi Minh City University
Ho Chi Minh City, Vietnam

K. Schmitt
Department of Mathematics
University of Utah
Salt Lake City, UT 84112, USA

L.K. Vy
Department of Computer Science
Ho Chi Minh City University
Ho Chi Minh City, Vietnam

# Analytic Semigroups: Applications to Inverse Problems for Flexible Structures

H.T. BANKS
D.A. REBNORD

Center for Applied Mathematical Sciences, University of Southern California
Department of Mathematics, Iowa State University

## 1. Introduction

In this presentation we consider abstract inverse problems in a least squares formulation for parameter dependent partial differential equations. We are interested in approximation ideas which lead to viable computational techniques for such problems. We pursue our investigations in the context of the general framework for convergence and stability developed by Banks and Ito in [BI]. Motivated by questions related to the use of accelerometer data to estimate parameters in flexible structures, we focus on second order (in time) systems with sufficient damping so that the system can be modeled by an analytic semigroup.

We state and prove a new approximation result (a Trotter-Kato type theorem) for analytic semigroups. This theorem gives conditions under which a family of approximating semigroups and all its time derivatives converges to a limit semigroup and all its time derivatives, respectively. These theoretical results are then stated in terms of simple, readily checked conditions on the sesquilinear forms defining "stiffness" and "damping" in the abstract second order systems.

---

This research was supported in part by the Air Force Office of Scientific Research under Contract F49620-86-C-0111, by the National Aeronautics and Space Administration under NASA Grant NAG-1-517, and by the National Science Foundation under NSF Grant DMS-8818530. Part of this research was carried out while the first author was a visiting scientist at the Institute for Computer Applications in Science and Engineering (ICASE), NASA Langley Research Center, Hampton, VA, which is operated under NASA Contract NAS1-18605.

We discuss several examples which indicate clearly the practical importance of these new convergence results. In two of the examples presented (the damped, cantilevered Euler-Bernoulli beam and the "RPL experiment" structure), we can apply the abstract theorem to substantially sharpen results already found in the research literature. In a third example involving a two dimensional grid structure, we note that the theory can be applied to obtain new results for acceleration data convergence in the least squares inverse problems.

## 2. Abstract inverse problems

In this section we formulate a class of inverse problems as abstract least squares optimization problems constrained by evolution equations in a Hilbert space and summarize some of our previous results for such problems [B], [BI], [BK].

We assume we are given a parameter dependent system

$$\dot{u}(t) = A(q)u(t) + F(t,q), \quad 0 < t < T,$$
$$u(0) = u_0(q) \tag{2.1}$$

for states $u(t;q)$ in a Hilbert space $H$. The parameters $q$ are to be chosen from an admissible parameter set $Q$ contained in a metric space $(Q_1, d)$. We assume throughout that $Q$ is a compact subset of $Q_1$ (a "regularization" assumption).

We are given observations $\{z_i\}$, a set of points in the observation space $Z$, along with an observation map $\mathcal{C} : C(0,T;H) \to Z$ from the states to the observations. The points $z_i$ are observations for $\mathcal{C}u(t_i;q), : 0 < t_i < T$, and it is this "data" to which we wish to fit the model by a best choice of the parameter $q$. Formally, the problem can be stated as

$$\text{Find } \bar{q} \in Q \text{ to minimize over } Q \text{ the}$$
$$\text{functional } J(q) = \sum_i |\mathcal{C}u(t_i;q) - z_i|_Z. \tag{$\mathcal{P}$}$$

The observation operator $\mathcal{C}$ is of fundamental importance to the discussions in this paper. For parabolic systems (2.1), a typical example for $\mathcal{C}$ arises from pointwise evaluation in the spatial variables, e.g., $\mathcal{C}u(t;q) = \{u(t,x_j;q)\}_{j=1}^\ell$ where $Z = R^\ell$. For problems involving structures such as beams or plates, several examples arise in practice. If one takes measurements with a laser vibrometer, then one obtains measurements at specific points in space for the velocities $u_t(t_i, x_j; q)$ so that the map $\mathcal{C}$ is a composite of time differentiation $\frac{\partial}{\partial t}$ followed by pointwise evaluation. If the measuring devices are accelerometers, one has observations for the accelerations $u_{tt}(t_i, x_j; q)$ and thus $\mathcal{C}$ is time differentiation (twice) $\frac{\partial^2}{\partial t^2}$ followed by pointwise evaluation.

The inverse problems outlined here are generally infinite dimensional in both the states $u(t) \in H$ and the parameters $q \in Q$ and, moreover, involve unbounded operators related to the states $(A(q))$ and the observations $(\mathcal{C})$. Thus to develop efficient computational methods, one must make finite dimensional approximations for both the state and parameter spaces, $H$ and $Q$, respectively. For the discussions of this paper, we shall restrict our considerations to approximations of the state space. In some cases (where the parameter sets $Q$ are finite dimensional, either naturally or through some a priori parameterization) no approximation of the parameter set is necessary. Methods for approximation of $Q$ by sets $Q^M$ have been fully discussed elsewhere (see [BK]) and these ideas could readily be incorporated in our presentation. Since this would add nothing to either the difficulties or their resolution that are the focus here, to avoid considerable notational tedium we do not consider such approximation ideas in the discussions below.

Thus, we consider approximate (to be made precise below) state spaces $H^N \subset H$ with associated approximate states $u^N(t; q) \in H^N$ satisfying the approximate systems

$$\dot{u}^N(t) = A^N(q) u^N(t) + P^N F(t, q)$$
$$u^N(0) = P^N u_0(q) \tag{2.2}$$

where $A^N$ is an approximation to $A$ and $P^N$ is the orthogonal projection of $H$ onto $H^N$. The corresponding approximate inverse problems are given by

Find $\bar{q}^N \in Q$ to minimize over $Q$ the
functional $J^N(q) = \sum_j |\mathcal{C} u^N(t_i; q) - z_i|_Z.$ $\quad (\mathcal{P}^N)$

Of course, there are a number of reasonable ways in which the approximations may be made to arrive at the problems $(\mathcal{P}^N)$. In analyzing different methods and their behavior in the context of inverse problems, a number of questions related to *parameter convergence* arise naturally. For example, given a fixed set of data $\tilde{z} = \{z_i\}$, do optimal parameters $\bar{q}^N$ of $(\mathcal{P}^N)$ converge in some sense to an optimal parameter $\bar{q}$ for $(\mathcal{P})$? More generally, one might also incorporate the continuous dependence of the estimates on the observations in the concept of *method stability* [B], [BK]. For this, one denotes by $\bar{q}^N(\tilde{z}^K)$ optimal parameters obtained from $(\mathcal{P}^N)$ for observations $\tilde{z}^K$ and requires that $\bar{q}^N(\tilde{z}^K) \to \bar{q}(\tilde{z}^0)$ in some sense as $N \to \infty$ and $\tilde{z}^K \to \tilde{z}^0$ in $Z$, where $\bar{q}(\tilde{z}^0)$ is an optimal parameter for $(\mathcal{P})$ corresponding to observations $\tilde{z}^0$. These issues are carefully discussed in [B], [BK] where it is shown that for both parameter convergence and method stability, it suffices to argue that

For arbitrary sequences $\{q^N\}$ in $Q$ converging
to $q \in Q$, we have $\mathcal{C} u^N(t; q^N) \to \mathcal{C} u(t; q)$ as $N \to \infty$ $\quad (2.3)$
for each $t \in (0, T)$.

Thus, certain fundamental aspects of approximation in inverse problems can be reduced to the convergence statement in (2.3) and we shall deal with conditions under which (2.3) can be guaranteed in the subsequent discussions of this paper.

In [BI], Banks and Ito developed a general framework for convergence and stability which we summarize here in an extended form to allow treatment of complex valued operators in complex state spaces (the sesquilinear forms introduced in [BI] were tacitly assumed to be generated by partial differential equations with real coefficients and the theory developed there was adequate for such examples). As in [BI], our goal is to isolate properties of the system (2.1) and the approximation families to insure that (2.3) holds. We state these properties in terms of conditions on a parameter dependent sesquilinear form arising from reformulating (2.1) in a weak or variational sense. To this end we consider the system as a variational equation holding for all $\phi \in V$:

$$(\dot{u}(t), \phi) + \sigma(q)(u(t), \phi) = \langle F(t, q), \phi \rangle,$$
$$u(0) = u_0(q). \tag{2.4}$$

Here $V$ and $H$ are complex Hilbert spaces with $V$ imbedded densely and continuously in the pivot space $H$; i.e. $V \hookrightarrow H = H^* \hookrightarrow V^*$. The sesquilinear form $\sigma(q) : V \times V \to \mathbb{C}$ is related to the operator $A(q)$ as described below, $\langle \cdot \rangle$ is the inner product in $H$, $: (\cdot, \cdot)$ is the duality product $(\cdot, \cdot)_{V^*, V}$ and equation (2.4) is interpreted in the $V^*$ sense. The conditions on $\sigma$ needed are:

**(A)** Continuity in $q$: For $q, \tilde{q} \in Q$, we have for all $\phi, \psi \in V$

$$|\sigma(q)(\phi, \psi) - \sigma(\tilde{q})(\phi, \psi)| \leq d(q, \tilde{q})|\phi|_V |\psi|_V.$$

**(B)** V-Coercivity: There exists $c_1 > 0$ and some real $\lambda_0$ such that for $q \in Q, \phi \in V$, we have

$$\operatorname{Re} \sigma(q)(\phi, \phi) + \lambda_0 |\phi|_H^2 \geq c_1 |\phi|_V^2.$$

**(C)** Boundedness in $q$: There exists $c_2 > 0$ such that, for $q \in Q, \phi, \psi \in V$, we have

$$|\sigma(q)(\phi, \psi)| \leq c_2 |\phi|_V |\psi|_V.$$

Several remarks concerning these conditions are useful. First, we can weaken (A) slightly (see [BRR]): From (C) it follows that one can define $A(q)$ in $\mathcal{L}(V, V^*)$ by $\sigma(q)(\phi, \psi) = (-A(q)\phi, \psi)_{V^*, V}$ and then one can replace condition (A) by the requirement that $q \to A(q)$ be continuous from $Q$ to $\mathcal{L}(V, V^*)$.

Moreover, for $\sigma : V \times V \to \mathbb{C}$ defined using real coefficients, i.e. real scalars, it suffices to replace (B) by the condition: $\sigma(q)(\phi, \phi) + \lambda_0 |\phi|_H^2 \geq c_1 |\phi|_V^2$ for all real valued functions $\phi$ in $V$ (this is condition (B) as stated in [BI]). To see this, we note that $\sigma(q)(\phi + i\psi, \phi + i\psi) = \sigma(q)(\phi, \phi) + \sigma(q)(\psi, \psi) + i[\sigma(q)(\psi, \phi) - \sigma(q)(\phi, \psi)]$ and thus $\operatorname{Re}\sigma(q)(\phi + i\psi, \phi + i\psi) = \sigma(q)(\phi, \phi) + \sigma(q)(\psi, \psi)$. Therefore, $\sigma(q)(\phi, \phi) + \lambda_0 |\phi|_H^2 \geq c_1 |\phi|_V^2$ for all real valued $\phi$ in $V$ implies immediately that condition (B) above holds.

Under conditions (B) and (C), $\sigma$ defines operators $A(q)$ such that $\sigma(q)(\phi, \psi) = \langle -A(q)\phi, \psi \rangle$ for $\phi \in \operatorname{dom}(A(q)), \psi \in V$ with $\operatorname{dom}(A(q))$ dense in $V$ (e.g., see [S]). Moreover, $A(q)$ is a sectorial operator with $(\lambda I - A(q))\operatorname{dom}(A(q)) = H$ for all $\lambda$

with Re$\lambda \geq \lambda_0$; indeed $R_\lambda(A(q)) = (\lambda I - A(q))^{-1}$ exists as a bounded operator on $H$ for all $\lambda$ in the complement of a sector with $\lambda_0$ as a vertex. The operator $A(q)$ generates an analytic semigroup $T(t;q)$ which can be used in defining mild solutions of the system (2.4) or (2.1); i.e.,

$$u(t;q) = T(t;q)u_0(q) + \int_0^t T(t-s;q)F(s,q)ds. \tag{2.5}$$

Galerkin approximations (e.g., systems (2.2)) for (2.5) can be developed in the context of sesquilinear forms satisfying conditions (B) and (C). Let $H^N$ be a family of finite dimensional subspaces of $H$ satisfying $H^N \subset V$ and the condition

For each $\phi \in V$, there exists $\hat{\phi}^N \in H^N$ such that
$|\phi - \hat{\phi}^N|_V \to 0$ as $N \to \infty$. (C1)

(We note that many popular approximation schemes - e.g., linear splines and more generally, many finite element schemes - satisfy condition (C1).) To define $A^N(q) : H^N \to H^N$, we restrict $\sigma(q)$ to $H^N \times H^N$ and denote this restriction by $\sigma^N(q)$. Then $\sigma^N$ satisfies conditions (A), (B), (C) on $H^N \times H^N$ which implies existence of a bounded linear operator $A^N(q)$ on $H^N$ satisfying $\sigma^N(q)(\phi,\psi) = \langle -A^N(q)\phi,\psi \rangle$ for all $\phi, \psi \in H^N$. From (B) and (C) it follows that the $A^N(q)$ are uniformly dissipative (sectorial) and generate analytic semigroups $T^N(t;q)$ on $H^N$. Solutions of the systems (2.2) are then given by

$$u^N(t;q) = T^N(t;q)P^N u_0(q) + \int_0^T T^N(t-s;q)P^N F(s,q)ds. \tag{2.6}$$

It can then be established that under conditions (A), (B), (C) and (C1), we have $u^N(t;q^N) \to u(t;q)$ in $V$ for arbitrary sequences $\{q^N\}$ with $q^N \to q$ in $Q$. Observe that this immediately yields (2.3) if the operator $\mathcal{C}$ possesses certain boundedness properties. The focus of this note involves cases where the operator $\mathcal{C}$ does not have such boundedness.

The arguments for the convergence $u^N(t;q^N) \to u(t;q)$ for the situation here (complex Hilbert spaces $V, H$ and complex valued sesquilinear form $\sigma$) are essentially the same as those in [BI]. Slight changes in the arguments for Theorem 2.2 of [BI] are necessary to treat the case of complex valued inner products and sesquilinear forms (essentially one need only use Re $\sigma(\cdot,\cdot)$ and Re$\langle \cdot,\cdot \rangle$ in some of the inequalities). In these arguments one does not use directly the analyticity properties of the semigroups $T(t;q), T^N(t;q)$. Rather one relies heavily on resolvent estimates of Tanabe [T] depending on the V-coercivity of $\sigma$ along with a resolvent convergence form of the Trotter-Kato approximation theorem of linear semigroup theory (see [BI] for details).

Our interest here is mainly in *second order* systems of the form

$$\ddot{u}(t) + B(q)\dot{u}(t) + A(q)u(t) = f(t,q),$$
$$u(0) = u_0, \tag{2.7}$$
$$\dot{u}(0) = v_0,$$

where $A(q)$ is a generalized stiffness operator and $B(q)$ is a generalized damping operator. Again we consider this equation in a weak or variational sense defined via parameter dependent sesquilinear forms in a complex Hilbert space $V \hookrightarrow H = H^* \hookrightarrow V^*$. We are given a stiffness sesquilinear form $\sigma_1(q) : V \times V \to \mathbf{C}$ that is symmetric and satisfies the boundedness condition (C). Then there exists $A(q) \in \mathcal{L}(V, V^*)$ such that $\sigma_1(q)(\phi, \psi) = (A(q)\phi, \psi)_{V^*, V}$. We also have a damping sesquilinear form $\sigma_2(q) : V \times V \to \mathbf{C}$ which satisfies (C) so that there exists $B(q) \in \mathcal{L}(V, V^*)$ with $\sigma_2(q)(\phi, \psi) = (B(q)\phi, \psi)_{V^*, V}$. We then reformulate the system (2.7) and seek solutions $u(t) \in V$ satisfying for all $\phi \in V$

$$\begin{aligned}(\ddot{u}(t), \phi) + \sigma_2(q)(\dot{u}(t), \phi) + \sigma_1(q)(u(t), \phi) &= \langle f(t, q), \phi \rangle, \\ u(0) = u_0, \dot{u}(0) &= v_0.\end{aligned} \qquad (2.8)$$

As is standard practice, we rewrite this in first order vector form on $\mathcal{H} = V \times H$ and $\mathcal{V} = V \times V$ in the coordinates $(u, \dot{u})$. To this end, define $\sigma(q) : \mathcal{V} \times \mathcal{V} \to \mathbf{C}$ by

$$\sigma(q)((u, v), (\phi, \psi)) = -\langle v, \phi \rangle_V + \sigma_1(q)(u, \psi) + \sigma_2(q)(v, \psi) \qquad (2.9)$$

so that (2.8) may be rewritten as

$$\begin{aligned}(\dot{w}(t), \chi) + \sigma(q)(w(t), \chi) &= \langle F(t, q), \chi \rangle_\mathcal{H}, \\ w(0) &= (u_0, v_0)\end{aligned} \qquad (2.10)$$

for $w(t) = (u(t), \dot{u}(t))$ and $\chi = (\phi, \psi)$ in $\mathcal{V}$ with $F(t, q) = (0, f(t, q))$. Or, if as in the usual practice, we abuse notation and do not distinguish between row and column vectors, we may write this in equivalent operator form

$$\begin{aligned}\dot{w}(t) &= \mathcal{A}(q)w(t) + F(t, q), \\ w(0) &= (u_0, v_0)\end{aligned}$$

where $\sigma(q)(\chi, \xi) = (-\mathcal{A}(q)\chi, \xi)$ with

$$\mathcal{A}(q) = \begin{bmatrix} 0 & I \\ -A(q) & -B(q) \end{bmatrix}.$$

For our treatment here (as for that in [BI]) we assume that $\sigma_1(q)$ satisfies conditions (A), (B), and (C). We further assume that $\sigma_2(q)$ satisfies conditions (A) and (C). Then the strength of the coercivity assumption on $\sigma_2(q)$ determines the properties of the semigroup generated by $\mathcal{A}(q)$. For example, if $\sigma_2(q)$ satisfies (B), then $\mathcal{A}(q)$ is strongly $\mathcal{V}$-coercive and generates an analytic semigroup on $\mathcal{H}$. (Actually, $\mathcal{A}(q)$ is $\mathcal{V}_q$-coercive with $\mathcal{V}_q = V_q \times V_q$ where $V_q$ is $V$ taken with the equivalent inner product $\sigma_1(q)(\cdot, \cdot)$— see [BI].) In [BI], it is assumed only that $\sigma_2$ is $H$-semicoercive: There exists $b \geq 0$ such that for all $\phi \in V$ we have Re $\sigma(q)(\phi, \phi) \geq b|\phi|_H^2$. In this case one can argue only that $\mathcal{A}(q)$ generates a strongly continuous semigroup on $\mathcal{H}$. The theory for second order systems with this weak damping is developed in [BI] in order to treat several forms of damping (spatial hysteresis, time hysteresis, bending rate damping) which are of physical interest and yet do not satisfy the strong $V$-coercivity assumption. The convergence theory obtained yields that $u^N(t; q^N) \to u(t; q)$ in $V$ norm, $\dot{u}^N(t; q^N) \to \dot{u}(t; q)$ in $H$ norm whenever $q^N \to q$.

As opposed to [BI], we wish to consider in this paper the case where $\sigma_2(q)$ does satisfy the strong $V$-coercivity condition (B) and hence $\mathcal{A}(q)$ is the infinitesimal generator of an analytic semigroup. We obtain immediately that $\sigma(q)$ of (2.9) is $\mathcal{V}$-coercive and the first order theory outlined above can be applied directly to the system (2.10). This yields the convergence statement in $\mathcal{V}$ - i.e., $u^N(t; q^N) \to u(t; q)$ in $\mathcal{V}$, $\dot{u}^N(t; q^N) \to \dot{u}(t; q)$ in $\mathcal{V}$. A natural question arises as to whether we can use the analyticity of the semigroups in this case to obtain stronger results. The next section is devoted to results that yield an affirmative answer to this question.

## 3. Analytic semigroups and approximation

We first give a general approximation theorem for analytic semigroups that is a generalization of the well-known Trotter-Kato theorem [BK].

**Theorem 3.1.** *Suppose we have complex Hilbert spaces $X$ and $X^N$, $N = 1, 2 \ldots$, with $X^N \subset X$. Let $P^N : X \to X^N$ denote the orthogonal projection of $X$ onto $X^N$ satisfying $P^N \to I$ strongly. Suppose that $A^N$ and $A$ are the infinitesimal generators of analytic semigroups $S^N(t)$ and $S(t)$ on $X^N$ and $X$ respectively that satisfy the following:*

*There exists a region $\sum = \sum_\delta = \{\lambda \in \mathbf{C} : |\arg(\lambda - \lambda_0)| < \frac{\pi}{2} + \delta\}$, where $\delta > 0$, such that $\sum \cup \{\lambda_0\} \subset \rho(A) \cap \bigcap_{N=1}^\infty \rho(A^N)$ and*

(i) *there exists a constant $M$ independent of $N$ such that*

$$|R_\lambda(A^N)| \leq \frac{M}{|\lambda - \lambda_0|}$$

*for all $\lambda \in \sum$ and $N = 1, 2, \ldots$;*

(ii) *for some $\lambda \in \sum$ and each $x \in X$ we have $R_\lambda(A^N)P^N x \to R_\lambda(A)x$.*

*Then we have*

(iii) *for each $x \in X$, $S^N(t)P^N x \to S(t)x$ uniformly in $t$ on compact subintervals of $[0, \infty)$;*

(iv) *for each $x \in X$ and integer $k \geq 1$, $(A^N)^k S^N(t) P^N x \to A^k S(t)x$ uniformly in $t$ on compact subintervals of $(0, \infty)$.*

**Proof.** The statement of (iii) under the given conditions is just a variant of the well-known Trotter-Kato theorem and follows immediately from Theorem II.1.14 of [BK]. To argue result (iv), we first note the convergence in (ii) for some $\lambda \in \sum$ implies that the convergence holds for *all* $\lambda \in \sum$. In light of the resolvent bounds of (i), this follows from the identity for $\mu, \lambda \in \sum$

$$R_\mu(A^N)P^N - R_\mu(A) = [I + (\lambda - \mu)R_\mu(A^N)P^N][R_\lambda(A^N)P^N - R_\lambda(A)] \\ \times [I + (\lambda - \mu)R_\mu(A)]$$

which is readily established using the standard resolvent identity $R_\lambda(A) - R_\mu(A) = (\mu - \lambda)R_\lambda(A)R_\mu(A)$.

The analyticity of the semigroups $S^N(t)$ and $S(t)$ allow us to write for $t > 0$ (e.g., see [P])

$$A^k S(t) = \frac{1}{2\pi i} \int_\Gamma \lambda^k e^{\lambda t} R_\lambda(A) d\lambda$$

and

$$(A^N)^k S^N(t) P^N = \frac{1}{2\pi i} \int_\Gamma \lambda^k e^{\lambda t} R_\lambda(A^N) P^N d\lambda.$$

Here $\Gamma$ is a positively oriented contour through $\lambda_0$ lying in $\sum$ with $\arg(\lambda - \lambda_0) = \pm \nu$ for $\lambda \neq \lambda_0$ where $\nu$ is fixed in $(\frac{\pi}{2} + \delta - \varepsilon, \frac{\pi}{2} + \delta)$. The desired convergence results follow immediately from the inequality

$$|(A^N)^k S^N(t) P^N x - A^k S(t) x| \leq \frac{1}{2\pi} \int_\Gamma |\lambda|^k |e^{\lambda t}| |R_\lambda(A^N) P^N x - R_\lambda(A) x| d\lambda$$

using the resolvent convergence of (ii) by noting that the integrand in this integral is dominated by an integrable function.

We now return to the second order systems of Section 2 - see (2.8), (2.9), (2.10) - and use Theorem 3.1 to obtain our main convergence results. As before we take $\mathcal{H} = V \times H$ and $\mathcal{V} = V \times V$. Let $\mathcal{H}^N = H^N \times H^N$ and $P^N$ be the orthogonal projection of $\mathcal{H}$ onto $\mathcal{H}^N$.

**Theorem 3.2.** *Let $\sigma_1(q)$ and $\sigma_2(q)$ in (2.8) satisfy conditions (A), (B) and (C) and let $H^N \subset V$ satisfy condition (C1). Let $\{q^N\}$ be arbitrary in $Q$ with $q^N \to q$. Then we have*

(i) *The sesquilinear form $\sigma(q)$ given by (2.9) satisfies conditions (A), (B), (C) in the norms of $\mathcal{V}$ and $\mathcal{H}$ and the operator $\mathcal{A}(q)$ defined via $\sigma(q)(\chi, \xi) = \langle -\mathcal{A}(q)\chi, \xi \rangle_\mathcal{H}$ for $\chi \in \text{dom}(\mathcal{A}(q))$ is the infinitesimal generator of an analytic semigroup $\mathcal{T}(t; q)$ on $\mathcal{H}$.*

(ii) *Let $\mathcal{A}^N(q)$ denote the operator obtained by restricting $\sigma(q)$ to $\mathcal{H}^N \times \mathcal{H}^N$ and let $\mathcal{T}^N(t; q)$ denote the corresponding analytic semigroups on $\mathcal{H}^N$. Then we have*

(a) *For each $\chi \in \mathcal{H}, : \mathcal{T}^N(t; q^N) P^N \chi \to \mathcal{T}(t; q)\chi$ in $\mathcal{H}$ uniformly in $t$ compact subintervals of $[0, \infty)$;*

(b) *For each $\chi \in \mathcal{H}$ and positive integer $k$, $\mathcal{A}^N(q^N)^k \mathcal{T}^N(t; q^N) P^N \chi \to \mathcal{A}(q)^k \mathcal{T}(t; q) \chi$ in $\mathcal{H}$ uniformly in $t$ on compact subintervals of $(0, \infty)$.*

The proof of this theorem is now rather straightforward. We first note that it is readily shown with routine calculations that $\sigma_1, \sigma_2$ satisfying (A), (B), (C) in the $V$ and $H$ norms imply that $\sigma$ satisfies (A), (B), (C) with the norms of $\mathcal{V} = V \times V$ and $\mathcal{H} = V \times H$. Result (i) of the theorem thus is established.

For result (ii), we apply Theorem 3.1 with $X = \mathcal{H} = V \times H$ and $X^N = \mathcal{H}^N = H^N \times H^N$ along with arguments of [BI]. Restricting $\sigma(q^N)$ to $\mathcal{H}^N \times \mathcal{H}^N$ to obtain the operators $A^N = \mathcal{A}^N(q^N)$, we have that condition (B) is satisfied in $\mathcal{H}^N$ with the constants uniform in $N$ - i.e., the uniform sector condition and the uniform resolvent bounds in (i) of Theorem 3.1 are readily seen to hold. For the resolvent convergence of (ii) we refer to Theorem 2.2 of [BI], noting that we have all the hypotheses of that theorem holding here in the sense of the $\mathcal{V}$ and $\mathcal{H}$ norms. Hence the same arguments (modified slightly as mentioned in Section 2 above to account for complex Hilbert spaces $\mathcal{V}$ and $\mathcal{H}$) given for Theorem 2.2 of [BI] can be used here to establish (ii) of Theorem 3.1. The convergence statements (a) and (b) of Theorem 3.2 then follow directly from (iii) and (iv) of Theorem 3.1.

We are now in a position to use the results of Theorem 3.2 to address the question of stronger convergence results for $u^N(t; q^N)$ to $u(t; q)$ raised in connection with the observation operator $\mathcal{C}$ in Section 2. We first consider the case of homogeneous systems (i.e., $f = 0$ in (2.8)) in which case $w^N(t; q^N) = (u^N(t; q^N), \dot{u}^N(t; q^N)) = \mathcal{T}^N(t; q^N) P^N \chi$ and $w(t; q) = (u(t; q), \dot{u}(t; q)) = \mathcal{T}(t; q)\chi$. For $\chi \in \text{dom}(\mathcal{A}(q))$ we have $\dot{w}^N(t) = \mathcal{A}^N(q^N)\mathcal{T}^N(t; q^N)P^N\chi$, $\ddot{w}^N(t) = \mathcal{A}^N(q^N)^2\mathcal{T}^N(t; q^N)P^N\chi$, etc., while $\dot{w}(t) = \mathcal{A}(q)\mathcal{T}(t; q)\chi$, $\ddot{w}(t) = \mathcal{A}(q)^2\mathcal{T}(t; q)\chi$, etc. Thus, from statement (b) of Theorem 3.2 we find $\ddot{w}^N(t; q^N) \to \ddot{w}(t; q)$ in $\mathcal{H}$, which yields $u^N_{tt}(t; q^N) \to u_{tt}(t; q)$ in the $V$ norm, uniformly in $t$ on compact subintervals of $(0, \infty)$. Indeed, for all $k = 1, 2 \ldots$, we obtain $\frac{\partial^k}{\partial t^k} u^N(t; q^N) \to \frac{\partial^k}{\partial t^k} u(t; q)$ in $V$.

Since, as we shall see in the next section, many interesting examples involve $V \subset H^2(\Omega)$ with $\Omega \subset \mathbb{R}^1$ or $\Omega \subset \mathbb{R}^2$, and since in this case $V$ imbeds continuously in $C(\bar{\Omega})$, the above results guarantee pointwise (in $t$ and $x$) convergence of $u^N(q^N)$ and all its time deviatives to $u(q)$ and its time deviatives, respectively. Thus, observation operators related to laser vibrometers ($u_t$) and accelerometers ($u_{tt}$) are included in the convergence and method stability theory for output least squares inverse problems.

Similar results are available for the nonhomogeneous equation (2.8) (or (2.10)) if one places appropriate regularity conditions on $f$. To obtain these, one considers the representation theorems (mild solutions)

$$w^N(t; q^N) = \mathcal{T}^N(t; q^N) P^N w_0 + \int_0^t \mathcal{T}^N(t - s; q^N) P^N F(s, q) ds,$$

$$w(t; q) = \mathcal{T}(t; q) w_0 + \int_0^t \mathcal{T}(t - s; q) F(s, q) ds$$

and uses regularity conditions on mild solutions in $\mathcal{H} = V \times H$. For example, if $f(\cdot, q)$ is in $C^1([0, T], H)$, one can differentiate once, if $f(\cdot, q)$ is in $C^2([0, T], H)$, one can differentiate twice, etc. (see [P]). Indeed, since we have analytic semigroups, we can weaken the conditions on $f$ needed for this procedure by considering special regularity theorems for mild solutions of Cauchy initial value problems in the case of analyticity (e.g., see Chapter 4.3 of [P]). For example, it suffices to have $f(\cdot, q) \in L^1(0, T; H)$ and locally Hölder continuous in order to differentiate $w^N$ and $w$ once in the above representations.

## 4. Examples

In this section we present briefly several examples to which the above theory can be readily applied. In some cases this sharpens the convergence and stability results currently available in the literature; in other cases it provides new results for the associated inverse problems.

**Example 4.1.** We consider a cantilevered Euler-Bernoulli beam with Kelvin-Voigt damping (i.e., stress proportional to a linear combination of the strain and the strain rate). The beam is assumed fixed at $x = 0$ and free at $x = \ell$, with the transverse displacement at time $t$ and position $x$ given by $u(t, x)$. Typical observations consist of acceleration $u_{tt}(t, x_j)$ or velocity $u_t(t, x_j)$ at several locations $x_j$. Balance of forces and moments yield the following system (we assume the linear mass density $\rho$ is normalized to unity):

$$\frac{\partial^2 u}{\partial t^2} + \frac{\partial^2}{\partial x^2}\{EI\frac{\partial^2 u}{\partial x^2} + c_D I\frac{\partial^3 u}{\partial x^2 \partial t}\} = f(t, x) \quad 0 < x < \ell,$$

$$u(t, 0) = \frac{\partial u}{\partial x}(t, 0) = 0, \qquad (4.1)$$

$$[EI\frac{\partial^2 u}{\partial x^2} + c_D I\frac{\partial^3 u}{\partial x^2 \partial t}]_{x=\ell} = 0,$$

$$[\frac{\partial}{\partial x}\{EI\frac{\partial^2 u}{\partial x^2} + c_D I\frac{\partial^3 u}{\partial x^2 \partial t}\}]_{x=\ell} = 0.$$

The parameters to be estimated in typical examples (see [BWIC], [BFWIC], [BIn]) are the stiffness and damping coefficients, $EI$ and $c_D I$, respectively. For a parameter set we choose $Q$ compact in

$$\tilde{Q}_\nu = \{q = (EI, c_D I) : q \in L^\infty(0, \ell) \times L^\infty(0, \ell), EI(x) \geq \nu > 0, c_D I(x) \geq \nu\}.$$

For the state spaces $H$ and $V$ we take $H = H^0(0, \ell)$, $V = H^2_L(0, \ell) \equiv \{\phi \in H^2(0, \ell) : \phi(0) = \phi'(0) = 0\}$. Then in writing the system (4.1) in the form (2.8) we take

$$\sigma_1(q)(\phi, \psi) = \langle EID^2\phi, D^2\psi\rangle_0,$$
$$\sigma_2(q)(\phi, \psi) = \langle c_D ID^2\phi, D^2\psi\rangle_0,$$

where $D^2 = \frac{\partial^2}{\partial x^2}$ and $\langle \cdot, \cdot \rangle_0$ is the inner product in $H^0(0,\ell) = L^2(0,\ell)$. For real valued $\phi$ in $V$ we have at once

$$\sigma_i(q)(\phi,\phi) \geq \nu |D^2\phi|_0 \geq c_1|\phi|_V^2$$

which, as we have noted, implies

$$\text{Re } \sigma_i(q)(\phi,\phi) \geq c_1|\phi|_V^2$$

for complex valued $\phi$ in $V$. It is equally trivial to argue that conditions (A) and (C) of Section 2 hold for this $\sigma_1(q), \sigma_2(q)$. For Galerkin schemes satisfying condition (C1) we thus have Theorem 3.2 applicable, and hence parameter convergence and method stability hold for the least squares problems ($\mathcal{P}$) and ($\mathcal{P}^N$) whenever one formulates these with pointwise observations of either velocity or acceleration.

**Example 4.2** For this example we return to the so-called "RPL experiment" discussed in some detail in [BGRW]. The focus of our attention is a cantilevered Euler-Bernoulli beam with a flexible gas hose and thruster nozzle attached to the free end as depicted in Figure 2.2 of [BGRW]. The structure is modeled as a uniform cantilevered beam with Kelvin-Voigt internal damping and tip mass with a mass-spring-dashpot assembly attached at the tip. Along with the usual damped Euler-Bernoulli beam equation

$$\rho \frac{\partial^2 u}{\partial t^2} + EI \frac{\partial^4 u}{\partial x^4} + c_D I \frac{\partial^5 u}{\partial x^4 \partial t} = 0, \ 0 < x < \ell, \tag{4.2}$$

we have the force balance equation at the tip

$$[m_T \frac{\partial^2 u}{\partial t^2} - c_D I \frac{\partial^4 u}{\partial x^3 \partial t} - EI \frac{\partial^3 u}{\partial x^3}]_{x=\ell}$$
$$= c_H(\dot{y}(t) - \frac{\partial u}{\partial t}(t,\ell)) + k_H(y(t) - u(t,\ell)) + f(t) \tag{4.3}$$

and the hose assembly state equation

$$m_H \ddot{y}(t) + c_H(\dot{y}(t) - \frac{\partial u}{\partial t}(t,\ell)) + k_H(y(t) - u(t,\ell)) = 0. \tag{4.4}$$

Here $\rho, EI,$ and $c_D I$ are the usual beam parameters whereas $m_T$ represents the tip mass, $m_H$ is the hose mass, $k_H$ is the hose stiffness (the "spring" or restoring force constant), $c_H$ is the hose damping coefficient and $f(t)$ represents an externally applied force at the tip (firing of the tip mounted thrusters). Boundary conditions for the coupled state equations (4.2) and (4.4) include the tip force balance equation (4.3), the tip moment balance equation (assuming that the hose assembly has negligible rotational inertia)

$$[EI \frac{\partial^2 u}{\partial x^2} + c_D I \frac{\partial^3 u}{\partial x^2 \partial t}]_{x=\ell} = 0, \tag{4.5}$$

and the zero displacement, zero slope conditions at the fixed end $x = 0$

$$u(t,0) = \frac{\partial u}{\partial x}(t,0) = 0. \tag{4.6}$$

The structure is assumed initially at rest so that initial conditions are given by

$$u(0,x) = \frac{\partial u}{\partial t}(0,x) = 0,$$

$$y(0) = \dot{y}(0) = 0.$$

The parameters to be estimated using accelerometer observations (see [BGRW]) include $q = (m_T, EI, c_D I, m_H, c_H, k_H)$, which is to be chosen from a compact subset $Q \subset \mathbb{R}_+^6$.

To write the system (4.2)-(4.6) in weak or variational form, we use the state variable $\hat{u}(t) = (y(t), u(t, \ell), u(t, \cdot))$ in the state space $H = \mathbb{R}^2 \times H^0(0, \ell)$ with inner product

$$\langle (\zeta, \eta, \phi), (\lambda, \mu, \psi) \rangle_H = \zeta\lambda + \eta\mu + \langle \phi, \psi \rangle_0.$$

For the space $V$ we choose $V = \{(\zeta, \eta, \phi) \in H : \phi \in H^2(0, \ell), \phi(0) = D\phi(0) = 0, \eta = \phi(\ell)\}$ with inner product

$$\langle (\zeta, \phi(\ell), \phi), (\lambda, \psi(\ell), \psi) \rangle_V = (\zeta - \phi(\ell))(\lambda - \psi(\ell)) + \langle D^2\phi, D^2\psi \rangle_0.$$

The stiffness and damping sesquilinear forms are given for $\hat{\phi} = (\zeta, \phi(\ell), \phi), \hat{\psi} = (\lambda, \psi(\ell), \psi)$ in $V$ by

$$\sigma_1(q)(\hat{\phi}, \hat{\psi}) = k_H(\zeta - \phi(\ell))(\lambda - \psi(\ell)) + EI \langle D^2\phi, D^2\psi \rangle_0$$

$$\sigma_2(q)(\hat{\phi}, \hat{\psi}) = c_H(\zeta - \phi(\ell))(\lambda - \psi(\ell)) + c_D I \langle D^2\phi, D^2\psi \rangle_0.$$

We also need an operator $\mathcal{M}(q) \in \mathcal{L}(H)$ given by

$$\mathcal{M}(q)(\zeta, \eta, \phi) = (m_H \zeta, m_T \eta, \rho\phi).$$

This operator can be extended to $\mathcal{L}(V^*)$ in an obvious manner. Then the system (4.2)-(4.6) can be written in variational form for the state $\hat{u}(t) \in V$ to satisfy for all $\hat{\phi} \in V$

$$\begin{aligned}(\mathcal{M}(q)\hat{u}_{tt}(t), \hat{\phi}) + \sigma_2(q)(\hat{u}_t(t), \hat{\phi}) + \sigma_1(q)(\hat{u}(t), \hat{\phi}) &= \langle F(t), \hat{\phi} \rangle, \\ \hat{u}(0) = \hat{u}_t(0) &= 0,\end{aligned} \tag{4.7}$$

where $F(t) = (0, f(t), 0)$. Since, for $m_H, m_T, \rho$ positive, the operator $\mathcal{M}(q)$ is invertible, equation (4.7) is obviously equivalent to an equation of the form (2.8). Thus the theory of Sections 2 and 3 is applicable if $\sigma_1$ and $\sigma_2$ satisfy the requisite hypotheses. If $Q$ is bounded below in $\mathbb{R}_+^6$, it is readily seen that both $\sigma_1$ and

$\sigma_2$ satisfy conditions (A), (B), and (C). For example, we see immediately for $\hat{\phi} = (\zeta, \phi(\ell), \pi) \in V$

$$\sigma_2(q)(\hat{\phi}, \hat{\phi}) = c_H(\zeta - \phi(\ell))^2 + c_D I |D^2\phi|_0^2$$
$$\geq c_1\{(\zeta - \phi(\ell))^2 + |D^2\phi|_0^2\} = c_1|\hat{\phi}|_V^2.$$

Similar arguments hold for $\sigma_1(q)$.

Application of the theory in Sections 2 and 3 substantially sharpens the results given in [BGRW]. In that paper the main results (see Lemma 3.1) yield

$$\int_0^T |\hat{u}_{tt}^N(t; q^N) - \hat{u}_{tt}(t; q)|_H dt \to 0 \tag{4.8}$$

as $N \to \infty$. The arguments are rather tedious and require the assumption $\hat{u}(q) \in H^2(0, T; V)$ on the limit function. Note also that (4.8) would require continuous time acceleration observations be used in the least squares criterion.

In contrast, the theory of this paper yields (since $H^2$ embeds compactly in C) $u^N(t, x; q^N) \to u(t, x; q)$, $\dot{u}^N(t, x; q^N) \to \dot{u}(t, x; q)$ uniformly in $x \in [0, \ell]$ for each $t \in [0, T]$ as well as $\hat{u}_{tt}^N(t; q^N) \to \hat{u}_{tt}(t; q^N)$ in $V$ for each $t$, which permits sampled time acceleration observations. The arguments are simple (given the theory developed above) and do not require the a priori smoothness assumption on $\hat{u}(q)$.

**Example 4.3** As a final example, we briefly describe the models for two dimensional grid structures ("plates with holes") developed and investigated in [R] and [BR]. We use Love-Kirchoff plate theory with Kelvin-Voigt damping. The "plate" is rectangular, e.g., on $(x, y) \in [0, \ell_1] \times [0, \ell_2]$, with rectangular holes periodically placed to produce a thin planar grid. It is assumed to be hanging vertically, clamped at the top, with the other three edges free. The basic equation for transverse displacements $w(t, x, y)$ is given by

$$\rho h \frac{\partial^2 w}{\partial t^2} + \frac{\partial^2 M^x}{\partial x^2} + 2\frac{\partial^2 M^{xy}}{\partial x \partial y} + \frac{\partial^2 M^y}{\partial y^2} = f$$

where $h$ is the thickness of the plate, the bending moments $M^x, M^y$ about the $x$ and $y$ axes, respectively, are given by

$$M^x = \frac{EI}{1-\nu^2}\left\{\frac{\partial^2 w}{\partial x^2} + \nu\frac{\partial^2 w}{\partial y^2}\right\} + \frac{c_D I}{1-\nu^2}\left\{\frac{\partial^3 w}{\partial x^2 \partial t} + \nu\frac{\partial^3 w}{\partial y^2 \partial t}\right\},$$

$$M^y = \frac{EI}{1-\nu^2}\left\{\frac{\partial^2 w}{\partial y^2} + \nu\frac{\partial^2 w}{\partial x^2}\right\} + \frac{c_D I}{1-\nu^2}\left\{\frac{\partial^3 w}{\partial y^2 \partial t} + \nu\frac{\partial^3 w}{\partial x^2 \partial t}\right\},$$

and the twisting moment is given by

$$M^{xy} = \frac{EI}{1-\nu^2}\left\{\frac{\partial^2 w}{\partial x \partial y} - \nu\frac{\partial^2 w}{\partial x \partial y}\right\} + \frac{c_D I}{1-\nu^2}\left\{\frac{\partial^3 w}{\partial x \partial y \partial t} - \nu\frac{\partial^3 w}{\partial x \partial y \partial t}\right\}.$$

Here $\nu$ is Poisson's ratio. For the boundary conditions at the top (clamped) we have the essential boundary conditions $w = \frac{\partial w}{\partial y} = 0$ (the top corresponds to the $x-$axis). The plate is free on the other outer edges and on the edges of the holes, where the natural boundary conditions of zero moment and zero shear are required. For example, on a free edge parallel to the $y-$axis this results in the conditions

$$M^x = 0, \ \frac{\partial M^x}{\partial x} + 2\frac{\partial M^{xy}}{\partial y} = 0.$$

As shown in [R], [BR], one can readily define the corresponding stiffness and damping sesquilinear forms $\sigma_1, \sigma_2$ on $V \times V$, where $V = \{\phi \in H^2(\Omega) : \phi = \frac{\partial \phi}{\partial y} = 0$ along $y = 0\}$, $\Omega$ is $[0, \ell_1] \times [0, \ell_2]$ less the holes, and argue the needed $V-$coerciveness along with conditions (A) and (C). In this case, the state space is $H = H^0(\Omega)$. The theory of Sections 3 and 4 can thus be shown to hold for this example.

# References

[B] Banks, H.T., *On a variational approach to some parameter estimation problems*, Distributed Parameter Systems, vol. 75, Springer L.N. in Control and Info. Sci., 1985, pp. 1–23.

[BGRW] Banks, H.T., Gates, S.S., Rosen, I.G. and Wang, Y., *The identification of a distributed parameter model for a flexible structure*, SIAM J. Control Opt. **26** (1988), 743–762.

[BI] Banks, H.T. and Ito, K., *A unified framework for approximation in inverse problems for distributed parameter systems*, Control-Theory and Adv. Tech. **4** (1988), 73–90.

[BIn] Banks, H.T. and Inman, D.J., *On damping mechanisms in beams*, CAMS Rep.# 89-3, University of Southern California, September 1989; ASME J. Applied Mech., submitted.

[BK] Banks, H.T. and Kunisch, K., *Estimation Techniques for Distributed Parameter Systems*, Birkhäuser, Boston, 1989.

[BR] Banks, H.T. and Rebnord, D.A., *Estimation of material parameters for grid structures* (to appear).

[BRR] Banks, H.T., Reich, S. and Rosen, I.G., *Estimation of nonlinear damping in second order distributed parameter systems*, CAMS Rep.#89-1, University of Southern California, September, 1989; Control-Theory and Adv. Tech., submitted.

[BFWIC] Banks, H.T., Fabiano, R.H., Wang, Y., Inman, D.J. and Cudney, H.H., *Spatial versus time hysteresis in damping mechanisms*, Proc. 27th IEEE Conf. Dec. Control, vol. 1988, pp. 1674–1677.

[BWIC] Banks, H.T., Wang, Y., Inman, D.J. and Cudney, H.H., *Parameter identification techniques for the estimation of damping in flexible structures experiments*, Proc. 26th IEEE Conf. Dec. Control, vol. 1987, pp. 1392–1395.

[P] Pazy, A., *Semigroups of Linear Operators and Applications to Partial Differential Equations*, Springer-Verlag, New York, 1983.

[R] Rebnord, D.A., *Parameter estimation for two-dimensional grid structures*, PhD. Thesis, Brown University, May, 1989.

[S] Showalter, R.E., *Hilbert Space Methods for Partial Differential Equations*, Pitman, London, 1979.

[T]     Tanabe, H., *Equations of Evolution*, Pitman, London, 1979.


H.T. Banks
Center for Applied Mathematical Sciences
University of Southern California
Los Angeles, CA 90089-1113, USA

D.A. Rebnord
Department of Mathematics
Iowa State University
Ames, IA 50011, USA

# A Maximum Principle for Semilinear Parabolic Network Equations

JOACHIM VON BELOW

Lehrstuhl für Biomathematik, Universität Tübingen

Reaction-diffusion equations and interaction phenomena on ramified networks with Kirchhoff type connecting operators have been investigated recently by several authors, cf. [1-11]. In this paper we present a strong maximum principle and an a priori estimate for semilinear parabolic network equations with excitatoric Kirchhoff laws in the ramification nodes. The results presented here extend those of [3, Chap.2] and [6] and include the proofs of the Lemma and Theorem 2 in [6]. Existence results for semilinear parabolic network equations can be found in [3] and [7].

Let $G$ denote a $C^2$-network with finite sets of vertices $E = \{E_i : 1 \leq i \leq n\}$ and edges $K = \{k_j : 1 \leq j \leq N\}$ as defined in [2, Chap.1]. Thus $G$ is the union of Jordan curves $k_j$ in $\mathbb{R}^m$ with arc length parametrizations $\pi_j \in C^2([0, l_j], \mathbb{R}^m)$. The arc length parameter of an edge $k_j$ is denoted by $x_j$. The topological graph $\Gamma$ belonging to $G$ is assumed to be simple and connected. Thus, by definition, $\Gamma = (E, K)$ consists in a collection of $N$ Jordan curves $k_j$ with the following properties: Each $k_j$ has its endpoints in the set $E$, any two vertices in $E$ can be connected by a path with arcs in $K$, and any two edges $k_j \neq k_h$ satisfy $k_j \cap k_h \subset E$ and $|k_j \cap k_h| \leq 1$. The valency of each vertex is denoted by $\gamma_i = \gamma(E_i)$. We distinguish the ramification nodes $\text{Int } E = \{E_i \in E : \gamma_i > 1\}$ from the boundary vertices $\partial G = \{E_i \in E : \gamma_i = 1\}$. The orientation of $\Gamma$ is given by the incidence matrix $D = (d_{ij})_{n \times N}$ with

$$d_{ij} = \begin{cases} 1 & \text{if } \pi_j(l_j) = E_i, \\ -1 & \text{if } \pi_j(0) = E_i, \\ 0 & \text{otherwise.} \end{cases}$$

Endowed with the induced topology $G$ is a connected and compact space in $\mathbb{R}^m$.

We introduce $t$ as the time variable and for $T > 0$

$$\Omega = G \times [0, T],$$
$$\Omega_j = [0, l_j] \times [0, T],$$
$$\Omega_p = (G \backslash \partial G) \times (0, T],$$
$$\Omega_{jp} = (0, l_j) \times (0, T],$$
$$\omega_p = (G \times \{0\}) \cup (\partial G \times (0, T]),$$
$$\Omega_{jp}^{\bullet} = \Omega_{jp} \cup \Big( \{\xi \in [0, l_j] | \pi_j(\xi) \in Int\, E\} \times (0, T] \Big),$$

and use the abbreviations $u_j = u \circ (\pi_j, id) : \Omega_j \to \mathbb{R}$ for $u : \Omega \to \mathbb{R}$ and $u_j(E_i, t) = u_j(\pi_j^{-1}(E_i), t)$ etc. Set $C^{2,1}(\Omega) = \{u \in C(\Omega) | u_j \in C^{2,1}(\Omega_j), 1 \leq j \leq N\}$.

On each edge $k_j$ we consider the semilinear equation

$$u_{jt} = a_j(x_j, t, u_j, u_{jx_j}) u_{jx_j x_j} + f_j(x_j, t, u_j, u_{jx_j}) =: D_j[u_j], \quad (1)$$

subject to the strict parabolicity condition

$$\exists \mu_1, \mu_2 \in \mathbb{R} \, \forall j \in \{1, \ldots, N\}, \, \forall (x, t, z, p) \in \Omega_j \times \mathbb{R}^2 :$$

$$0 < \mu_1 \leq a_j(x, t, z, p) \leq \mu_2. \quad (2)$$

At the ramification nodes $E_i$ we impose continuity conditions and a classical Kirchhoff law

$$\sum_{j=1}^{N} d_{ij} c_{ij}(t) u_{jx_j}(E_i, t) = 0 \quad \text{for } \gamma_i \geq 2 \text{ and } 1 \leq i \leq n \quad \text{(K)}$$

with positive conductivity functions $c_{ij}$, or more generally the excitatoric Kirchhoff condition with coefficient functions $c_{ij}, \rho_i, \sigma_i : [0, T] \to \mathbb{R}$

$$V_i(u, t) := \rho_i(t) u(E_i, t) - \sum_{j=1}^{N} d_{ij} c_{ij}(t) u_{jx_j}(E_i, t) - \sigma_i(t) u_t(E_i, t) = 0 \quad \text{(GK)}$$

$$\text{for } \gamma_i \geq 2 \text{ and } 1 \leq i \leq n$$

subject to

$$c_{ij}(t) > 0 \text{ for } t \in [0,T], i \in \{1,\ldots,n\} \text{ and } j \in \{1,\ldots,N\}$$

and

$$\sigma_i(t) \geq 0 \quad \text{for} \quad t \in [0,T] \quad \text{and} \quad i \in \{1,\ldots,n\}. \tag{3}$$

Note that (3) plays the rôle of a parabolicity condition in $Int > E$. A maximum principle with respect to the parabolic boundary $\omega_p$ cannot hold, when (2) is violated. An easy example is given by the solution $u \in C^{2,1}(\Omega)$ with $u_j(x_j,t) = \exp(t-x_j)$ of $u_{jt} = u_{jx_jx_j}$ on the edges of a star graph $\Gamma$ with $\{E_1\} = IntE$, $N > 1$, $d_{1j} = -1$ for all edges, and the condition $(GK)$ $Nu_t(E_1,t) - \sum_{j=1}^N d_{1j}u_{jx_j}(E_1,t) = 0$. Furthermore, the coefficient of the flow term $\sum d_{ij}c_{ij}u_{jx_j}$ can be set equal to 1, since the condition $\sigma_i u_t = \rho_i u$ corresponds to a Dirichlet boundary condition at $\gamma_i$ boundary vertices. For $\sigma_i \neq 0$, $(GK)$ is not well-posed in sense of Solonnikov, cf. [5, Chap.8], but, of course, it has a natural physical interpretation, cf. [4], [8]. For function spaces on $\Omega' \subseteq \Omega$ let the indices $K$ and $GK$ indicate the validity of $K$ and $GK$ in $\Omega'$, respectively.

The basic technique, analogous to the classical one for domains established in [12, Chap.24], is comprehended in the following

**Lemma.** *Let $\phi, \psi \in C(\Omega) \cap C^{2,1}(\Omega_p)$ satisfy*

$$V_i(\psi,t) \leq V_i(\varphi,t) \quad \text{for all} \quad t \in (0,T] \quad \text{and for all} \quad E_i \in IntE, \tag{4}$$

*and the test point implication:*

$$\begin{array}{c} \text{If } \varphi_j = \psi_j, \quad \varphi_{jx_j} = \psi_{jx_j}, \quad \varphi_{jx_jx_j} \leq \psi_{jx_jx_j} \text{ at a point in } \Omega_p, \text{ then} \\ \varphi_{jt} < \psi_{jt} \text{ at this point.} \end{array} \tag{5}$$

*Then precisely one of the following cases holds:*

(a) $\varphi < \psi$ in $\Omega_p$
(b) *There exists a maximal $t^* \in [0,T)$ such that $\varphi < \psi$ in $(G \setminus \partial G) \times (0,t^*]$. Thus there is a sequence $\{(y_k,t_k)|k \in \mathbb{N}\} \subset \Omega_p$ with $t_k > t^*$ and $\varphi(y_k,t_k) \geq \psi(y_k,t_k)$ for all $k \in \mathbb{N}$ with $\lim_{k\to\infty} t_k = t^*$ and $\lim_{k\to\infty}(y_k,t_k) \in \omega_p$.*

**Proof.** In the case $IntE = \emptyset$ the assertion is a special case of [12. Lemma 24.I]. Next, we consider the case

$$\Gamma \text{ is a star graph with} \quad |\partial G| = N \quad \text{and} \quad IntE = \{e\}. \tag{*}$$

Introduce

$$t^* = sup\{\tau \in [0,T]: \phi < \psi \text{ on } (G \setminus \partial G) \times (0,\tau)\}.$$

Then $\phi \leq \psi$ on $H := \Omega_p \cap (G \times \{t^*\})$. (Note that $H = \emptyset$ is possible.) Suppose that $\phi(\tilde{x}, t^*) = \psi(\tilde{x}, t^*)$ for some $(\tilde{x}, t^*) \in H$. Then $\psi - \phi$ attains a minimum at $\tilde{x}$ as a function on $H$ and, by definition of $t^*$,

$$0 \geq \lim_{\theta \to t^*, \theta < t^*} \frac{\psi(\tilde{x}, t^*) - \phi(\tilde{x}, t^*) - \psi(\tilde{x}, \theta) + \phi(\tilde{x}, \theta)}{t^* - \theta} = \psi_t(\tilde{x}, t^*) - \phi_t(\tilde{x}, t^*).$$

If $\tilde{x}$ is an interior point of some edge $k_j$, then $\phi_{jt}(\tilde{x}, t) < \psi_{jt}(\tilde{x}, t)$ by condition (5), which is a contradiction.

For $\tilde{x} = e = E_i$ we have $d_{ij}(\psi_{jx_j}(e, t^*) - \phi_{jx_j}(e, t^*)) \leq 0$ and by (4)

$$\psi_t(e, t^*) = \phi_t(e, t^*) \quad \text{and} \quad \psi_{jx_j}(e, t^*) = \varphi_{jx_j}(e, t^*) \quad \text{for all } j.$$

Regarding $\psi - \varphi$ as a function on each pair of incident edges, $\psi - \varphi$ is continuously differentiable at $\tilde{x} = e$ and $\varphi_{jx_jx_j}(e, t^*) \leq \psi_{jx_jx_j}(e, t^*)$ for all $j$. By (5) we conclude $\varphi_t(e, t^*) < \psi_t(e, t^*)$, which again is a contradiction.

Thus we have shown that $\varphi < \psi$ on $H$. Due to the maximality of $t^*$ either case (a) or case (b) holds corresponding to $t^* = T$ or $t^* < T$.

In the case $|IntE| \geq 2$, we consider the stars

$$S_i = \bigcup_{d_{ij} \neq 0} k_j \quad \text{for} \quad E_i \in IntE.$$

Applying the case (*) to $S_i, \Delta_i := S_i \times [0, T], \varphi \mid_{\Delta_i}$ and $\psi \mid_{\Delta_i}$, we find numbers $t_i^*$ as obtained in (*) and set

$$t^* = \min\{t_i^* : \gamma_i \geq 2\}.$$

In the case $t^* = T$, (a) holds. For $t^* < T$ there is a vertex $E_s \in IntE$ and a sequence $\{(y_k, t_k) : k \in \mathbb{N}\} \subset \Delta_{sp}$ with

$$\lim_{k \to \infty} t_k = t_s^* = t^* \quad \text{and} \quad t_k > t^*, \phi(y_k, t^*) \geq \psi(y_k, t^*) \quad \text{for all} \quad k \in \mathbb{N},$$

such that $\lim_{k \to \infty}(y_k, t_k) = (y, t^*)$ lies on the parabolic boundary $\delta_{sp}$ of $\Delta_s$. If $t^* > 0$, then the construction of $H$ and the continuity of $\varphi$ and $\psi$ imply that $y$ cannot be a ramification node. Thus $y$ has to be a boundary vertex of $\Gamma$. If $t^* = 0$, then $(y, 0) \in \omega_p$. In both cases the assertion (b) is shown.

Immediate applications are the comparison principle [6, Theorem 1] for general parabolic network equations with condition (7) below and weak maximum-minimum principles with respect to the parabolic boundary $\omega_p$. Here we show the following strong maximum-minimum principle.

**Theorem 1.** *Suppose there is some constant $c > 0$ such that all $f_j : \Omega_j \times \mathbb{R}^2 \to \mathbb{R}$ satisfy*

$$f_j(x,t,z,p) \leq c|p| \quad \left(f_j(x,t,z,p) \geq -c|p|\right) \quad \text{in} \quad \Omega_j \times \mathbb{R}^2. \tag{6}$$

*Assume that all $\rho_i$ in (GK) vanish. Let $u \in C(\Omega) \cap C^{2,1}_{GK}(\Omega_p)$ be a solution of*

$$u_{jt} \leq D_j[u_j] \quad \text{in} \quad \Omega_{jp} \quad \left(u_{jt} \geq D_j[u_j] \quad \text{in} \quad \Omega_{jp}\right)$$

*for all $j \in \{1,\ldots,N\}$. Then*

$$\max_\Omega u = \max_{\omega_p} u \quad \left(\min_\Omega u = \min_{\omega_p} u\right)$$

*More precisely: If $u$ attains its maximum $M$ (its minimum $m$) at some point $(x_o, t_o)$ in $\Omega_p$, then $u = M$ $(u = m)$ in $G \times [0, t_0]$.*

**Proof.** Suppose $u(x_0, t_0) = M$ at some point $(x_0, t_0) \in \Omega_p$.
*Case 1:* $x_0 \notin E$. Thus $x_0$ is an interior point of some edge $k_j$. Then by the strong maximum principle for domains [12, Chap. 26] we find

$$u_j(x_j, t) = M \quad \text{for all} \quad (x_j, t) \in [0, l_j] \times [0, t_0].$$

For an arbitrary edge $k_h$ with $k_h \cap k_j = \{E_i\} \subset E$ and for $t \in (0, t_0]$ we conclude

(α) $u_h(E_i, t) = M$, since $u \in C(\Omega)$.
(β) $u_{hx_h}(E_i, t) = u_{jx_j}(E_i, t) = 0$ by (GK).
(γ) $u_{hx_hx_h}(E_i, t) \leq 0$ (By (α) and (β) $u$ is continuously differentiable at $(E_i, t)$ as a function on $k_j \cup k_h$.)
(δ) $u_{hx_hx_h}(E_i, t) = 0$. (Condition (6), $u \in C^{2,1}(\Omega_p)$ and $u_{ht} \leq D_h[u_h]$ imply $\lim_{\xi \to 0} u_{hx_hx_h}\left(l_h \frac{1+d_{ih}}{2} - \xi d_{ih}, t\right) \geq 0$.)

Define

$$z(\xi, t) = \begin{cases} u_j(l_j \frac{1+d_{ij}}{2} - \xi d_{ij}, t) & \text{for } 0 \leq \xi \leq l_j, \\ u_h(l_h \frac{1+d_{ih}}{2} + \xi d_{ih}, t) & \text{for } -l_h \leq \xi \leq 0. \end{cases}$$

Then $z \in C([-l_h, l_j] \times [0, T]) \cap C^{2,1}((-l_h, l_j) \times (0, T))$ is a solution of the differential inequality

$$z_t \leq a(\xi, t, z, z_\xi) z_{\xi\xi} + f(\xi, t, z, z_\xi),$$

where $a$ and $f$ are defined with the same argument transformation in $x$ by

$$a = \begin{cases} a_j & \text{for } \xi > 0, \\ a_h & \text{for } \xi < 0, \\ \mu_1 & \text{for } \xi = 0, \end{cases} \qquad f = \begin{cases} f_j & \text{for } \xi > 0, \\ f_h & \text{for } \xi < 0, \\ c & \text{for } \xi = 0. \end{cases}$$

(Note that $p$ is replaced by $-d_{ij}p$ or $d_{ih}p$.)
At $(0, t_0)$, $z$ attains its maximum, so that $u_h = M$ for $t \leq t_0$ by the classical strong maximum principle. Since $\Gamma$ is connected, the assertion is shown in the case 1.

*Case 2:* $x_0 = E_i \in Int E$. If $u \equiv M$ in $\{E_i\} \times [0, t_0]$ then we can proceed as in the case 1. Thus we may assume that there is $t_1 \in (0, t_0)$ such that $u(x_0, t_1) < M$. Using the case 1 we conclude that $u < M$ in $\cup_{j=1}^N \Omega_{jp}$. We may assume $d_{ij} = -1$ for all edges $k_j$ with $E_i \in k_j$. Choose $a > 0$ with $a < \min_j l_j$ and introduce

$$\Sigma = \{(\pi_j(x_j), t) \in \Omega : t_1 \leq t \leq t_0, \quad x_j \leq a, \quad d_{ij} \neq 0, \quad 1 \leq j \leq N\},$$
$$\Sigma_p = \{(\pi_j(x_j), t) \in \Omega : t_1 < t \leq t_0, \quad x_j < a, \quad d_{ij} \neq 0, \quad 1 \leq j \leq N\},$$
$$\sigma_p = \Sigma \setminus \Sigma_p,$$
$$A(\xi) = a^2 - \xi^2, \quad \alpha = \max\{\mu_2, c\},$$
$$w(\xi, t) = A^2(\xi) e^{-Bt}$$

with $B > \max\{1, 4a^{-2}\alpha\}$ sufficiently large such that

$$8\mu_1 \xi^2 + BA^2(\xi) - 4A(\xi)\alpha(1 + \xi) > 0 \quad \text{for} \quad 0 \leq \xi \leq a.$$

Then $w$ satisfies the strict differential inequality

$$w_t < \alpha w_{\xi\xi} - \alpha |w_\xi| \quad \text{in} \quad [0, a] \times [0, T]. \tag{$\varepsilon$}$$

Define $\varphi = u \mid_\Sigma$ and $\psi : \Sigma \to \mathbb{R}$ via

$$\psi_j(x_j, t) = M + \varepsilon - \zeta w(x_j, t - t_1).$$

Then $\psi \in C_K^{2,1}(\Sigma)$ and $d_{ij}\psi_{jx_j} = 0$ in $\{E_i\} \times [t_1, t_0]$. Since $u < M$ on $\sigma_p$, we can choose $\zeta > 0$ sufficiently small such that

$$\psi > M - \zeta w \geq u \quad \text{on} \quad \sigma_p.$$

Then $\varphi$ and $\psi$ satisfy (4)

$$V_i(\psi, t) = -B\sigma_i(t) a^4 e^{-B(t-t_1)} \leq 0 = V_i(\varphi, t) \quad \text{for all} \quad t \in [t_1, t_0].$$

Next suppose $\varphi_j = \psi_j, \varphi_{jx_j} = \psi_{jx_j}$ and $\varphi_{jx_jx_j} \leq \psi_{jx_jx_j}$ at some point $(y, s) \in \Sigma_p$. If $y \neq 0$ then by $(\varepsilon)$

$$\varphi_{jt}(y, s) \leq D_j[u_j](y, s) \leq \mu_2 \psi_{jx_jx_j}(y, s) + c|\psi_{jx_j}(y, s)|$$
$$= -\zeta(\mu_2 w_{\xi\xi}(y, s - t_1) - c|w_\xi(y, s - t_1)|)$$
$$< -\zeta w_t(y, s - t_1) = \psi_{jt}(y, s).$$

For $y = 0$ we have
$$\psi_{jt}(E_i, s) = B\zeta a^4 e^{-B(s-t_1)} > \mu_2 \zeta 4a^2 e^{-B(s-t_1)} = \mu_2 \psi_{jx_j x_j}(E_i, s) > 0.$$
In the nontrivial case $0 < \varphi_{jt}(E_i, s), \varphi_{jx_j x_j}(x_j, s)$ has to be positive near $E_i$ due to $\varphi \in C^{2,1}(\Sigma_p), \varphi_{jx_j}(E_i, s) = 0$ and
$$\varphi_{jt}(x_j, s) \le \mu_2 \varphi_{jx_j x_j}(x_j, s) + \alpha|\varphi_{jx_j}(x_j, s)| \quad \text{for all} \quad x_j \in (0, l_j),$$
and furthermore,
$$\varphi_{jt}(E_i, s) \le \alpha \varphi_{jx_j x_j}(E_i, s) \le \alpha \psi_{jx_j x_j}(E_i, s) < \psi_{jt}(E_i, s).$$
Thus $\varphi$ and $\psi$ satisfy the test point implication (5). Applying the Lemma to $\varphi, \psi$, and $\Sigma$ yields $\varphi < \psi$ in $\Sigma_p$, and letting $\varepsilon$ tend to zero shows
$$M - \zeta w \ge u \quad \text{in} \quad \Sigma_p.$$
But at $(E_i, t_0)$ this leads to the contradiction $M - \zeta a^4 e^{-B(t_0-t_1)} \ge M$. Therefore $u$ cannot attain its maximum $M$ at $(x_o, t_0)$ unless $u \equiv M$ in $G \times [0, t_0]$.

The minimum case is shown similarly with $\varphi = m - \varepsilon + \zeta w$.
$\diamond$

In the case $\rho_i(t_0) > 0$ $(\rho_i(t_0) < 0)$, $u \in C_{GK}^{2,1}(\Omega_p)$ cannot have a negative maximum or a positive minimum (a positive maximum or a negative minimum) at $(E_i, t_0)$. More precisely: The following example shows that a maximum principle with respect to $\omega_p$ cannot hold when $\rho_i(t_0) > 0$, no matter what $\sigma_i(t_0)$ or $\rho_i(t_0)(\sigma_i(t_0))^{-1}$ amount to. On a star graph $\Gamma$ with $\text{Int} E = \{E_1\}$, $N > 1$, $d_{1j} \equiv -1$ consider $u \in C^{2,1}(\Omega)$ with $u_j(x_j, t) = \exp(\lambda t - \nu x_j)$. On the edges $u$ satisfies $u_{jt} = \lambda \nu^{-2} u_{jx_j x_j}$, and at $E_1$ the condition
$$0 = (N\nu + \varepsilon)u(E_1, t) - \sum_{j=1}^{N} d_{1j} u_{jx_j}(E_1, t) - \frac{\varepsilon}{\lambda} u_t(E_1, t)$$
for arbitrary $\lambda > 0, \nu > 0$ and $\varepsilon \ge 0$. Nevertheless, for $\rho_i(t_0) \le 0$, the conclusions $(\alpha) - (\delta)$ remain valid in the case 1, and $V_i(\psi, t) \le 0$ still holds in the case 2. Thus we can prove in the same way the following result.

**Corollary.** *Under the assumptions of Theorem 1 except that all $\rho_i$ need only to be nonpositive, $u$ cannot attain its positive maximum $M$ (its negative minimum $m$) at some point $(x_0, t_0) \in \Omega_p$, unless all $\rho_i(t_0) = 0$. In that case $u = M$ ($u = m$) in $G \times [0, t_0]$.*

Another classical estimate can be carried over to parabolic network equations with $(GK)$ type connecting operators. Introduce the condition
$$\exists b_3 \ge 0 \, \forall E_i \in Int E \, \forall t \in (0, T] : \rho_i(t) \le b_3 \sigma_i(t). \tag{7}$$

**Theorem 2.** Let $u \in C(\Omega) \cap C^{2,1}_{GK}(\Omega_p)$ be a solution of

$$u_{jt} = a_j(x_j, t, u_j, u_{jx_j}) u_{jx_jx_j} + f_j(x_j, t, u_j, u_{jx_j}) \text{ in } \Omega^\bullet_{jp}$$

for all $j \in \{1, \ldots, N\}$, where all $f_j : \Omega_j \times \mathbb{R}^2 \to \mathbb{R}$ satisfy the Osgood condition

$$\exists\, b_1, b_2 \geq 0\ \forall\, j \in \{1, \ldots, N\}\ \forall\, (x, t, z) \in \Omega^\bullet_{jp} \times \mathbb{R} : \qquad (8)$$

$$z f_j(x, t, z, 0) \leq b_1 z^2 + b_2.$$

Let condition (7) be fulfilled. Then the following estimate holds

$$\max_\Omega |u| \leq \inf_{\lambda > b_1, \lambda \geq b_3} \left( e^{\lambda T} \max \left\{ \max_{\omega_p} |u|, \sqrt{\frac{b_2}{\lambda - b_1}} \right\} \right).$$

**Proof.** Apply the Lemma to $\varphi = u$ ($\varphi = -u$) and

$$\psi = \psi(x_j, t) = (1 + \varepsilon) e^{\lambda t} \max \left\{ \max_{\omega_p} |u|, \sqrt{\frac{b_2}{\lambda - b_1}} \right\}$$

for $\lambda > b_1$ and $\lambda \geq b_3$ and $\varepsilon > 0$ in order to obtain the estimate from above (from below). By (7) we obtain $V_i(\psi, t) \leq 0 = V_i(\varphi, t)$ in $(0, T]$ and $IntE$. Using the differential equation, condition (8) and the inequality $\psi^2 \leq e^{2\lambda t} b_2 (\lambda - b_1)^{-1}$ we conclude at a test point $(x_0, t_0)$ as defined in (5)

$$\psi_t - \varphi_{jt} = \lambda \psi - a_j(x_0, t_0, \varphi_j, 0) \varphi_{jx_jx_j} - f_j(x_0, t_0, \varphi_j, 0)$$

$$\geq \psi(\lambda - b_1 - \frac{b_2}{\psi^2}) \geq \psi(1 - e^{-2\lambda t})(\lambda - b_1) > 0.$$

Since $\varepsilon$ was arbitrary and $\varphi < \psi$ on $\omega_p$, the desired estimate follows.

$\diamond$

The proof shows that the assertion of Theorem 2 remains valid under the condition of weak parabolicity, i.e. $\mu_1 \geq 0$ in (2).

## References

[1] F. Ali Mehmeti, *Linear and nonlinear transmission and interaction problems*, Semesterbericht Funktionalanalysis Tübingen **13** (1987/88).
[2] J. v. Below, *A characteristic equation associated to an eigenvalue problem on $c^2$-networks*, Lin. Alg. Appl. **71** (1985), 309–325.
[3] J. v. Below, *Diffusion und Reaktion auf Netzwerken*, Dissertation, Tübingen 1984.
[4] J. v. Below, *Sturm-Liouville eigenvalue problems on networks*, Math. Meth. Applied Sciences **10** (1988), 383–395.
[5] J. v. Below, *Classical Solvability of linear parabolic equations on networks*, J. Differential Equ. **72** (1988), 316–337.

[6] J. v. Below, *Comparison and maximum principles for parabolic network equations*, Semesterbericht Funktionalanalysis Tübingen **14** (1988), 33–37.

[7] J. v. Below, *An existence result for semilinear parabolic network equations*, Semesterbericht Funktionalanalysis Tübingen **15** (1988/89), 33–41.

[8] H. Camerer, *Die elektrotonische Spannungsausbreitung im Soma, Dendritenbaum und Axon von Nervenzellen*, Dissertation, Tübingen 1980.

[9] G. Lumer, *Connecting of local operators and evolution equation on networks*, Potential Theory Copenhagen 1979, vol. 787, Springer Lect. Notes Math., 1980.

[10] S. Nicaise, *Diffusion sur les espaces ramifiés*, Thèse du Doctorat, Mons 1986.

[11] S. Nicaise, *Spectre des réseaux topologiques finis*, Bull. Sc. Math. $2^e$ Série **111** (1987), 401–413.

[12] W. Walter, *Differential and integral inequalities*, Springer-Verlag Berlin.

Joachim von Below
Lehrstuhl für Biomathematik
Universität Tübingen
Auf der Morgenstelle 10
D-7400 Tübingen 1, FRG

# Pair Formation in Structured Populations

CARLOS CASTILLO-CHAVEZ
STAVROS BUSENBERG
KEN GEROW

Biometric Unit/Center for Applied Mathematics, Cornell University
Department of Mathematics, Harvey Mudd College
Biometric Unit, Cornell University

## 1. Introduction

The grim scenario created by the AIDS epidemic has driven researchers to develop mathematical models to improve our understanding for the mechanisms responsible for HIV (the etiological agent for AIDS) transmission and of the evaluation of possible intervention measures. Recent reviews of the literature on models include those of [A1,A2], [CC1,CC2], and [SchCCH]. Some of the important conclusions generated by mathematical models include the clear identification of three key mechanisms which have the greatest effect on HIV transmission at the population level: variable infectivity, mixing or pair formation, and long, variable periods of infectiousness. For an extensive in depth study of some of the most recent mathematical and statistical work in these and other areas to AIDS epidemiology see [CC2].

This paper is organized as follows: in Section 2, we outline a unified axiomatic approach to the problem of mixing which extends and generalizes the one–sex framework of [BlCC1] and [CCB1] and provide an expression for the general solution due to [BuCC1], as well as some numerical illustrations of particular mixing functions; in Section 3, we formulate a two–sex mixing or pair formation framework that is a natural generalization of the one–sex framework, and construct some explicit solutions; in Section 4 we formulate a demographic model that follows pairs and provide some preliminary analysis of this model.

## 2. Mixing framework

The formulation described in this section can be used in the modeling of social or sexual mixing interactions. The mixing or pair formation function can describe the proportion of "dates" between individuals in distinct groups, or it can represent the proportion of sexual partnerships or sexual contacts between these individuals. In addition, the mixing function can be generalized to include the geographical distribution or the geographical movement of individuals through the use of "localized" mixing functions, i.e., functions that represent the proportion of partnerships formed between individuals from clearly defined groups (social, demographic, etc.) at a particular geographical location. The local geographical heterogeneities can then be linked through the specification of migration or movement matrices (see [Sa1,Sa2] and [SaSi]). Therefore, our approach allows from the specification of a spatial mixing framework. In this paper, however, we concentrate in the study of localized mixing functions.

Since our work has been motivated by HIV dynamics, we concentrate on the study of mixing functions in the context of SIR models where $S$ represents the class of susceptible individuals, $I$ the class of infected individuals, and $R$ the class of removed or recovered individuals. We consider first the interactions of a single, socially–homogeneous group of individuals who are structured according to the following variables: $a$ = age; $\tau$ = time (or age) since infection, $r$ = activity or risk level. We let $N(r, a, \tau, t)$ denote the total population density per unit age, activity, and time since infection, at time $t$. This population is divided into the following epidemiological classes: $S$ = susceptible; $I$ = asymptotic or slightly symptomatic infective; $A$ = highly symptomatic infective. This classification is fairly general and includes implicitly the traditional exposed, but not infected, class $E$ (see [BuCC1]). In our dicussion, $\tau$ is a hidden internal variable that does not distinguish individuals other than through their level of infectivity, and perhaps mortality. When modeling the sexual transmission of AIDS, we assume that $A$–individuals (i.e. individuals with severe symptoms or "full–blown" AIDS) are sexually inactive (i.e. this class represents the "removed" individuals) and hence that

$$T(r, a, t) = S(r, a, t) + \int_0^\infty I(r, a, \tau, t) d\tau$$

represents the total age and activity–level density of a population active in disease transmission contacts. Sexual mixing (or pair formation) is defined through the mixing function $p$. Specifically,

$p(r, a, r', a', t)$ = the proportion of partners of an $(r, a)$ individual
(i.e., a person of activity level $r$ at age $a$), with $(r', a')$
individuals at time $t$.

$C(r, a, t)$ = the expected or average number of partners per unit time
of an $(r, a)$ individual given at time $t$. We assume $C \geq 0$.

The following natural conditions characterize the mixing function:

(i) $\rho \geq 0$,
(ii) $\int_0^\infty \int_0^\infty \rho(r, a, r', a', t) dr' da' = 1$,
(iii) $\rho(r, a, r', a', t) C(r, a, t) T(r, a, t) = \rho(r', a', r, a, t) C(r', a', t) T(r', a', t)$,
(iv) $C(r, a, t) T(r, a, t) C(r', a', t) T(r', a', t) = 0 \Rightarrow \rho(r, a, r', a', t) = 0$.

Condition (ii) simply says that $\rho$ is a proportion. Condition (iii) states that the total number of pairs of $(r, a)$ individuals with $(r', a')$ individuals equals the total number of pairs of $(r', a')$ individuals with $(r, a)$ individuals (all this is per unit time, age, and time since infection). Condition (iv) says that there is no mixing in the age and activity levels at which there are not active individuals; i.e.; on the set

$$S = \{(r, a, r', a') : C(r, a, t) T(r, a, t) C(r', a', t) T(r', a', t) = 0\},$$

where there is no mixing. Condition (iv) arises naturally in the study of the solutions of the above framework (see [BuCC2]).

In some situations it is necessary to consider mixing functions $\rho$, which are Dirac delta functions or, more generally, distributions or generalized functions. Hence, we are forced to consider solutions to the axiomatic framework in the space of distributions or generalized functions (see [Schw], [GS]). To accommodate this possibility the following modification to the interpretation of axioms (i) and (iv) is necessary:

(i') $\rho \geq 0$ in the sense of distributions; i.e.,

$$\iint_0^\infty \rho(r, a, r', a', t) f(r', a', t) dr' da' \geq 0 \quad \text{for all} \quad f \geq 0, \quad \text{and}$$

(iv') $\rho(r, a, r', a', t) = 0$ on a set $F$, means

$$\iint_F \rho(r, a, r', a', t) f(r, a', t) dr da' = 0 \quad \text{for all} \quad f.$$

Pair formations can involve selectivity by individuals according to age or activity level, they can be random pairings without regard to these variables, or they can be any combination or mixture of the two extremes. A detailed discussion of these possibilities and of the restrictions they place on the mixing function $\rho$ is found in [BuCC2].

A solution of critical importance to the mixing framework is that of total (i.e. in age and risk) proportionate mixing:

$$\bar{\rho}(r, a, r', a', t) = \frac{C(r', a', t) T(r', a', t)}{\iint_0^\infty C(r', a', t) T(r', a', t) da' dr'}. \tag{1}$$

This solution plays an important role in the determination of all possible solutions to the mixing framework (i) – (iv). Note that proportionate mixing vacuously satisfies condition (iv). This condition prevents us from accidentally dividing by zero, and hence prevents us from arbitrarily defining a mixing function for subpopulations that either are not sexually active or that have been depleted of individuals by disease dynamics. Further examples of specific mixing functions can be found in [BlCC1] and [BuCC2]. We further observe that convex linear combinations of mixing functions are mixing functions. Specifically, if $\alpha_1, \ldots, \alpha_N$ are positive constants such that $\sum_{i=1}^{N} \alpha_i = 1$ and $\rho_1, \ldots, \rho_N$ are mixing functions, then $\sum_{i=1}^{N} \alpha_i \rho_i$ is a mixing function. This last observation provides a recipe for the construction of a variety of mixing functions. Furthermore, it clearly shows that preferred mixing (a convex combination of two mixing functions), contrary to the suggestions of some researchers, does not contain all reasonable possibilities. Specifically, (omitting age) preferred mixing is given by

$$\rho(s,r) = (1-\alpha) \frac{C(r)T(r)}{\int_0^\infty C(u)T(u)du} + \alpha \delta(s-r), \qquad (2)$$

where $\delta$ denotes the Dirac delta (see [BlCC1], i.e., it is the convex linear combination of the Dirac delta (a mixing function) and proportionate mixing. The two extreme points of this particular convex linear combination (when $\alpha = 0$ or 1) do not obviously represent sociological or mathematical mixing extremes – this was pointed out to us by S. Gupta and R. Anderson.

A mixing function $\rho$ is called *separable* if it can be written in the form

$$\rho(r, a, r', a', t) = \rho_1(r, a, t) \rho_2(r', a', t). \qquad (3)$$

The total proportionate mixing function $\bar{\rho}$ is separable, and our first result states that there are no other separable pairing functions.

**Theorem 2.1.** *The only separable pairing function $\rho$ satisfying conditions (i) – (ii) – (iii) is the total proportionate mixing function $\bar{\rho}$ given by (1).*

**Proof.** This result can be easily obtained by direct substitution of (3) into the mixing axioms. Since the proof of this result to that of Theorem 3.1 (included later) we omit the details.

All other solutions to the mixing framework are given by multiplicative perturbations of total proportionate mixing. The nature of the perturbations is specified in the following theorem:

**Theorem 2.2.** *Let $\phi : \mathbb{R}_+^4 \to \mathbb{R}$ be measurable and jointly symmetric: $\phi(r, a, r', a') = \phi(r', a', r, a)$, and suppose that*

$$\iint_0^\infty \bar{\rho}(r', a') \phi(r, a, r', a') dr' da' \leq 1,$$

and

$$\iint_0^\infty \bar{\rho}(r, a) \left( \iint_0^\infty \bar{\rho}(r', a') \phi(r, a, r', a') dr' da' \right) dr da < 1.$$

Let
$$\rho_1(r,a) = 1 - \iint_0^\infty \bar{\rho}(r',a')\phi(r,a,r',a')dr'da', \qquad (4)$$

so that

$$\rho(r,a,r',a') = \bar{\rho}(r',a')\left[\frac{\rho_1(r,a)\rho_1(r',a')}{\iint_0^\infty \bar{\rho}(r',a')\phi_1(r',a')dr'da'} + \phi(r,a,r',a')\right] \qquad (5)$$

is a mixing function. Conversely, for every mixing function $\rho$ there exists a $\phi$ that satisfies the hypotheses of the theorem such that $\rho$ is given by (5) with $\rho_1$ defined by (4).

**Proof.** That the expression given by equation (5) is a mixing function is immediate. For the proof of converse, see [BuCC2].

The function $\phi$ provides us with a measure of the deviation from proportionate mixing and therefore it is a measure of preference. We call this perturbation the structural covariance of preference function (note that this covariance is always positive). To illustrate the effects of $\phi$ on the shape of the mixing or pair formation function, we look at some examples for situations in which the mixing function is only a function of the age or risk (related to frequency and type of sexual activity) of the individuals but not of both. The version that is illustrated in our numerical corresponds to the following version of Theorem 2.2:

**Theorem 2.3.** Let $\phi : \mathbb{R}_+^2 \to \mathbb{R}^+$ be a measurable and jointly symmetric function, and suppose that

$$\int_0^\infty \bar{\rho}(r)\phi(r,r')dr \leq 1 \quad \text{and} \quad \int_0^\infty \bar{\rho}(r)\left\{\bar{\rho}\int_0^\infty (u)\phi(u)du\right\} < 1.$$

Defining $\rho_1(r)$ by

$$\rho_1(r) = 1 - \int_0^\infty \bar{\rho}(u)\phi(r,u)du \qquad (6)$$

we obtain the following representation formula for a two dimensional one–sex mixing function:

$$\rho(r,r') = \bar{\rho}(r)\left[\frac{\rho_1(r)\rho_1(r')}{\int_0^\infty \bar{\rho}(r)\rho_1(r)dr} + \phi(r,r')\right], \qquad (7)$$

where

$$\bar{\rho}(r) = \frac{C(r)T(r)}{\int_0^\infty C(u)T(u)du}, \qquad (8)$$

i.e., we have a multiplicative perturbation of proportionate mixing. Also for every mixing function $\rho$, there exists a structural covariance or preference function

$\phi$ satisfying the hypotheses of the theorem such that $\rho$ is given by (7) with $\rho_1$ defined by (6).

We now proceed to illustrate the effects of preference on the shape of the mixing function. As a model for the distribution of activity levels in a population, the lognormal distribution has appeal due to its flexibility. Formally, if $\ln(R)$ has a normal distribution with mean $\mu$ and variance $\sigma^2$, then $R$ has a two–parameter lognormal distribution with parameters $\mu$, and $\sigma$. For convenience, define $b = e^\mu$. The probability density function for the lognormal may be written as

$$T(r) = \frac{1}{\sigma r \sqrt{2\pi}} \exp\left[-\frac{1}{2\sigma^2}(\ln(\frac{r}{b}))^2\right] = \text{Prob}\left[\ln(R) = r,\right], r > 0.$$

The mean and variance of $R$ are

$$E(R) = \mu_R = b \exp\left(\frac{\sigma^2}{2}\right),$$

and

$$\text{Var}(R) = \sigma_R^2 = b^2 a^{\sigma^2}(e^{\sigma^2} - 1).$$

A more natural parameterization for our modeling purposes is to describe the distribution in terms of $\mu_R$ and $\sigma_R^2$. Given values of these two population parameters (either arbitrarily, or as suggested by data), we can easily determine that

$$\sigma^2 = \ln[\sigma_R^2 \mu_R^{-1} + 1],$$

and

$$b = \mu_R \exp(-0.5\sigma^2).$$

We may further simplify our model prescription if we accept the empirical "power law" of [AM]:

$$\sigma_R^2 = 0.555 \mu_R^{3.231}, \tag{9}$$

whence

$$\sigma^2 = \ln[0.555 \mu_R^{2.231} + 1],$$

and

$$b = \mu_R \exp(-0.5\sigma^2).$$

With $C(r) = r$, then (8) becomes

$$\bar{\rho}(r') = \frac{r' T(r')}{\int_0^\infty u T(u) du},$$

where the denominator is the expected value of a lognormal random variable, i.e.,

$$= \frac{1}{\sigma b \sqrt{2\pi}} \exp - \left[\frac{1}{2\sigma^2}(\ln(\frac{r'}{b})) + (\frac{\sigma^2}{2})\right].$$

$T(r)$ is really a function of $t$, i.e. $T = T(r,t)$, and its behavior is governed by an appropriate partial differential equation (see [BuCC1]). Note however, that the "power law" of [AM] suggests that the mean and variance of $T(r,t)$ (regardless of how we model it) has to satisfy equation (9): Further, since our purpose is to *illustrate* the effects of the structural covariance or preference function on the shape of $\rho(r,r',t)$, we "bypass" the dynamic model and concentrate on the effects of $\phi$ on $\rho$ when $\mu_R$ and $\sigma_R$ satisfy (9) (for a more orthodox procedure to test mixing frameworks is found in [CCBl]. We observe that a population that is experiencing a decrease in sexual activity will have to do it in a restricted fashion, i.e., by moving down along the line (in log–log scale) defined by equation (9). Finally, we remark that the "power law" of Anderson and May can be explained through the processes of pair formation and dissolution (for details see [BlCC2].

In our numerical illustrations we take a fairly general $\phi$, namely:

$$\phi(r,r') = \exp\left[-(c_1(r^2 + r'^2) + c_2 rr'))\right]. \tag{10}$$

Recall that $\rho_1(r) = 1 - \int_0^\infty \bar{\rho}(r,r')dr'$ must be $\leq 1$; and note that for our current choices for $T$ and $\phi$, this condition is met for a wide range of $\mu_R$, $c_1$ and $c_2$, including values which may be reasonable for human populations. The denominator term of $\rho(r,r')$ is fairly cumbersome, but can be cleaned up a little:

$$\int_0^\infty \bar{\rho}(r)\rho_1(r)dr = \int_0^\infty \frac{1}{\sigma b \sqrt{2\pi}} \exp\left[\frac{-(\ln(r'/b))^2}{2\sigma^2} - \left(\frac{\sigma^2}{2}\right)\right] dr$$
$$- \int_0^\infty \int_0^\infty \frac{c_1}{2\sigma^2 b^2 \pi} \exp - \left[\frac{-(\ln(r'/b))^2}{\sigma^2} + \sigma^2 + c_1(r^2 + r'^2) + c_2 rr'\right] dr' dr.$$

The second term can be reduced to a one-dimensional integral by use of the change of variable defined by $u = \sqrt{2c_1} r + \frac{c_2 r'}{\sqrt{2c_1}}$. Then we have

$$\int_0^\infty \frac{c_1}{2\sigma^2 b^2 \sqrt{c_1 \pi}} \exp\left[-\left(\frac{(\ln(r'/b))^2}{\sigma}\right)^2 - \sigma^2 - r'^2\left(c_1 - \frac{c_2^2}{4c_1}\right)\right] (1 - \Phi(\frac{c_2 r'}{\sqrt{2c_1}})) dr',$$

where $\Phi(\cdot)$ is the standardized Gaussian cumulative distribution function.

In our set of simulations, we use all six combinations of two choices for $T(r)$ (determined by the Anderson and May's (1988) power law, with values of 2 and 8 for $\mu_R$) and three choices for $\phi$: $\phi_1$ with a well–defined narrow ridge along the line $r = r'$ (determined by the pair $(c_1, c_2) = (0.3, -0.6)$), $\phi_2$ with a somewhat broader profile $((c_1, c_2) = (0.05, -0.08))$, and $\phi_3 = 0$, representing proportional mixing. We have plots also of the corresponding structural covariance functions $\phi_1$ and $\phi_2$.

The plots* illustrate the interaction between the structural covariance function and the degree to which the population exhibits proportionate mixing. For a given mean activity level (2 and 8 in these simulations) the preference function

---
*See pp. 61-65.

exhibiting the sharpest degree of preference ($\phi_1$, plot 1) has the mixing function which is (visually, at least) furthest removed from proportionate mixing (plots 2,3). As the preference function gets less sharply peaked ($\phi_2$, plot 4), the mixing function (plots 5,6) is more similar to proportionate mixing (plot 7,8). Also, for a given $\phi$, as the population mean activity level increases, the mixing function look more and more like a simple additive combination of $\phi$ with proportionate mixing (plots $2 \to 3$, $5 \to 6$).

## 3. Two–sex mixing framework

In this section we provide an outline of our two–sex framework. Since an extensive account will be provided later (see [CCBu]), we look exclusively at our mixing framework in the context of a two–sex age–structured population. We further concentrate on a framework suitable for a two–sex demographic model. The modifications needed to transform this demographic model into an epidemiological model for sexually-transmitted diseases are straightforward and can be found in [CCBu].

We let $M(a,t)$ denote the density of males of age $a$ who are not in pairs at time $t$; and let $F(a',t)$ denote the density of females of age $a'$ who are not in pairs at time $t$. Pairing is defined through the mixing functions:

$p(a,a',t) =$ proportion of partnerships of males of age $a$ with females of age $a'$ at time $t$,

$q(a,a',t) =$ proportion of partnerships of females of age $a'$ with males of age $a$ at time $t$,

and we let

$C(a,t) =$ expected or average number of partners of a male of age $a$ at time $t$ per unit time,

$D(a',t) =$ expected or average number of partners of a female of age $a'$ at time $t$ per unit time.

The following natural conditions characterize these mixing functions:

(a) $p, q \geq 0$,
(b) $\int_0^\infty p(a,a',t)da' = \int_0^\infty p(a',a,t)da = 1$,
(c) $p(a,a',t)C(a,t)M(a,t) = q(a',a,t)D(a',t)F(a',t)$,
(d) $C(a,t)M(a,t)D(a',t)F(a',t) = 0 \Rightarrow p(a,a',t) = q(a',a,t) = 0$.

Condition (ii) is due to the fact that $p$ and $q$ are proportions. Condition (iii) simply states that the total number of pairs of males of age $a$ with females of age $a'$ equals the total number of pairs of females of age $a'$ with males of age $a$ (all per unit time and age). Condition (iv) says that there is no mixing in the age and activity levels where there are no active individuals, i.e., on the set $S(t) = \{(a, a', t) : C(r, a, t)M(a, t)D(a', t)F(a', t) = 0\}$.

The pair $(p, q)$ is called a *two-sex mixing function* iff it satisfies axioms (a-d). Further, a two-sex mixing function is called *separable* iff

$$p(a, a', t) = p_1(a, t)p_2(a', t) \quad \text{and} \quad q(a, a', t) = q_1(a, t)q_2(a', t).$$

If we let
$$h_p(a, t) = C(a, t)M(a, t) \tag{11}$$

and
$$h_q(a, t) = D(a, t)F(a, t), \tag{12}$$

then, omitting $t$ to simplify the notation, we establish the following result:

**Theorem 3.1.** *The only two-sex separable mixing function satisfying conditions (a-d) is given by $(\bar{p}, \bar{q})$, where*

$$\bar{p}(a') = \frac{h_q(a')}{\int_0^\infty h_p(u)du}, \tag{13}$$

$$\bar{q}(a) = \frac{h_p(a)}{\int_0^\infty h_q(u)du}. \tag{14}$$

**Proof.** It is clear that the expressions given by equations (13) – (14) satisfy the axioms (a – d), and hence, $(p, q)$ is a two-sex mixing function. Let's now assume that $(p, q)$ is separable, then using axiom (b), we see that

$$p_1(a) = \frac{1}{\int_0^\infty p_2(u)du} = \ell, \quad \text{(a constant)}$$

and
$$q(a') = \frac{1}{\int_0^\infty q_2(u)du} = k, \quad \text{(a constant)};$$

therefore,
$$p(a, a') = \ell p_2(a') \quad \text{and} \quad q(a, a') = k q_2(a).$$

If we substitute the above expressions into axiom (c) and integrate over all ages $a$, then we arrive at

$$\ell p_2(a') \int_0^\infty h_p(n)dn = h_q(a)$$

from which (13) follows. Equation (14) is obtained similarly.

Castillo-Chavez and Busenberg [CCBu] have established that all two-sex mixing functions are multiplicative perturbations (with appropriate structural covariance functions) of the only separable two-sex mixing function given by (13)

– (14). Although, the general solution may prove to be quite useful in theoretical considerations, it is still of practical importance to provide modelers and theoreticians with flexible families of two–sex mixing functions. The following two–sex biased mixing familiy for $N$–interacting subpopulations may fulfill this need. To introduce it, we let $u_i$ ($v_i$) denote the proportion of partnerships by males (females) of group $i$ reserved for mixing with females (males) in group $i$; necessarily $0 \leq u_i, v_i \leq 1$. If $F_i(t)$ ($M_i(t)$) denote the number of males (females) in group $i$ at time $t$, and $C_i$ ($D_i$) denote the average number of female (male) sexual partners of males (females) in group $i$, and $p_{ij}(t)$ ($q_{ij}(t)$) denote the proportion of partnerships of males (females) in group $i$ with females (males) in group $j$. Then

$$p_{ij}(t) = u_i \delta_{ij} + (1 - u_i)\frac{(1-v_j)D_j F_j}{\sum_{k=1}^{N}(1-v_k)D_k F_k}, \qquad (15)$$

$$q_{ji}(t) = v_j \delta_{ji} + (1 - v_i)\frac{(1-u_i)C_i M_i}{\sum_{k=1}^{N}(1-u_k)C_k M_k}, \qquad (16)$$

where

$$\delta_{ij} = \begin{cases} 1 & \text{if } i = j \\ 0 & \text{if } i \neq j \end{cases}.$$

The above family of two–sex biased mixing functions is easily incorporated into classical epidemiological models as well as into models that follow pairs. This is the topic of the next section where we introduce the simplest demographic model that follows pairs and that makes use of the framework of this section.

## 4. Demographic pair formation models

Demographic models that consider pairs and follow the dynamics of pairs have been studied by [K], [F], [DH], [D], [H1], [H2] and [W]. Their approach is based on the use of a nonlinear function $\psi$ to model the process (rate) of pair formation. This mixing/pair formation function is assumed to satisfy the Fredrickson/McFarland ([F],[McF]) properties:

(e) $\psi(0, F) = \psi(M, 0) = 0$.
   In the absence of either males or females there will be no heterosexual pair formation.
(f) $\psi(\alpha M, \alpha F) = \alpha \psi(M, F)$ for all $\alpha, M, F \geq 0$.
   If the sex ratio remains constant, then the increase in the rate of pair formation is assumed to be proportional to total population size.
(g) $\psi(M + u, F + v) \geq \psi(M, F)$ for all $u, v, F, M \geq 0$.
   Increases in the number of males and/or females does not decrease the rate of pair formation.

Condition (f) implies that all mixing functions are of the form

$$\psi(M, F) = Mg(\frac{F}{M}) = Fh(\frac{M}{F}),$$

where $h$ and $g$ are functions of one-variable.

Mixing functions satisfying the above axioms, and that have been used in demographic studies, include:

$$\psi(M, F) = k\min(M, F), \quad k \text{ is a constant}$$

$$\psi(M, F) = k\sqrt{MF},$$

and

$$\psi(M, F) = 2k\frac{MF}{M + F}.$$

Let $\sigma$ denote the rate of pair dissolution, $\mu$ denote the natural mortality rate, $\Lambda$ denote the "recruitment" rate, and $W$ denote the number of (heterosexual) pairs. Then a simple demographic model is given by the following set of equations:

$$\frac{dM}{dt} = \Lambda - \mu M + (\sigma + \mu)W - \psi(M, F)$$

$$\frac{dF}{dt} = \Lambda - \mu F + (\sigma + \mu)W - \psi(M, F)$$

$$\frac{dW}{dt} = -(\sigma + 2\mu)W + \psi(M, F).$$

If $\Lambda, \mu,$ and $\sigma,$ are constant, then there is always a globally stationary solution $(M, F, W)$ and $W$ is determined by the equation

$$\psi\left(\frac{\Lambda}{\mu} - W, \frac{\Lambda}{\mu} + W\right) = (\sigma + 2\mu)W.$$

(for references to this and related results see [W]).

If we now let $f(a', t)$ and $m(a, t)$ denote the age-specific densities for single males and single females respectively, and assume that $D$ (as defined in Section 3) and $\mu_m$ and $\mu_f$ are functions of age (the mortality rates for males and females), and assume that $W(a, a', t)$ denotes the age-specific density of heterosexual pairs (where $a$ denotes the age of the male and $a'$ the age of the female), then using the two-sex mixing functions $p$ and $q$ of Section 3, we arrive at the following demographic model for heterosexual populations:

$$\frac{\partial m}{\partial t} + \frac{\partial m}{\partial a} = -C(a)m(a, t)\int_0^\infty p(a, a', t)da' \\ - \mu_m(a)m(a, t) + \int_0^\infty [\mu_f(a') + \sigma]W(a, a', t)da', \quad (17)$$

$$\frac{\partial f}{\partial t} + \frac{\partial f}{\partial a'} = -D(a')f(a', t)\int_0^\infty q(a', a, t)da \\ - \mu_f(a')f(a', t) + \int_0^\infty [\mu_m(a) + \sigma]W(a, a', t)da, \quad (18)$$

$$\frac{\partial W}{\partial t} + \frac{\partial W}{\partial a} + \frac{\partial W}{\partial a'} = D(a')f(a't)q(a,a',t) \\ - [\mu_f(a') + \mu_m(a) + \sigma]W(a,a',t). \tag{19}$$

To complete this model we need to specify the initial and boundary conditions. To this effect we let $\lambda_m$ and $\lambda_f$ denote the female–age–specific fertility rates, and let $m_0$, $f_0$, and $w_0$ denote the initial age densities. Hence, the initial and boundary conditions are given by

$$m(0,t) = \int_0^\infty \lambda_m(a')N_f(a',t)da', \tag{20}$$

$$f(0,t) = \int_0^\infty \lambda_f(a')N_f(a',t)da', \tag{21a}$$

$$W(0,0,t) = 0, \tag{21b}$$

$$f(a,0) = f_0(a), \ m(a,0) = m_0(a), \ W(a,a',0) = W_0(a,a'), \tag{22}$$

where

$$N_f(a',t) = \int_0^\infty W(a,a',t)da.$$

Further, we observe that $N_f$ and $f + N_f$ satisfy the following set of equations:

$$\left(\frac{\partial}{\partial t} + \frac{\partial}{\partial a'}\right)(f + N_f) = -\mu_f(a')[f + N_f] \tag{23}$$

and

$$\left(\frac{\partial}{\partial t} + \frac{\partial}{\partial a'}\right)N_f = D(a')f(a',t) - [\mu_f(a') + \sigma]N_f \\ - \int_0^\infty \mu_m(a)W(a,a',t)da. \tag{24}$$

Note that if we let $\sigma \to \infty$ (while fixing $a$, $t$, and $\mu_m(a)$ contant) then $N_f(a',t) \to 0^+$ and (formally) equation (23) approaches the classical MacKendrick/Von Foerster model. Further note that since in the model given by equations (17) – (22) only pairs reproduce we do not recover the classical boundary condition (nevertheless the boundary condition is consistent).

If we assume that $\mu_m$, $\mu_f$ and $D$ are constants and look for solutions of the form

$$N_f(a',t) = e^{rt}\tilde{N}(a'),$$

and

$$f(a',t) = e^{rt}\tilde{f}(a), \tag{25}$$

then in the usual fashion, we arrive at a characteristic equation for $r$ of the form

$$H(r) = 1$$

where

$$H(r) = \int_0^\infty \lambda_f(a) \frac{D}{\mu_m + D + \sigma} e^{(-r+\mu_f)a}[1 - e^{(-\mu_m+\sigma+D)a}]da. \quad (26)$$

Since $H(r)$ is decreasing, then (25) has a unique real root $r_*$. Clearly

$$H(0) > 1 \Leftrightarrow r_* > 0,$$

and if $r = \alpha + i\beta$ is a complex root then one easily sees that $\alpha < r_*$. We observe further that the above analysis is independent of $\rho$, however, note that in order to recover $W$ or to study the stability of product solutions we need to have specific knowledge of the mixing function $\rho$.

Although further analysis is possible, we will not present it here, as one of the main purposes of this article is to show an alternative approach to that of Fredrickson/McFarland/Dietz/Hadeler for the formulation of demographic models that follow the dynamics of pairs. Epidemiological models that fit into our framework are easily formulated and the appropriate details will be discussed elsewhere.

## 5. Conclusions

In this article we have presented a general solution to the one sex mixing/pair formation problem. Our representation theorem states that any mixing function can be represented as a multiplicative perturbation of proportionate mixing. This perturbation, through its structural covariance or preference function, provides us with a measure of divergence from proportionate mixing. Simulations based on the "power law" of [AM] were provided to illustrate the role of preference in the shape of the mixing function. Our discussion of the simulation results, combined with our previous studies (see [BlCC1, CCBl, BuCC1]), show that to understand the role of preference in disease dynamics we need to develop methods of estimating the effects to the structural covariance function on the shape of the basic mixing function (i.e. proportionate mixing). Knowledge of "realistic" mixing structures is needed in the evaluation of possible intervention programs aimed at disease prevention.

We have also introduced a two-sex mixing framework and constructed a variety of solutions that may prove useful in applications. We have introduced a demographic model that follows pairs based on our mixing/pair-formation framwork, and have shown that this model has nontrivial solutions. Further analysis of this model is being carried out and will be published elsewhere.

Finally, we remark that in order for these results to be useful in applications we need to critically examine the assumptions behind our model (see [BlCC2],[BlCC4], [SaCC]), we need to be able to determine ways of estimating parameters from data (see [BlCCCa]), and we need to be able to explain the available data (see [BlCC3]).

**Acknowledgments.** This research has been partially supported by the Center for Applied Mathematics at Cornell University and NSF grant DMS-8703631 to Stavros Busenberg, and by NSF grant DMS-8906580, NIAID Grant R01 A129178-01 and Hatch project grant NYC 151-409, USDA to Carlos Castillo-Chavez. We thank S.P. Blythe for his valuable comments.

## References

[A1] Anderson, R.M., *The role of mathematical models in the study of HIV transmission and the epidemiology of AIDS*, Journal of AIDS **1** (1988), 241–256.

[A2] Anderson, R.M., *Editorial Review: Mathematical and Statistical Studies of the epidemiology of HIV*, AIDS **3** (1989), 333–346.

[AM] Anderson, R.M. and R.M. May, *Epidemiological parameters of HIV transmission*, Nature **333** (1988), 514–519.

[ABGK] Anderson, R.M., S.P. Blythe, S. Gupta and E. Konings, *The transmission dynamics of the Human Immunodeficiency Virus Type 1 in the male homosexual community in the United Kingdom: the influence of changes in sexual behavior*, Manuscript.

[BlCC1] Blythe, S.P. and C. Castillo–Chavez, *Like–with–like preference and sexual mixing models*, Math. Biosci **96** (1989), 221–238.

[BlCC2] Blythe, S.P. and C. Castillo–Chavez, *Scaling of sexual activity*, Nature **344** (1990), 202.

[BlCC3] Blythe, S.P. and C. Castillo–Chavez, *The one-sex mixing problem: a choice of solutions?*, submitted.

[BlCC4] Blythe, S.P. and C. Castillo–Chavez, *Like–with–like mixing and sexually transmitted disease epidemics in one-sex populations*, submitted.

[BlCCCa] Blythe, S.P, C. Castillo-Chavez, and G. Casella, *Empirical methods for the estimation of the mixing probabilities for socially structured populations from a single survey sample*, submitted to J. of AIDS.

[BuCC1] Busenberg, S. and C. Castillo-Chavez, *Interaction, pair formation and force of infection terms in sexually transmitted diseases*, Mathematical and Statistical Approaches to AIDS Epidemiology, (C. Castillo-Chavez, Ed.), Lecture Notes in Biomathematics **83**, Springer–Verlag, Berlin, Heidelberg, New York, London, Paris, Tokyo, Hong Kong (1989), pp. 289–300.

[BuCC2] Busenberg, S. and C. Castillo-Chavez, *On the role of preference in the solution of the mixing problem, and its application to risk– and age-structured epidemic models*, submitted to IMA J. of Math. Applic. to Med. and Biol.

[CC1] Castillo-Chavez, C., *Review of recent models of HIV/AIDS transmission*, Applied Mathematical Ecology, (S.A. Levin, T.G. Hallam, and L.J. Gross, Eds.), Biomathematics **18**, Springer–Verlag, Berlin, Heidelberg, New York, London, Paris, Tokyo, Hong Kong (1989), pp. 253–262.

[CC2] Castillo-Chavez, C.(Ed.), *Mathematical and Statistical Approaches to AIDS Epidemiology*, Lecture Notes in Biomathematics, **83**, Springer–Verlag, Berlin, Heidelberg, New York, London, Paris, Tokyo, Hong Kong (1989), pp. 275–288.

[CCBl] Castillo-Chavez, C. and S.P. Blythe, *A "test-bed" procedure for evaluating one-sex mixing frameworks*, submitted.

[CCBu] Castillo-Chavez, C. and S. Busenberg, *On the general solution of the two-sex pair-formation problem*, submitted.

[D] Dietz, K., *On the transmission dynamics of HIV*, Math. Biosci **90** (1988), 397–414.

[DH] Dietz, K. and H.P. Hadeler, *Epidemiological models for sexually transmitted diseases*, J. Math. Biol. **26** (1988), 1–25.

[F] Fredickson, A.G., *A mathematical theory of age structure in sexual populations: Random mating and monogamous marriage models*, Math. Biosci **20** (1971), 117–143.

[GS]      Gel'fand, I.M. and G.E. Shilov, *Generalized Functions, Vol. 1*, Academic press, New York, London, 1964.

[H1]      Hadeler, K.P., *Pair formation in age-structured populations*, Acta Applicandae Mathematicae **14** (1989), 91–102.

[H2]      Hadeler, K.P., *Modeling AIDS in structured populations*, Manuscript.

[K]       Kendall, D.G., *Stochastic processes and population growth*, Roy. Statist. Soc. **Ser. B2** (1949), 230–264.

[McF]     McFarland, D.D., *Comparison of alternative marriage models*, Population Dynamics, (Greville, T.N.E. Ed.), Academic Press, New York, London, 1972, pp. 107–135.

[Sa1]     Sattenspiel, L., *Population structure and the spread of disease*, Human Biology **59** (1987), 41–438.

[Sa2]     Sattenspiel, L., *Epidemics in nonrandomly mixing populations: a simulation*, American Journal of Physical Anthropology **73** (1987), 251–265.

[SaSi]    Sattenspiel, L. and C.P. Simon, *The spread and persistence of infectious diseases in structured populations*, Math. Biosci **90** (1988), 341–366.

[SaCC]    Sattenspiel, L. and C. Castillo-Chavez, *Environmental context, social interactions, and the spread of HIV*, American Journal of Human Biology, Volume 2, Number 4 (in press).

[SchCCH]  Schwager, S.J., C. Castillo-Chavez, and H. Hethcote, *Statistical and mathematical approaches in HIV/AIDS modeling: a review*, Mathematical and Statistical Approaches to AIDS Epidemiology, (C. Castillo-Chavez, Ed.), Lecture Notes in Biomathematics **83**, Springer-Verlag, Berlin, Heidelberg, New York, London, Paris, Tokyo, Hong Kong, 1989, pp. 2–37.

[Schw]    Schwartz, L., *Théorie des distributions*, Hermann, Paris, 1966.

[W]       Waldstätter, R., *Pair formation in sexually transmitted diseases*, Mathematical and Statistical Approaches to AIDS Epidemiology, (C. Castillo-Chavez, Ed.), Lecture Notes in Biomathematics **83**, Springer-Verlag, Berlin, Heidelberg, New York, London, Paris, Tokyo, Hong Kong, 1989, pp. 260–274.

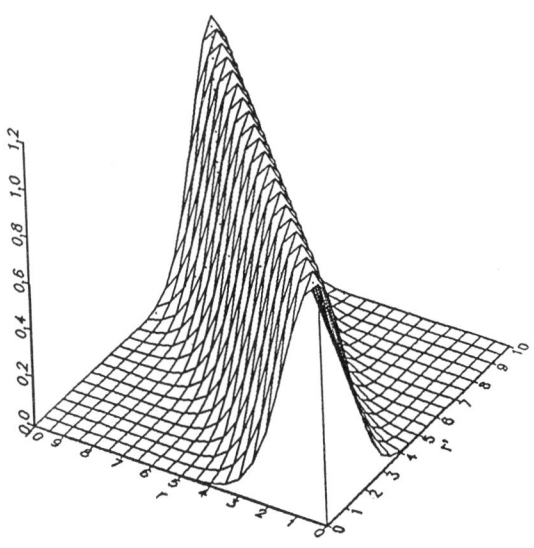

PLOT 1: $\phi$, with $c_1 = 0.3$, $c_2 = -0.6$.

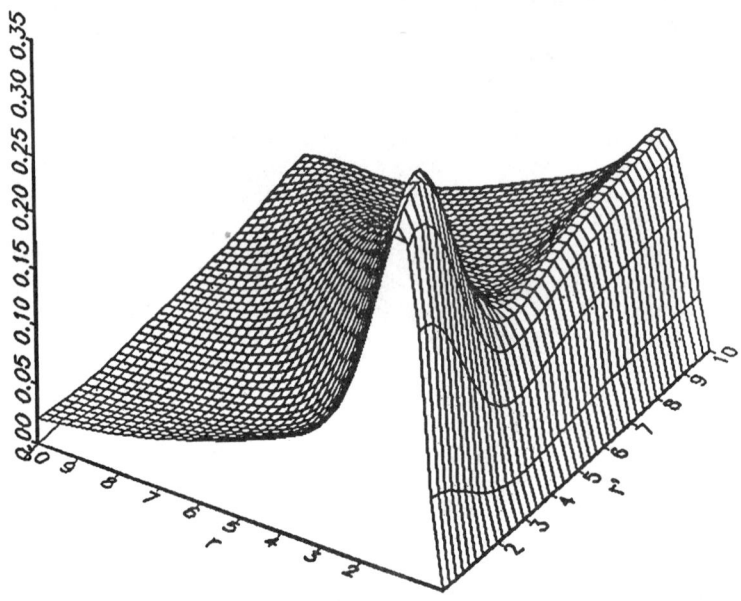

PLOT 2: $\rho$, with $\mu_R = 2$, $c_1 = 0.3$, $c_2 = -0.6$.

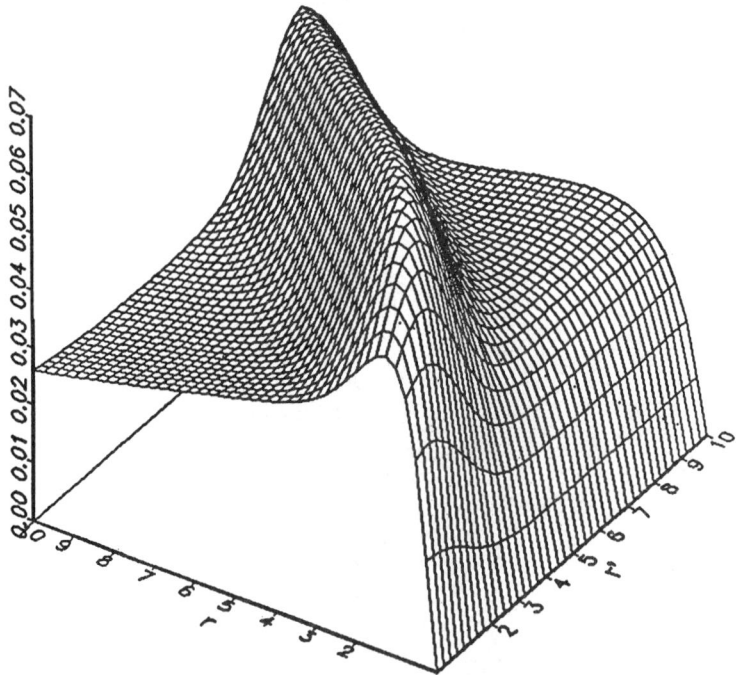

PLOT 3: $\rho$, with $\mu_R = 8$, $c_1 = 0.3$, $c_2 = -0.6$.

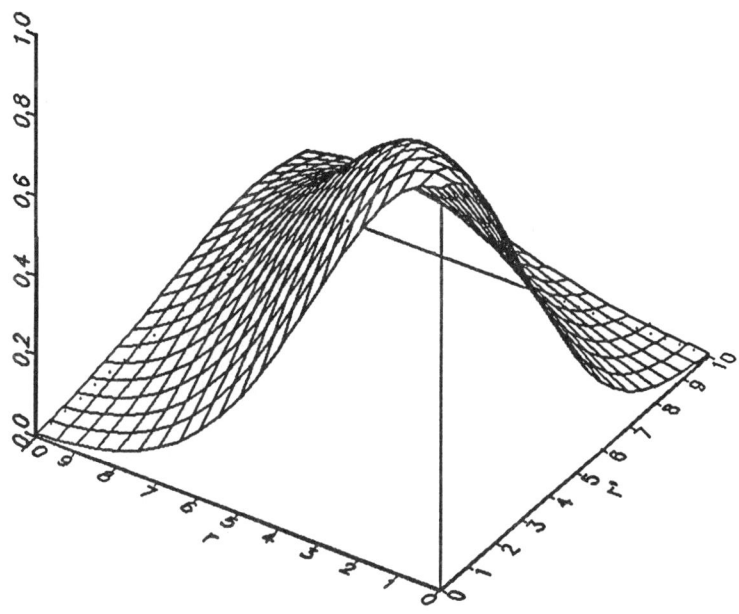

PLOT 4: $\phi$, with $c_1 = 0.05$, $c_2 = -0.08$.

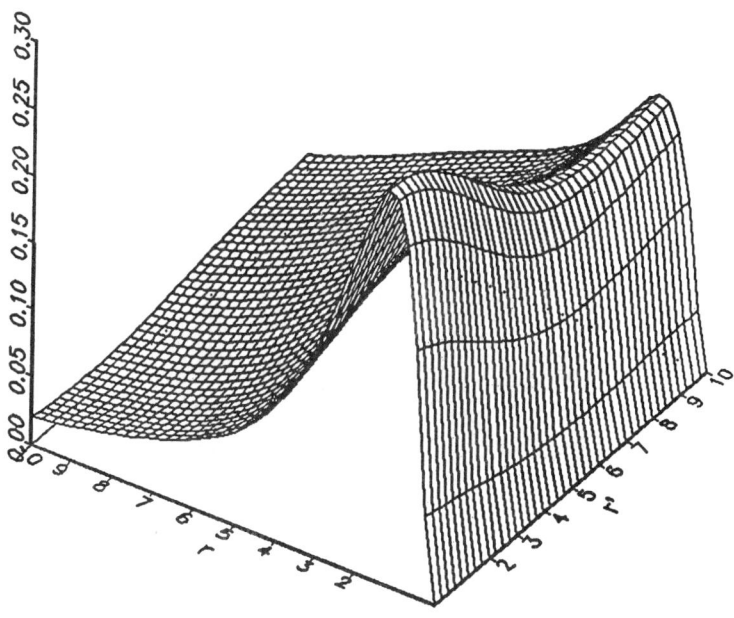

PLOT 5: $\rho$, with $\mu_R = 2$, $c_1 = 0.05$, $c_2 = -0.08$.

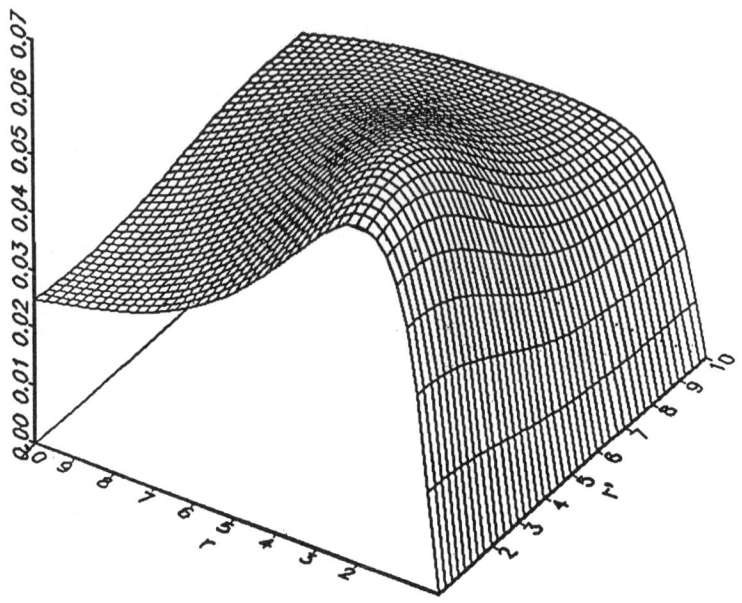

PLOT 6: $\rho$, with $\mu_R = 8$, $c_1 = 0.05$, $c_2 = -0.08$.

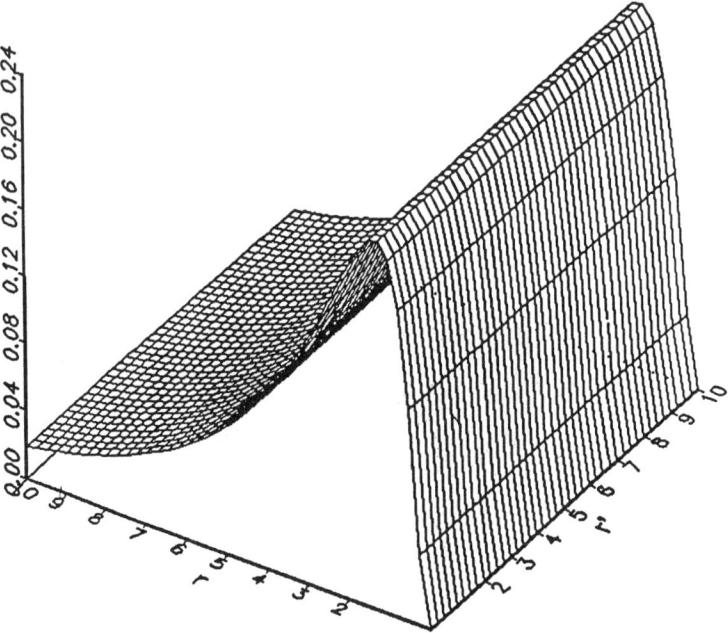

PLOT 7: $\rho$, with $\mu_R = 2$, $\phi = 0$.

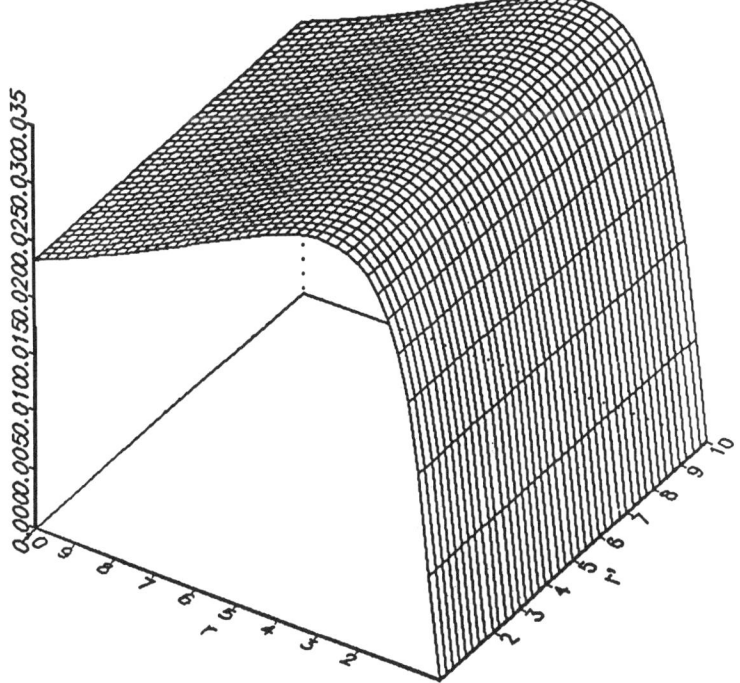

PLOT 8: $\rho$, with $\mu_R = 8$, $\phi = 0$.

Carlos Castillo–Chavez
Biometric Unit/Center for Applied Mathematics
Cornell University
341 Warren Hall
Ithaca, NY 14853–7801, USA

Stavros Busenberg
Department of Mathematics
Harvey Mudd College
Claremont, CA 91711, USA

Ken Gerow
Biometric Unit
Cornell University
341 Warren Hall
Ithaca, NY 14853–7801, USA

# Positivity for Operator Matrices

PANOS CHARISSIADIS
RAINER NAGEL

Mathematisches Institut, Universität Tübingen

## 1. Introduction

Systems of linear evolution equations can be written as

$$\dot{u}(t) = \mathcal{A}u(t), \quad u(0) = u_0, \qquad (*)$$

where $u(\cdot) = \begin{pmatrix} u_1(\cdot) \\ \vdots \\ u_n(\cdot) \end{pmatrix}$ takes values in a product $\mathcal{E} = E_1 \times \ldots \times E_n$ of Banach spaces $E_i$ and $\mathcal{A} = (A_{ij})_{n \times n}$ is a matrix whose entries $A_{ij}$ are linear operators from $E_j$ into $E_i$ (see [Be] or [Na 1], Sect. 3). In the one-dimensional case, i.e., if $E_i = \mathbb{C}$ for all $i$, we obtain a complex matrix $\mathcal{A}$ and linear algebra yields the appropriate tools for a detailed analysis of the initial value problem $(*)$. In the general case, i.e., for arbitrary Banach spaces $E_i$, the equation $(*)$ seems to be a notational trick with only few of the classical matrix results remaining true. In fact, since we have to allow (for resonable applications) unbounded and non-commuting entries $A_{ij}$ in the operator matrix $\mathcal{A}$ strange and difficult phenomena occur:

— It is not clear how to define an appropriate domain $D(\mathcal{A})$ of $\mathcal{A}$.
— There is no way to define some of the basic concepts from matrix theory such as *"determinant"*, *"trace"*, etc.

In contrast to these difficulties, the more general framework gives new flexibility and allows to restrict attention to $2 \times 2$-matrices. In fact, the $n \times n$-matrix

$\mathcal{A} = (A_{ij})_{n \times n}$ can be succesively built up by submatrices $\mathcal{A}_k := (A_{ij})_{k \times k}$ each of which is a $2 \times 2$-matrix

$$\mathcal{A}_k = \begin{pmatrix} \mathcal{A}_{k-1} & \mathcal{B}_k \\ \mathcal{C}_k & A_{kk} \end{pmatrix},$$

where $\mathcal{B}_k = \begin{pmatrix} A_{1k} \\ \vdots \\ A_{k-1\,k} \end{pmatrix}$, $\mathcal{C}_k = \begin{pmatrix} A_{k1} & \ldots & A_{k\,k-1} \end{pmatrix}$ and $E_1 \times \ldots \times E_k$ is considered as the product $(E_1 \times \ldots \times E_{k-1}) \times E_k$. Unfortunately many important properties of matrices are not inherited or characterized by submatrices.

**1.1. Example.** Let $A$ be the generator of a strongly continuous semigroup on a Banach space $E$ with spectrum equal to the left halfplane in $\mathbb{C}$ (e.g., $A := \frac{d}{dx}$ on $E := L^2(\mathbb{R}_+)$). Then

$$\mathcal{A} = \begin{pmatrix} (1+i)A & -\sqrt{\frac{3}{2}}A \\ -\sqrt{\frac{3}{2}}A & (1-i)A \end{pmatrix}$$

is an operator matrix generating a strongly continuous semigroup on $E \times E$ (see [Na 1], Thm. 2.3 or [E 1]). But none of its submatrices (i.e., entries) is a generator on $E$.

In this paper we will show that *positivity* properties are extremly helpful in order to overcome this drawback. They allow to a large extend (and in particular for the investigation of stability properties) to concentrate on $2 \times 2$-matrices for which the so-called "Schur complements" yield appropriate tools even in the unbounded, non-commutative case. But let us first review the corresponding results for scalar matrices.

## 2. Positivity in the scalar case

We consider a complex matrix $\mathcal{A} = (a_{ij})_{n \times n}$ and its generated semigroup $(T(t))_{t \geq 0} = (e^{t\mathcal{A}})_{t \geq 0}$. The following results can be found in most books on matrix theory (e.g., [Ga], [B-P], [L-T], [M]).

**2.1. Characterization Theorem.** *The following assertions are equivalent:*

(a) $e^{t\mathcal{A}} \geq 0$ for all $t \geq 0$.

(b) $a_{ii} \in \mathbb{R}$ and $a_{ij} \geq 0$ for $i \neq j$.

In order to state the basic spectral properties of positive matrix semigroups and their generators we recall that the *resolvent* of $\mathcal{A}$ is denoted by $R(\lambda, \mathcal{A}) := (\lambda - \mathcal{A})^{-1}$ for $\lambda$ not in the *spectrum* $\sigma(\mathcal{A})$ while the *spectral bound* of $\mathcal{A}$ is

$$s(\mathcal{A}) := \sup\{\operatorname{Re}\lambda : \lambda \in \sigma(\mathcal{A})\}.$$

**2.2. Perron-Frobenius Theorem.** *If the matrix $\mathcal{A}$ generates a positive semigroup then its spectral bound satisfies*

$$s(\mathcal{A}) = \inf\{\lambda \in \mathbb{R} : \lambda \notin \sigma(\mathcal{A}) \text{ and } R(\lambda, \mathcal{A}) \geq 0\} \in \sigma(\mathcal{A}).$$

The property most important for application is *stability* of the semigroup, i.e.,

$$\lim_{t \to \infty} e^{t\mathcal{A}} = 0.$$

By Liapunov's theorem this is characterized by the negativity of the spectral bound $s(\mathcal{A})$. But the special structure of positive matrices and their spectral properties mentioned above allow various other, quite useful characterizations.

**2.3. Stability Theorem.** *If the matrix $\mathcal{A}$ generates a positive semigroup then the following assertions are equivalent:*

(a) *There exists $\epsilon > 0$ such that $\lim_{t \to \infty} e^{\epsilon t} e^{t\mathcal{A}} = 0$.*

(b) $s(\mathcal{A}) < 0$.

(c) $0 \notin \sigma(\mathcal{A})$ *and* $\mathcal{A}^{-1} \leq 0$.

(d) $(-1)^{k+1} \det \mathcal{A}_k < 0$ *for each $1 \leq k \leq n$ and each submatrix $\mathcal{A}_k$.*

We remark that in that case $-\mathcal{A}$ is called an M-matrix (see [M], Chap. VI, Sect. 6.4).

## 3. The characterization problem

We now tackle the problem of extending Theorem 2.1 to the infinite dimensional situation. To that purpose we consider Banach lattices $E_1, \ldots, E_n$ (such as $\mathcal{C}(X)$-or $L^p$-spaces, see [S]) and an operator matrix $\mathcal{A} = (A_{ij})_{n \times n}$ with (possibly unbounded) linear operators $A_{ij}$ from $E_j$ into $E_i$. In addition we assume the reader to be familiar with the theory of strongly continuous semigroups of linear (see [Go]) and positive (see [Na 2]) operators. Our approach will be based on perturbation theory. More precisely we write $\mathcal{A}$ as the sum of the diagonal matrix $\mathcal{A}_0 := \text{diag}(A_{ii})$ and the matrix $\mathcal{B}$ containing the off-diagonal elements of $\mathcal{A}$. Then $\mathcal{A}_0$ is simple to understand while we need appropriate assumptions on $\mathcal{B}$ in order to discuss the sum

$$\mathcal{A} = \mathcal{A}_0 + \mathcal{B}.$$

Since these assumptions prevail throughout this paper we state them explicitly.

**3.1. Standard assumptions.**

($A_1$) The diagonal operators $A_{jj}$ (with dense domain $D(A_{jj})$) generate strongly continuous semigroups $(T_j(t))_{t \geq 0}$ on $E_i$.

($A_2$) The off-diagonal operators $A_{ij}$, $i \neq j$, are relatively bounded with respect to $A_{jj}$, i.e. $D(A_{ij}) \supset D(A_{jj})$ and $A_{ij}R(\lambda, A_{jj})$ is bounded for $\lambda > s(A_{jj})$.

($A_3$) The off-diagonal matrix $\mathcal{B}$ is "small" with respect to the diagonal $\mathcal{A}_0$. More precisely, there exists $\lambda_0 > s(\mathcal{A}_0)$ ($= \max\{s(A_{jj}) : 1 \leq j \leq n\}$) such that the (bounded) operator $\mathcal{B}R(\lambda_0, \mathcal{A}_0)$ has spectral radius $r(\mathcal{B}R(\lambda_0, \mathcal{A}_0)) < 1$.

A few comments on these assumptions might be helpful.

### 3.2. Comments.

($C_1$) The diagonal matrix $\mathcal{A}_0$ with domain $D(\mathcal{A}_0) := D(A_{11}) \times \ldots \times D(A_{nn})$ generates the semigroup $(\mathcal{T}_0(t))_{t \geq 0} := (\text{diag}(T_j(t)))_{t \geq 0}$ on $\mathcal{E}$.

($C_2$) The assumption ($A_3$) implies that $\mathcal{A}$ is a closed operator on the domain $D(\mathcal{A}) = D(\mathcal{A}_0)$. This follows from the decomposition

$$\lambda_0 - \mathcal{A} = (1 - \mathcal{B}R(\lambda_0, \mathcal{A}_0))(\lambda_0 - \mathcal{A}_0)$$

and the fact that a product of an invertible bounded operator with a closed operator remains closed.

($C_3$) If a certain off-diagonal matrix $\mathcal{B}$ satisfies ($A_3$) then clearly all multiples $\epsilon \mathcal{B}$ for $|\epsilon| \leq 1$ do the same. The easiest example is obtained from bounded $\mathcal{B}$, i.e., if all entries $A_{ij}$, $i \neq j$, are bounded.

($C_4$) The assumptions ($A_1$)–($A_3$) do not automatically imply that $\mathcal{A}$ generates a strongly continuous semigroup. For that one needs some extra condition (e.g. see [Go], Chap. I, Thm. 6.1) and we refer particularly to the results of Voigt [Vo] and Arendt-Rhandi [Ar-R] on perturbation of positive semigroups.

If we assume that $\mathcal{A} = \mathcal{A}_0 + \mathcal{B}$ generates a semigroup on $\mathcal{E}$ we are now ready to give a characterization for the positivity of the generated semigroup in terms of the entries of $\mathcal{A}$.

### 3.3. Characterization Theorem.
Let $\mathcal{A} = (A_{ij})_{n \times n}$ be an operator matrix on $\mathcal{E} = E_1 \times \ldots \times E_n$ satisfying ($A_1$)–($A_3$) and generating a strongly continuous semigroup $(\mathcal{T}(t))_{t \geq 0}$. Then the following conditions are equivalent:

(a) The semigroup $(\mathcal{T}(t))_{t \geq 0}$ is positive.

(b) (i) The semigroups $(T_j(t))_{t \geq 0}$ generated by the diagonal operators $A_{jj}$ are positive for $1 \leq j \leq n$.

(ii) The off-diagonal operators $A_{ij}$, $i \neq j$, are positive from $D(A_{jj})$ into $E_i$.

**Proof.** (b)$\Longrightarrow$(a): By decomposing $\mathcal{A}$ into the sum of $\mathcal{A}_0$ and $\mathcal{B}$ we see that $\mathcal{A}_0$ generates a positive semigroup while $\mathcal{B}$ is a positive perturbation. It follows from assumption ($A_3$) and the identity in ($C_2$) that $R(\lambda_0, \mathcal{A})$ is positive. As in [Ar], Thm. 3.1, this implies $R(\lambda, \mathcal{A})$ to be positive for all $\lambda \geq \lambda_0$, hence $\mathcal{A}$ has a positive resolvent and therefore generates a positive semigroup by [Na 2], B-II, Prop. 1.1.

(a)$\Longrightarrow$(b): We show first that (i) holds. For $1 \leq i \leq n$ consider the family $F_i(t)$, $t \geq 0$, of bounded positive operators on $E_i$ given by

$$F_i(t)f := P_i e^{t\mathcal{A}} J_i f, \quad f \in E_i,$$

where $P_i : \mathcal{E} \longrightarrow E_i$ (resp., $J_i : E_i \longrightarrow \mathcal{E}$) is the projection (resp., embedding). For $f \in D(A_{ii})$ we have $F_i'(0)f = A_{ii}f$. On the other hand, from the estimate

$$0 \leq F_i(t)^2 f = P_i e^{t\mathcal{A}} (J_i P_i) e^{t\mathcal{A}} J_i f \leq P_i e^{2t\mathcal{A}} J_i f \quad \text{for} \quad 0 \leq f \in E_i$$

we obtain

$$\|F_i(t)^2\| \leq \|e^{2t\mathcal{A}}\|.$$

Proceeding by induction this yields

$$\|F_i(t)^k\| \leq \|e^{kt\mathcal{A}}\| \quad \text{for} \quad k = 1, 2, \ldots \quad \text{and} \quad t \geq 0.$$

Thus $\|F_i(t)^k\| \leq M e^{kt\omega}$ for constants $M$, $\omega$ and all $k = 1, 2, \ldots$ and we have shown that the family $F_i(t)$, $t \geq 0$, satisfies the conditions of [Pa], Chap. III, Cor. 5.4. Therefore

$$\lim_{n \to \infty} F_i(\tfrac{t}{n})^n f = e^{tA_{ii}} f \quad \text{for all} \quad f \in E_i.$$

Since each $F_i(\tfrac{t}{n})^n$, $n = 1, 2, \ldots$, is positive the limit $e^{tA_{ii}}$ is also positive and statement (i) in (b) is proved.

We now proceed with part (ii). For each $f \in D(A_{ii})_+ := D(A_{ii}) \cap (E_i)_+$ we consider differentiable functions $\eta_f, \xi_f : \mathbb{R}_+ \longrightarrow \mathcal{E}_+$ defined by

$$\xi_f(t) := e^{t\mathcal{A}} J_i f \quad \text{and} \quad \eta_f(t) := F_i(t)f.$$

Obviously, $\eta_f(0) = \xi_f(0)$ and – since $(e^{t\mathcal{A}})_{t \geq 0}$ is positive – $\xi_f(t) \geq \eta_f(t)$ for $t \geq 0$ and $1 \leq i \leq n$. Consequently we obtain the following estimate for the derivatives:

$$\begin{pmatrix} A_{1i}f \\ \vdots \\ A_{ni}f \end{pmatrix} = \mathcal{A} \begin{pmatrix} 0 \\ \vdots \\ f \\ \vdots \\ 0 \end{pmatrix} = \xi_f'(0) \geq \eta_f'(0) = \begin{pmatrix} 0 \\ \vdots \\ A_{ii}f \\ \vdots \\ 0 \end{pmatrix}.$$

Since the order on $\mathcal{E}$ is defined coordinatewise we conclude $A_{ji}f \geq 0$ for $f \in D(A_{ii})_+$ and $j \neq i$ and the proof is complete. ∎

The above theorem clearly generalizes in a natural and satisfactory way the scalar Characterization Theorem 2.1. Two more examples are presented to show its broad range of applicability.

**3.4. Examples.** (1) In the semigroup approach to Volterra integro-differential equations (see [D-S], [D-G-S]) the operator matrix

$$\mathcal{A} := \begin{pmatrix} a & \delta_0 \\ c & D \end{pmatrix}$$

is considered on $\mathcal{E} := \mathbb{R} \times L^2(\mathbb{R}_+)$ where $a \in \mathbb{R}$, $D := \frac{d}{dx}$, $D(D) := W^1(\mathbb{R}_+)$, $\delta_0$ the Dirac measure in 0 and $c \in L^2(\mathbb{R}_+)$ is understood as an operator from $\mathbb{R}$ into $L^2(\mathbb{R}_+)$. The diagonal operators $a$ and $D$ generate positive semigroups. From $R(\lambda, D)f(t) = \int_t^\infty e^{-\lambda(s-t)} f(s) ds$ it follows that $\|\delta_0 \circ R(\lambda, D)\| = (2\lambda)^{-\frac{1}{2}}$ becomes small for large $\lambda$. Hence $\mathcal{A}$ satisfies all the assumptions $(A_1)$–$(A_3)$ and, in addition, generates a strongly continuous semigroup by [Na 3], Sect. 3, Ex. 1. Our theorem shows that the following statements are equivalent:

(a) The semigroup generated by $\mathcal{A}$ is positive.

(b) The function $c \in L^2(\mathbb{R}_+)$ is positive.

(2) Consider $A$ to be an elliptic differential operator (such as the Laplacian) on $E = L^p(\Omega)$, $1 \leq p < \infty$, $\Omega \subset \mathbb{R}^m$. For many natural boundary conditions this operator generates a positive semigroup on $E$ (see [Na 2], C-III, Ex. 2.14) but clearly is never positive itself. Take now the system given by the operator matrix

$$\mathcal{A} = (a_{ij}A)_{n \times n}$$

on $E^n$ with complex coefficient matrix $(a_{ij})_{n \times n}$. From the results in [Na 1] (or more generally [E 1] and [E 2]) it can be decided when $\mathcal{A}$ is a generator on $E^n$. In that case this semigroup is positive if and only if $a_{ij} = 0$ for $i \neq j$ and $a_{ii} \geq 0$ for $1 \leq i \leq n$. This explains the special case treated by [Ka], Thm. 2.1.

## 4. The stability problem

In this section we will show how the stability of the solutions of the Cauchy problem

$$\dot{u}(t) = \mathcal{A}u(t), \quad u(0) = u_0, \qquad (*)$$

for an $n \times n$-operator matrix $\mathcal{A}$ on a product space $\mathcal{E} := E_1 \times \ldots \times E_n$ generating a positive semigroup can be characterized by conditions for certain operators on the factor spaces $E_i$. Our approach combines results from the abstract theory of positive semigroups on Banach lattices (see [Na 2], C-IV) with the analogue of the Schur-complement for operator matrices (see [Na 3], Sect. 2).

We first recall the stability result for arbitrary positive semigroups on Banach lattices based on the (infinite dimensional) Perron-Frobenius theory as developed in [Na 2] (compare Thm. 2.2, 2.3 above).

**4.1. Stability criterion** (see [Na 2], C-III, Thm. 1.1). *Let $A$ with domain $D(A)$ be generator of a strongly continuous semigroup $(T(t))_{t\geq 0}$ of positive operators on some Banach lattice $E$. Then the spectral bound $s(A) := \sup\{\text{Re}\lambda : \lambda \in \sigma(A)\}$ satisfies*

$$s(A) = \inf\{\lambda \in \mathbb{R} : R(\lambda, A) \geq 0\} \in \sigma(A). \qquad (\star)$$

*As a consequence* (use [Na 2], C-IV, Cor. 1.4) *the following assertions are equivalent:*
  (a) *There exists $\epsilon > 0$ such that $\lim_{t\to\infty} e^{\epsilon t}T(t)f = 0$ for every $f \in D(A)$, i.e., $(T(t))_{t\geq 0}$ is exponentially stable.*
  (b) *The spectral bound satisfies $s(A) < 0$.*
  (c) *The generator $A$ is invertible and its inverse $A^{-1}$ is a negative operator.*

We are now interested in adding to this list an equivalent condition (d) taking into account the matrix structure and generalizing condition 2.3.d. To that purpose we again assume $\mathcal{A} = (A_{ij})_{n\times n}$ to satisfy the assumptions $(A_1)$–$(A_3)$ and, in addition, to generate a positive semigroup $(\mathcal{T}(t))_{t\geq 0}$ on the Banach lattice $\mathcal{E} := E_1 \times \ldots \times E_n$. Since we want to apply the concept of Schur complements for $2\times 2$-matrices we show first that these assumptions are preserved by considering certain submatrices.

**4.2. Lemma.** *Let $(\mathcal{T}(t))_{t\geq 0}$, resp. $(\mathcal{T}_0(t))_{t\geq 0}$ be the positive semigroups on $\mathcal{E}$ generated by $\mathcal{A} = (A_{ij})_{n\times n}$, resp. by $\mathcal{A}_0 := \text{diag}(A_{ii})_{n\times n}$. Write $\mathcal{A} = \mathcal{A}_0 + \mathcal{B}$ and take a matrix $\tilde{\mathcal{B}}$ satisfying $0 \leq \tilde{\mathcal{B}}f \leq \mathcal{B}f$ for all $f \in D(\mathcal{A})_+$. Then the operator matrix $\tilde{\mathcal{A}} := \mathcal{A}_0 + \tilde{\mathcal{B}}$ generates a positive semigroup $(\tilde{\mathcal{T}}(t))_{t\geq 0}$ such that $0 \leq \mathcal{T}_0(t) \leq \tilde{\mathcal{T}}(t) \leq \mathcal{T}(t)$ for all $t \geq 0$.*

**Proof.** We have seen in $(C_2)$ how to compute the resolvent of the perturbed operator $\mathcal{A}_0 + \tilde{\mathcal{B}}$. Our assumption on $\tilde{\mathcal{B}}$ implies $\tilde{\mathcal{B}}R(\lambda, \mathcal{A}_0) \leq \mathcal{B}R(\lambda, \mathcal{A}_0)$ for $\lambda > \lambda_0$. Therefore $r(\tilde{\mathcal{B}}R(\lambda, \mathcal{A}_0)) \leq r(\mathcal{B}R(\lambda, \mathcal{A}_0)) < 1$ and $0 \leq R(\lambda, \tilde{\mathcal{A}}) \leq R(\lambda, \mathcal{A})$ for large $\lambda$. But the domination of resolvents implies the domination of the semigroups by [Na 2], C-II, Prop. 4.1 . ∎

This lemma shows in particular that $\mathcal{A}$ generates an exponentially stable, positive semigroup on $\mathcal{E}$ (i.e. $s(\mathcal{A}) < 0$) if and only if each submatrix $\mathcal{A}_k := (A_{ij})_{k\times k}$ on $\mathcal{E}_k := E_1 \times \ldots \times E_k$, $1 \leq k \leq n$, generates an exponentially stable, positive semigroup. In fact one has $s(\mathcal{A}) \geq s(\mathcal{A}_k) \geq s(\mathcal{A}_{k-1})$ for $k = 2, \ldots, n$. Hence stability for positive semigroups is a property inherited by submatrices (compare Sect. 1) and we are able to concentrate on 2 × 2-matrices by decomposing each $\mathcal{A}_k$ into a 2 × 2- matrix of the form

$$\mathcal{A}_k = \begin{pmatrix} \mathcal{A}_{k-1} & \mathcal{B}_k \\ \mathcal{C}_k & A_{kk} \end{pmatrix},$$

where $\mathcal{B}_k = \begin{pmatrix} A_{1k} \\ \vdots \\ A_{k-1k} \end{pmatrix}$, $\mathcal{C}_k = \begin{pmatrix} A_{k1} & \cdots & A_{kk-1} \end{pmatrix}$ and $\mathcal{E}_k = (E_1 \times \ldots \times E_{k-1}) \times E_k$.

To all these 2 × 2-matrices we shall apply the invertibility criterion developed in [Na 3], Sect. 2 using (infinite dimensional) *Schur complements*. For the readers convenience we repeat the relevant formulas from [Na 3], Lemma 2.1 and Thm. 2.4.

**4.3. Lemma.** *Let $E$, $F$ be Banach spaces and* $\mathcal{A} := \begin{pmatrix} A & B \\ C & D \end{pmatrix}$ *an operator matrix on* $\mathcal{E} := E \times F$ *satisfying the assumptions (analogous to)* $(A_1)$- $(A_3)$.

*(1) If $A$ is invertible on $E$ the following assertions are equivalent:*

$(a_1)$ *The operator matrix $\mathcal{A}$ is invertible on $\mathcal{E}$.*

$(b_1)$ *The Schur complement $D - CA^{-1}B$ is invertible on $F$.*

In that case the inverse of $\mathcal{A}$ is obtained as

$$\mathcal{A}^{-1} = \begin{pmatrix} A^{-1}(Id + B(D - CA^{-1}B)^{-1}CA^{-1}) & -A^{-1}C(D - CA^{-1}B)^{-1} \\ -(D - CA^{-1}B)^{-1}CA^{-1} & (D - CA^{-1}B)^{-1} \end{pmatrix}.$$

*(2) These facts applied to the matrix* $\lambda - \mathcal{A} = \begin{pmatrix} \lambda - A & -B \\ -C & \lambda - D \end{pmatrix}$ *for $\lambda \notin \sigma(A)$ yield the following equivalence:*

$(a_2)$ $\lambda \notin \sigma(\mathcal{A})$.

$(b_2)$ $\lambda \notin \sigma(D + CR(\lambda, A)B)$.

In that case the resolvent of $\mathcal{A}$ is obtained as

$$R(\lambda, \mathcal{A}) = \begin{pmatrix} R(\lambda, A + BR(\lambda, D)C) & R(\lambda, A + BR(\lambda, D)C)BR(\lambda, D) \\ R(\lambda, D + CR(\lambda, A)B)CR(\lambda, A) & R(\lambda, D + CR(\lambda, A)B) \end{pmatrix}.$$

With these concepts, applied to the submatrices $\mathcal{A}_k$, we are now able to characterize the operator matrices generating exponentially stable, positive semigroups.

**4.4. Theorem.** Let $\mathcal{A} = (A_{ij})_{n \times n}$ be an operator matrix satisfying the assumptions $(A_1)$–$(A_3)$ and generating a positive semigroup $(\mathcal{T}(t))_{t \geq 0}$ on $\mathcal{E} := E_1 \times \ldots \times E_n$, $E_i$ Banach lattices. The following assertions are equivalent:

(a) The semigroup $(\mathcal{T}(t))_{t \geq 0}$ is exponentially stable.

(b) The spectral bound of $\mathcal{A}$ satisfies $s(\mathcal{A}) < 0$.

(c) The operator matrix $\mathcal{A}$ is invertible with negative inverse on $\mathcal{E}$.

(d) The spectral bounds of $A_{11}$ and $A_{kk} - \mathcal{C}_k \mathcal{A}_{k-1}^{-1} \mathcal{B}_k$ satisfy $s(A_{11}) < 0$ and $s(A_{kk} - \mathcal{C}_k \mathcal{A}_{k-1}^{-1} \mathcal{B}_k) < 0$ for $k = 2, \ldots, n$.

(d') The operators $A_{11}$ and $A_{kk} - \mathcal{C}_k \mathcal{A}_{k-1}^{-1} \mathcal{B}_k$ are invertible with negative inverses on $E_k$ for $k = 2, \ldots, n$.

**Proof.** The equivalences $(a) \iff (b) \iff (c)$ hold for each positive semigroup on any Banach lattice (see Stability Criterion 4.1).

In the next preparatory step we show that for any $\mu > s(\mathcal{A}) \geq s(\mathcal{A}_k) \geq s(\mathcal{A}_{k-1})$ the operators $A_{kk} + \mathcal{C}_k R(\mu, \mathcal{A}_{k-1}) \mathcal{B}_k$ appearing in the Schur complement representation of $R(\mu, \mathcal{A}_k)$ (compare Lemma 4.3 (2)) are resolvent positive (see [Ar]). All the matrices $\mathcal{A}_{k\epsilon} := \begin{pmatrix} \mathcal{A}_{k-1} & \epsilon \mathcal{B}_k \\ \mathcal{C}_k & A_{kk} \end{pmatrix}$, $0 < \epsilon \leq 1$, satisfy the assumptions $(A_1)$–$(A_3)$ and generate positive semigroups dominated by the semigroup $(\mathcal{T}_k(t))_{t \geq 0}$ generated by $\mathcal{A}_k$. The invertibility of $\mu - \mathcal{A}_{k-1}$ implies by Lemma 4.3.2 that the operators $\mu - A_{kk} - \epsilon \mathcal{C}_k R(\mu, \mathcal{A}_{k-1}) \mathcal{B}_k$ are invertible for all $0 < \epsilon \leq 1$. Moreover, the identity

$$\mu - A_{kk} - \epsilon \mathcal{C}_k R(\mu, \mathcal{A}_{k-1}) \mathcal{B}_k = [1 - \epsilon \mathcal{C}_k R(\mu, \mathcal{A}_{k-1}) \mathcal{B}_k R(\mu, A_{kk})](\mu - A_{kk}) \quad (*)$$

shows that $\frac{1}{\epsilon}$ is contained in the resolvent set of the positive (bounded) operator $\mathcal{C}_k R(\mu, \mathcal{A}_{k-1}) \mathcal{B}_k R(\mu, A_{kk})$ for all $0 < \epsilon \leq 1$. Since the spectral radius of a positive operator is contained in its spectrum (see [S], Chap. V, Prop. 4.1) this implies that the spectral radius $r(\mathcal{C}_k R(\mu, \mathcal{A}_{k-1}) \mathcal{B}_k R(\mu, A_{kk}))$ is smaller than 1. For $\lambda > \mu$ we have $R(\lambda, A_{kk}) \leq R(\mu, A_{kk})$ (observe that $A_{kk}$ generates a positive semigroup and use [Ar], Sect. 2), hence

$$0 \leq \mathcal{C}_k R(\mu, \mathcal{A}_{k-1}) \mathcal{B}_k R(\lambda, A_{kk}) \leq \mathcal{C}_k R(\mu, \mathcal{A}_{k-1}) \mathcal{B}_k R(\mu, A_{kk}) \quad (**)$$

and a similar inequality for the spectral radii. We now decompose $\lambda - A_{kk} - \mathcal{C}_k R(\mu, \mathcal{A}_{k-1}) \mathcal{B}_k$ as in $(*)$ and obtain for its inverse

$$0 \leq R(\lambda, A_{kk} + \mathcal{C}_k R(\mu, \mathcal{A}_{k-1}) \mathcal{B}_k) \leq R(\mu, A_{kk} + \mathcal{C}_k R(\mu, \mathcal{A}_{k-1}) \mathcal{B}_k) \quad (***)$$

for $\lambda > \mu$, i.e. $A_{kk} + \mathcal{C}_k R(\mu, \mathcal{A}_{k-1}) \mathcal{B}_k$ has a positive resolvent (see [Ar]). Since the assertion $(\star)$ and hence the equivalence of $(b)$ and $(c)$ in the Stability Criterion

4.1 hold for any resolvent positive operator (see [Na 2], C-III, Remark 2.15), we have shown that $(d)$ and $(d')$ are equivalent.

$(b) \Longrightarrow (d')$. By assumption the operator matrix $\mathcal{A}$ and all submatrices $\mathcal{A}_k$ are invertible with negative inverses. If we take the matrix representation

$$\mathcal{A}_k = \begin{pmatrix} \mathcal{A}_{k-1} & \mathcal{B}_k \\ \mathcal{C}_k & A_{kk} \end{pmatrix}$$

we obtain that the Schur complement $A_{kk} - \mathcal{C}_k \mathcal{A}_{k-1}^{-1} \mathcal{B}_k$ is invertible. Its inverse appears as the lower right entry of $\mathcal{A}_k^{-1}$ (use Lemma 4.3.1), hence is a negative operator on $E_k$.

$(d') \Longrightarrow (c)$. Again it follows successively from Lemma 4.3.1 that each submatrix $\mathcal{A}_k$ is invertible. A careful inspection shows that this inverse has only negative entries. For $k = n$ we obtain the assertion. ∎

The above theorem extends [Na 4], Thm. 3.3 and Cor. 3.4 to operator matrices with unbounded, off-diagonal entries. But clearly it contains the scalar Stability Theorem 2.3 as a special case (use the identity

$$\det \mathcal{A}_k = (a_{kk} - \mathcal{C}_k \mathcal{A}_{k-1}^{-1} \mathcal{B}_k) \det \mathcal{A}_{k-1}$$

valid for scalar matrices $\mathcal{A}_k = (a_{ij})_{k \times k}$).

More examples demonstrating the usefulness of the above theorem can be found in [Na 4], Sect. 4. Here we discuss an additional example with unbounded, off-diagonal entries.

**4.5. Example.** Let us go back to the matrix

$$\mathcal{A} = \begin{pmatrix} a & \delta_0 \\ c & D \end{pmatrix}$$

on $\mathbb{R} \times L^2(\mathbb{R}_+)$ introduced in Example 3.4 (1). In order to obtain a positive semigroup we assume $c \in L^2(\mathbb{R}_+)$ to be a positive function. Since $D$ generates a contraction semigroup on $L^2(\mathbb{R}_+)$ with $s(D) = 0$ it is clear that $\mathcal{A}$ will not satisfy $s(\mathcal{A}) < 0$. Therefore we consider the matrix

$$\mathcal{A}_\mu := \begin{pmatrix} a - \mu & \delta_0 \\ c & D - \mu \end{pmatrix}$$

for $\mu > 0$. Then $s(D - \mu) = -\mu < 0$ and condition 4.4.d yields stability if and only if

$$0 > s(a - \mu + \delta_0 \circ R(0, D - \mu) \circ c)$$
$$= a - \mu + \delta_0(R(\mu, D)c) = a - \mu + \int_0^\infty e^{-\mu s} c(s) ds, \quad \text{or}$$
$$\mu > a + \int_0^\infty e^{-\mu s} c(s) ds.$$

In terms of the original operator $\mathcal{A}$ this means that for any $\mu > 0$ the following statements are equivalent:

(a) $s(\mathcal{A}) < \mu$.
(b) $a + \int_0^\infty e^{-\mu s} c(s) ds < \mu$.

This extremely simple criterion allows estimates for the growth of the solutions of the corresponding Volterra integro-differential equation

$$\dot{x}(t) = ax(t) + \int_0^t c(s)x(s)ds$$

(see [D-G-S], [Na 3]). Moreover it can be generalized easily to vector valued and even Banach space valued equations.

**4.6. Concluding remark.** Theorem 4.4 reduces the stability problem for an operator on a product space $\mathcal{E} = E_1 \times \ldots \times E_n$ to $n$ stability problems on the factor spaces $E_1, \ldots, E_n$. This is essentially based on the positivity of the generated semigroups. If this assumption is not satisfied one might change $\mathcal{A}$ into a matrix $\mathcal{A}^\sharp$ generating a *dominating* semigroup (see [Na 2], C-II, Sect. 4). Since the stability of this dominating semigroup implies the stability of the original semigroup our Theorem 4.4 is applicable by, e.g., verifying conditions 4.4.d for $\mathcal{A}^\sharp$. This will be worked out in detail in [Ch].

## References

[Ar]    W. Arendt, *Resolvent positive operators*, Proc. London Math. Soc. **54** (1987), 321–349.
[Ar-R]  W. Arendt and A. Rhandi, *Perturbation of positive semigroups* (to appear).
[Be]    A. Belleni-Morante, *Applied semigroups and evolution equations*, Oxford University Press, 1979.
[B-P]   A. Berman and R. Plemmons, *Nonnegative matrices in the mathematical sciences*, Academic Press, 1979.
[Ch]    P. Charissiadis, *Beiträge zu einer Perron-Frobenius Theorie für Operatormatrizen*, Dissertation, Tübingen 1990.
[D-G-S] W. Desch, R. Grimmer and W. Schappacher, *Wellposedness and wave propagation for a class of integrodifferential equations in Banach space*, J. Diff. Eq. **74** (1988), 391–441.
[D-S]   W. Desch and W. Schappacher, *A semigroup approach to integrodifferential equations on Banach spaces*, J. Int. Eq. **10** (1985), 99–110.
[E 1]   K.-J. Engel, *Polynomial operator matrices as semigroup generators: the $2 \times 2$ case*, Math. Ann. **284** (1989), 563–576.
[E 2]   K.-J. Engel, *Polynomial operator matrices as semigroup generators: the general case*, Integral Equations and Operator Theory (to appear).
[Ga]    F. R. Gantmacher, *Matrizentheorie*, Springer-Verlag, 1986.
[Go]    J. A. Goldstein, *Semigroups of Linear Operators and Applications*, Oxford University Press, 1985.
[Ka]    C. Kahane, *On the nonnegativity of solutions of reaction diffusion equations*, Rocky Mountain J. Math. **17** (1987), 491–498.
[L-T]   P. Lancaster and M. Tismenetsky, *The Theory of Matrices*, Academic Press, 1985.
[M]     H. Minc, *Nonnegative Matrices*, Wiley Interscience, 1988.

[Na 1]   R. Nagel, *Well-posedness and positivity for systems of linear evolution equations*, Conferenze del Seminario di Mathematica **203** (1985), Bari.

[Na 2]   R. Nagel (ed.), *One-parameter Semigroups of Positive Operators*, Lecture Notes Math., vol. 1184, Springer-Verlag, 1986.

[Na 3]   R. Nagel, *Towards a "matrix theory" for unbounded operator matrices*, Math. Z. **201** (1989), 57–68.

[Na 4]   R. Nagel, *On the stability of positive semigroups generated by operator matrices*, Semesterbericht Funktionalanalysis Tübingen, Sommersemester 1988.

[Pa]     A. Pazy, *Semigroups of linear operators and applications to partial differential equations*, Springer-Verlag, 1983.

[S]      H. H. Schaefer, *Banach Lattices and Positive Operators*, Springer-Verlag, 1974.

[Vo]     J. Voigt, *On resolvent positive operators and positive $C_0$-semigroups on AL-spaces*, Semigroup Forum **38** (1989), 263–266.


Panos Charissiadis
Rainer Nagel
Mathematisches Institut
Universität Tübingen
Auf der Morgenstelle 10
D-7400 Tübingen, FRG
Electronic address:
MINA001@DTUZDV5A.bitnet

# Time Dependent Differential Equations in Non Reflexive Banach Spaces

GIUSEPPE DA PRATO
EUGENIO SINESTRARI

Scuola Normale Superiore di Pisa
Dipartimento di Matematica, Università di Roma

## 1. Introduction

We consider two Banach spaces, $(E, |\cdot|)$ and $(D, \|\cdot\|)$ with $D$ continuously imbedded in $E$ and denote by $\overline{D}$ the closure of $D$ in $E$.

If $L : D(L) \subset E \to E$ is a linear operator in $E$, we denote by $\rho(L)$ the resolvent set of $L$, and set for $\lambda \in \rho(L), R(\lambda, L) = (\lambda - L)^{-1}$. Moreover $\mathcal{L}(E)$ (resp.$\mathcal{L}(D)$), resp. $\mathcal{L}(D; E)$) represent the Banach space of all linear continuous operators from $E$ into $E$ (resp. from $D$ into $D$, resp. from $D$ into $E$). The norms in $\mathcal{L}(E)$, $\mathcal{L}(D)$ and $\mathcal{L}(D; E)$ are denoted by $\|\cdot\|$; if some confusion could happen we shall write the suitable sub-scripts.

Given a mapping :

$$A : [0, T] \to \mathcal{L}(D; E), t \to A(t)$$

and chosen $\tau \in [0, T[$ , we consider the Cauchy problem

$$\begin{aligned} u'(t) &= A(t)u(t) + f(t); \ \tau \leq t \leq T, \\ u(\tau) &= x, \end{aligned} \qquad (1.1)$$

where $f \in L^1(\tau, T; E)$ and $x \in E$.

If $A$ is independent of $t$ and $D$ is dense in $E$, problem (1.1) can be studied by the classical semigroups theory (see for instance [8]). If $\overline{D} \neq E$ this theory was generalized in [3],[4]. When $A$ depends on $t$, problem (1.1) was extensively studied by T. Kato ([5]-[7]), (see also [2]), under the hypothesis of density of $D$.

The goal of this paper is to extend these results to the case in which $D$ is not dense in $E$.

Our main hypotheses ($\mathcal{H}$) are the following :
  (i) There exists $C > 0$ such that $\frac{1}{C}\|x\| \leq |x| + |A(t)x| \leq C\|x\|$ for all $x \in D$ and $t \in [0, T]$.
  (ii) There exist $\omega < 0$ and $M > 0$ such that $\rho(A(t)) \supset ]\omega, \infty[$, for all $t \in [0, T]$. In addition for $n \in \mathbf{N}$

$$\left\|\prod_{i=1}^{n} R(\lambda, A(t_i))\right\| \leq \frac{M}{(\lambda - \omega)^n},$$

when $T \geq t_1 \geq \cdots \geq t_n \geq 0, \lambda > \omega$.

When necessary we shall write $(\mathcal{H})_{C,\omega,M}$ instead of $(\mathcal{H})$. We shall also suppose that (at least).
  (iii) $A \in C([0,T]; \mathcal{L}(D; E))$.

But to get some of our existence results (see Theorems 3.3 and 4.1) we will require that $A$ can be suitably approximated by more regular operators $A_k$: in future papers dedicated to the applications of this theory to partial differential equations we will show how this condition can be verified under mild regularity assumptions on the coefficients of the differential operator.

We say that u is a *strict solution* of (1.1) if

$$\begin{aligned} &u \in C^1([\tau, T]; E) \cap C([\tau, T]; D), \\ &u'(t) = A(t)u(t) + f(t); \; t \in [\tau, T] \; ; \; u(\tau) = x. \end{aligned} \tag{1.2}$$

Let $W^{1,1}(\tau, T; E)$ denote the usual Sobolev space: $u \in L^1(\tau, T; E)$ is said to be a *strong solution* of (1.1) if there exists a sequence $\{u_k\}$ such that :

  (i)   $u_k \in W^{1,1}(\tau, T; E) \cap L^1(\tau, T; D)$.
  (ii)  $u_k(\cdot) \to u(\cdot)$ and $u_k'(\cdot) - A(\cdot)u_k(\cdot) \to f$ in $L^1(\tau, T; E)$. (1.3)
  (iii) $u_k(\tau) \to x$ in $E$.

## 2. Uniqueness

It is useful to introduce two linear operators in $L^1(\tau, T; E)$, by setting

$$\begin{aligned} &Bu = -u'; \\ &D(B) = W_0^{1,1}(\tau, T; E) = \{u \in W^{1,1}(\tau, T; E); \; u(\tau) = 0\} \end{aligned} \tag{2.1}$$

and

$$\begin{aligned} &(\Lambda u)(t) = A(t)u(t); \; t \in [\tau, T]; \\ &D(\Lambda) = L^1(\tau, T; D). \end{aligned} \tag{2.2}$$

From $(\mathcal{H})$ we deduce that

$$L^1(\tau,T;D) = \{u \in L^1(\tau,T;E);\ u(t) \in D, t \in [\tau,T]\ \text{a.e.} \\ \text{and}\ A(\cdot)u(\cdot) \in L^1(\tau,T;E)\} \quad (2.3)$$

$\rho(\Lambda) \supset ]\omega,\infty[$ and :

$$(R(\lambda,\Lambda)u)(t) = R(\lambda,A(t))u(t),\ t \in [\tau,T],\ u \in L^1(\tau,T;E). \quad (2.4)$$

It is known that $B$ is the infinitesimal generator of a contractions semi-group in $L^1(\tau,T;E)$. We shall consider the approximating problem

$$B_n(u_n - x) + \Lambda u_n + f = 0, \quad (2.5)$$

where $B_n = nBR(n,B) = n^2 R(n,B) - n$ are the Yosida approximations of $B$.
Equation (2.5) can be written as an integral equation :

$$n^2 \int_\tau^t e^{-n(t-s)}(u_n(s) - x)ds + n(x - u_n(t)) + A(t)u_n(t) + f(t) = 0. \quad (2.6)$$

The following result can be found e.g. in Chapter 3 of [1].

**Lemma 2.1.** *Let* $s \in [0,T[$, $K \in C([s,T];\mathcal{L}(E))$, $\varphi \in L^1(s,T;E), x \in E$. *Then the initial value problem:*

$$\begin{aligned} z'(t) &= K(t)z(t) + \varphi(t);\ s \leq t \leq T \\ z(s) &= x \end{aligned} \quad (2.7)$$

*has a unique solution* $z \in W^{1,1}(s,T;E)$, *given by the formula* :

$$z(t) = U(t,s)x + \int_s^t U(t,\sigma)\varphi(\sigma)d\sigma \quad (2.8)$$

*where* :

$$U(t,s) = I + \sum_{k=1}^\infty \int_{\Delta_k(s,t)} K(t_k)\ldots K(t_1)dt_k\ldots dt_1 \quad (2.9)$$

*and* :

$$\Delta_k(s,t) = \{(t_1,\ldots,t_k) \in \mathbb{R}^k;\ s \leq t_1 \leq t_2 \leq \cdots \leq t_k \leq t\}. \quad (2.10)$$

*If, in addition,* $\varphi \in C^k([s,T];E), K \in C^k([s,T];\mathcal{L}(E))$ *then* $z \in C^{k+1}([s,T];E)$.

We can now prove the following result :

**Proposition 2.2.** Let $x \in E$ and $L^1(s,T;E)$. Then problem (2.5) has a unique solution $u_n \in L^1(s,T;D)$ and the following estimate holds for $t \in [\tau,T]$, a.e. :

$$|u_n(t)| \leq \frac{Mn}{n-\omega} e^{\frac{n\omega}{n-\omega}(t-\tau)}|x|$$
$$+ \frac{Mn^2}{(n-\omega)^2} \int_\tau^t e^{\frac{n\omega}{n-\omega}(t-s)} |f(s)|ds + \frac{M}{n-\omega}|f(t)|. \quad (2.11)$$

Moreover if $A \in C^1([\tau,T]; \mathcal{L}(D;E))$ and $f \in W^{1,1}(\tau,T;E)$, then $u_n \in W^{1,1}(\tau,T;D)$ and

$$u_n(\tau) = R(n,A(\tau))f(\tau) + nR(n,A(\tau))x. \quad (2.12)$$

**Proof.** We follow here [4]. If $u_n \in L^1([\tau,T];D)$ is a solution of equation (2.5), we deduce from (2.6) :

$$n^2 R(n,A(t)) \int_\tau^t e^{-n(t-s)} u_n(s)ds + ne^{-n(t-\tau)} R(n,A(t))x - u_n(t) \\ + R(n,A(t))f(t) = 0 \quad (2.13)$$

and this implies (2.12) and

$$n^2 R(n,A(t)) \int_\tau^t e^{ns} u_n(s)ds + ne^{n\tau} R(n,A(t))x + e^{nt} R(n,A(t))f(t) = e^{nt} u_n(t).$$

Setting :

$$w_n(t) = \int_\tau^t e^{ns} u_n(s)ds \quad (2.14)$$

we have, $w_n \in W^{1,1}(\tau,T;D)$ and :

$$w'_n(t) = K_n(t)w_n(t) + \varphi_n(t) \\ w_n(\tau) = 0 \quad (2.15)$$

where :

$$\varphi_n(t) = R(n,A(t))(ne^{n\tau}x + e^{nt}f(t)) \quad K_n(t) = n^2 R(n,A(t)). \quad (2.16)$$

Thus, by Lemma 2.1, we know that $w_n$ must be given by the formula :

$$w_n(t) = \int_\tau^t U(t,\sigma)R(n,A(\sigma))(ne^{n\tau}x + e^{n\sigma}f(\sigma))d\sigma \quad (2.17)$$

where

$$U(t,s) = I + \sum_{k=1}^{\infty} \int_{\Delta_k(s,t)} n^{2k} R(n, A(\sigma_k)) \ldots R(n, A(\sigma_1)) d\sigma_k \ldots d\sigma_1 \quad (2.18)$$

and $\Delta_k(s,t)$ is defined by (2.10). It follows that:

$$w_n(t) = \int_\tau^t \{[R(n, A(s))$$
$$+ \sum_{k=1}^{\infty} \int_{\Delta_k(s,t)} n^{2k} R(n, A(\sigma_k)) R(n, A(\sigma_1)) d\sigma_k \ldots d\sigma_1 R(n, A(s))]\} \quad (2.19)$$
$$\times [n e^{n\tau} x + e^{ns} f(s)] ds$$

which implies :

$$u_n(t) = n^2 e^{-nt} R(n, A(t)) w_n(t) + n R(n, A(t)) e^{-n(t-\tau)} x + R(n, A(t)) f(t)$$
$$= n R(n, A(t)) e^{-n(t-\tau)} x + R(n, A(t)) f(t) + n^2 e^{-nt} \int_\tau^t \{ R(n, A(t)) R(n, A(s))$$
$$+ \sum_{k=1}^{\infty} \int_{\Delta_k(s,t)} n^{2k} R(n, A(t)) R(n, A(\sigma_k)) \ldots R(n, A(\sigma_1)) d\sigma_k \ldots d\sigma_1 R(n, A(s)) \}$$
$$\times [n e^{n\tau} x + e^{ns} f(s)] ds. \quad (2.20)$$

Since the measure of $\Delta_k(s,t)$ is $\frac{(t-s)^k}{k!}$, we deduce from $(\mathcal{H})$

$$|u_n(t)| \le \frac{Mn}{n-\omega} e^{-n(t-\tau)} |x| + \frac{M}{n-\omega} |f(t)| + n^2 e^{-nt} \quad (2.21)$$
$$\times \int_\tau^t \left\{ \left[ \frac{M}{(n-\omega)^2} + \sum_{k=1}^{\infty} \frac{(t-s)^k}{k!} n^{2k} \frac{M}{(n-\omega)^{k+2}} \right] [n e^{n\tau} |x| + e^{ns} |f(s)|] \right\} ds$$

and (2.11) follows. Conversely, given $x \in E$ and $f \in L^1(\tau, T; E)$ we get from Lemma 2.1 a solution $w_n \in W^{1,1}(\tau, T; E)$ of (2.15)-(2.16): then $u_n(t) = e^{-nt} w'_n(t)$ is a solution of (2.5). $\square$

We give now an a priori estimate for the solutions of problem (1.1).

**Proposition 2.3.** *Let $x \in E, f \in L^1(\tau, T; E)$, and let $u$ be a strong solution of (1.1). Then the following estimate holds :*

$$|u(t)| \le M e^{\omega(t-\tau)} |x| + M \int_\tau^t e^{\omega(t-s)} |f(s)| ds. \quad (2.22)$$

*Moreover, (possibly modifying u in a set of Lebesgue measure zero) we have*

$$u \in C([\tau,T]; E), \ u(\tau) = x \quad (2.23)$$

*and $u(t) \in \overline{D}$ for $t \in [\tau, T]$.*

**Proof.** We first assume that u is a strict solution of (1.1). Then we have

$$B_n(u-x) + \Lambda u = h_n \quad (2.24)$$

where

$$h_n = (B_n - B)(u - x) - f \quad (2.25)$$

By (2.11) it follows :

$$|u(t)| \leq \frac{Mn}{n-\omega} e^{\frac{n\omega}{n-\omega}(t-\tau)} |x| + \frac{Mn^2}{(n-\omega)^2} \int_\tau^t e^{\frac{n\omega}{n-\omega}(t-s)} |h_n(s)| ds \\ + \frac{M}{n-\omega} |h_n(s)|. \quad (2.26)$$

Since $u - x \in D(B)$, $\{h_n\}$ is bounded in $L^1(\tau, T; E)$. Thus (2.22) follows from (2.26) letting $n$ tend to infinity.

Let now $u$ be a strong solution of (1.1) and let $\{u_k\}$ be a sequence such that (1.3) hold. Setting

$$f_k = u'_k - \Lambda u_k, \ x_k = u_k(\tau) \quad (2.27)$$

$u_h - u_k$ is a strict solution of the problem :

$$\begin{aligned}(u_h - u_k)' - \Lambda(u_h - u_k) &= f_h - f_k, \\ (u_h - u_k)(\tau) &= x_h - x_k \end{aligned} \quad (2.28)$$

and by (2.22) it follows

$$|u_h(t) - u_k(t)| \leq M \left\{ \int_\tau^t e^{\omega(t-s)} |f_h(s) - f_k(s)| ds + e^{\omega(t-\tau)} |x_h - x_k| \right\}. \quad (2.29)$$

Thus $\{u_k\}$ is a Cauchy sequence in $C(\tau, T; E)$ and so there exists $\underline{u} \in C(\tau, T; E)$ such that $u_k \to \underline{u}$ in $C(\tau, T; E)$. Since, on the other hand, $u_k \to \underline{u}$ in $L^1(\tau, T; E)$ we have $u(t) = \underline{u}(t)$ a.e. and $x_k \to \underline{u}(\tau) = x$. This proves (2.23). As (2.22) is true for each $u_k$, it holds also for $u$. □

As an immediate consequence of Proposition 2.3 we obtain the main result of this section.

**Theorem 2.4.** *Problem (1.1) has at most one strong solution.*

The following result can be proved as theorem 4.2 of [3].

**Theorem 2.5.** *If $u$ is a strong solution of (1.1) then $\lim_{n \to \infty} u_n = u$ in $L^1(\tau, T; E)$, where $u_n$ is the solution of (2.5).*

## 3. Existence of the strong solution

To prove the existence of the strong solution of problem (1.1), we will first make stronger assumptions, then we prove the general result.

We set for $n \in \mathbb{N}$

$$C_0^n([\tau, T]; E) = \{f \in C^n[\tau, T]; E); \ f^{(k)}(\tau) = 0, k = 0, 1, ..., n\}.$$

**Lemma 3.1.** *Assume that* $A \in C^3([0, T]; \mathcal{L}(D; E))$ *and* $f \in C_0^3([0, T]; E))$ . *Let* $u_n$ *be the solution of the equation*

$$B_n u_n + \Lambda u_n + f = 0. \tag{3.1}$$

*Then* $u_n \in C_0^3([\tau, T]; D)$ *and, for* $k = 1, 2, 3$, *there exists* $n_k \in \mathbb{N}$ *and* $C_k > 0$, *depending on* $C, M, \omega, T - \tau$ *and* $\|A\|_{C^k([\tau, T]; \mathcal{L}(D; E))}$ *such that for* $n \geq n_k$ :

$$\|u_n^{(k)}\|_{C^k([\tau, T]; E)} \leq C_k \|f\|_{C^k([\tau, T]; E)}. \tag{3.2}$$

**Proof.** From the proof of Proposition 2.2 one deduces that $u_n \in C_0^3([\tau, T]; D)$. We set now $v_n = u_n'$; since $u_n(\tau) = 0$ we can write (3.1) as

$$-nR(n, B)v_n + \Lambda u_n + f = 0 \tag{3.3}$$

so that

$$u_n = n\Lambda^{-1} R(n, B)v_n - \Lambda^{-1} f. \tag{3.4}$$

Now, differentiating (3.1) with respect to $t$ and setting $(\Lambda' u)(t) = A'(t)u(t)$ we have

$$B_n v_n + \Lambda v_n + \Lambda' u_n + f' = 0. \tag{3.5}$$

Substituting $u_n$ (given by (3.4)) into (3.5) we find :

$$B_n v_n + \Lambda v_n + T_n v_n + g = 0 \tag{3.6}$$

where

$$T_n = n\Lambda' \Lambda^{-1} R(n, B) \tag{3.7}$$

and

$$g = f' - \Lambda' \Lambda^{-1} f. \tag{3.8}$$

By (2.11) we have

$$|v_n(t)| \leq M \int_\tau^t \{|g(s)| + |(T_n v_n)(s)|\} \, ds + \frac{M}{n} \{|g(t)| + |(T_n v_n)(t)|\} \tag{3.9}$$

and so, if we denote by $C'$ and $C_1$ functions of $C, M, \omega, T - \tau$ and $\|A\|_{C^1([\tau,T];\mathcal{L}(D;E))}$, we get

$$|v_n(t)| \leq C' \left\{ \int_\tau^t \left[ \|v_n\|_{C([\tau,s];E)} + \|f\|_{C^1([\tau,s];E)} \right] ds \right. \\ \left. + \frac{1}{n} \left( \|v_n\|_{C([\tau,s];E)} + \|f\|_{W^{1,1}(\tau,T;E)} \right) \right\} \quad (3.10)$$

From Gronwall's lemma we deduce the existence of $n_1 \in \mathbb{N}$ such that

$$\|u_n'\|_{C([\tau,T];E)} \leq C_1 \|f\|_{C^1([\tau,T];E)} \quad (3.11)$$

for $n \geq n_1$. By repeating the same argument for $u_n''$ and $u_n'''$ we obtain the other estimates of (3.2). □

**Lemma 3.2.** *If $A \in C^3([0,T];\mathcal{L}(D;E))$ and $f \in C_0^3([\tau,T];E)$ then problem*

$$u'(t) = A(t)u(t) + f(t); \quad \tau \leq t \leq T; \quad u(\tau) = 0 \quad (3.12)$$

*has a strict solution $u$ and*

$$\|u\|_{C^1([\tau,T];E)} + \|u\|_{C([\tau,T];D)} \leq C_4 \|f\|_{C^3([\tau,T];E)} \quad (3.13)$$

*with $C_4$ depending on the same variables of $C_1$.*

**Proof.** If $u_n$ denotes the solution of (3.1) given by lemma 3.1 we deduce from (33) for $n, m \geq n_1$:

$$B_n(u_n - u_m) + \Lambda(u_n - u_m) + (B_n - B_m)u_m = 0. \quad (3.14)$$

Hence from (14) we obtain

$$\|(u_n - u_m)\|_{C([\tau,T];E)} \leq C \|(B_n - B_m)u_m\|_{C([\tau,T];E)} \quad (3.15)$$

where $C$ is independent on $n$ and $m$. As $u_m \in D(B^2)$, by virtue of (3.2) we have

$$\|(B_n - B_m)u_m\|_{C([\tau,T];E)} = \|(m-n)R(n,B)R(m,B)B^2 u_m\|_{C([\tau,T];E)} \\ \leq M^2 C_3 \left| \frac{m-n}{(m-\omega)(n-\omega)} \right| \|f\|_{C^3([\tau,T];E)} \quad (3.16)$$

and so there exists $u \in C([\tau,T];E)$ such that

$$\lim_{n \to \infty} \|u - u_n\|_{C([\tau,T];E)} = 0. \quad (3.17)$$

Starting from (3.6), with the same procedure we deduce the convergence of $\{u_n'\}$ in $C([\tau,T];E)$ and so $u \in C^1([\tau,T];E)$.

As $A(t)$ is a closed operator, $u$ satisfies the equation (3.12) and so we have also $u \in C([\tau,T];D)$. By using the convergence of $u_n$ to $u$ in $C^1([\tau,T];E)$ we deduce (3.13). □

**Theorem 3.3.** *Let us suppose that (iii) holds and that there exists $A_k \in C^3([\tau, T]; \mathcal{L}(D; E))$ verifying $(\mathcal{H})_{C,\omega,M}$ and such that*

$$\lim_{k \to \infty} \|A - A_k\|_{C([\tau,T];\mathcal{L}(D;E))} = 0, \ \sup_{k \in \mathbb{N}} \|A'_k\|_{C([\tau,T];\mathcal{L}(D;E))} < \infty.$$

*Then problem (1.1) has a unique strong solution for each given $f \in L^1(\tau, T; E)$ and $x \in \overline{D}$.*

**Proof.** Let us first suppose that $f \in C_0^3([\tau, T]; E)$ and $x = 0$ : from lemma 3.2 we deduce the existence of $u_k \in C^1([\tau, T]; E) \cap C([\tau, T]; D), (k = 1, 2, \ldots)$ solution of

$$\begin{aligned} u'_k(t) &= A_k(t)u_k(t) + f(t); \ \tau \leq t \leq T \\ u_k(\tau) &= 0. \end{aligned} \quad (3.18)$$

Let us set

$$f_k(t) = u'_k(t) - A(t)u_k(t) = (A_k(t) - A(t))u_k(t) + f(t).$$

From estimate (3.13) we deduce that $\{u_k\}$ is bounded in $C([\tau, T]; D)$: and so

$$\lim_{k \to \infty} \|f - f_k\|_{C([\tau,T];E)} = 0 \quad (3.19)$$

By using (2.22) we deduce also the existence of $u \in C([\tau, T]; E)$ such that

$$\lim_{k \to \infty} \|u - u_k\|_{C([\tau,T];E)} = 0. \quad (3.20)$$

In conclusion $u$ is a strong solution of problem (1.1). This result obtained for the case $f \in C_0^3([\tau, T]; E)$ and $x = 0$ can be generalized by density when $f \in L^1(\tau, T; E)$ and $x = 0$; then we can pass to the case $f \in L^1(\tau, T; E)$ and $x \in D$ by substituting $f$ with $f - A(\cdot)x$ and finally by density it is possible to prove the general result corresponding to $f \in L^1(\tau, T; E)$ and $x \in \overline{D}$. $\square$

## 4. Existence of the strict solution

We are now in position of proving the existence and uniqueness of the strict solution of problem (1.1) under suitable regularity assumptions ; this result is a generalization of an analogous theorem proved in [3] for the autonomous case.

**Theorem 4.1.** *Let us suppose that $A \in C^1([\tau, T]; \mathcal{L}(D; E))$ and that there exist $A_k \in C^4([\tau, T]; \mathcal{L}(D; E))$ verifying $(\mathcal{H})_{C,\omega,M}$ and such that*

$$\lim_{k \to \infty} \|A - A_k\|_{C^1([\tau,T];\mathcal{L}(D;E))} = 0.$$

*If $f \in W^{1,1}(\tau, T; E)$ and $x \in D$ satisfy the compatibility condition*

$$x_1 = A(\tau)x + f(\tau) \in \overline{D} \quad (4.1)$$

then there exists a unique strict solution of (1.1).

**Proof.** Le $u$ be the strong solution of (1.1) given by Theorem 3.3 and $u_n \in W^{1,1}(\tau, T; D)$ the solution of

$$B_n(u_n - x) + \Lambda u_n + f = 0. \tag{4.3}$$

We know from Theorem 2.5 that

$$\lim_{n \to \infty} \|u - u_n\|_{L^1(\tau, T; E)} = 0. \tag{4.4}$$

By using (2.12) we deduce from (4.3)

$$nR(n, B)u_n'(t) + ne^{-n(t-\tau)}R(n, A(\tau))x_1 - A(t)u_n(t) - f(t) = 0 \tag{4.5}$$

and so

$$\begin{aligned} u_n(t) &= A^{-1}(t)(nR(n,B)u_n')(t) \\ &\quad + ne^{-n(t-\tau)}A^{-1}(t)R(n, A(\tau))x_1 - A^{-1}(t)f(t). \end{aligned} \tag{4.6}$$

Differentiating (4.3) we get

$$B_n(u_n' - x_1)(t) + A(t)u_n'(t) + C(t)(nR(n,B)u_n')(t) + h_n(t) = 0 \tag{4.7}$$

where we have set

$$\begin{aligned} C(t) &= A'(t)A^{-1}(t) \\ h(t) &= f'(t) - C(t)f(t) \\ h_n(t) &= e^{-n(t-\tau)}n\left[A(\tau)R(n, A(\tau))x_1 + C(t)R(n, A(\tau))x_1\right] + h(t). \end{aligned} \tag{4.8}$$

As $C \in C([\tau, T]; \mathcal{L}(E))$, by using Theorem 5.2.3 of [8] we deduce that problem

$$\begin{aligned} z'(t) &= (A(t) + C(t))z(t) + h(t), \\ z(\tau) &= x_1 \end{aligned} \tag{4.9}$$

has a strong solution and Theorem 2.5 implies

$$\lim_{n \to \infty} \|z - z_n\|_{L^1(\tau, T; E)} = 0 \tag{4.10}$$

where $z_n$ is the solution of

$$B_n(z_n - x_1) + \Lambda z_n + C(\cdot)z_n + h = 0. \tag{4.11}$$

From (4.7) and (4.11) we obtain

$$\begin{aligned} &B_n(u_n' - z_n) + \Lambda(u_n' - z_n) + C(\cdot)[nR(n,B)(u_n' - z_n) \\ &+ nR(n,B)(z_n - z) + nR(n,B)z - z_n] + h_n - h = 0. \end{aligned} \tag{4.12}$$

An application of the Gronwall's lemma (as in the proof of lemma 3.1) and the use of condition (4.1) let us deduce

$$\lim_{n\to\infty} \|u'_n - z_n\|_{L^1(\tau,T;E)} = 0. \tag{4.13}$$

Now (4.4), (4.10) and (4.13) imply $u \in W^{1,1}(\tau,T;E)$ and $u' = z$ : but $z$ is continuous (see Proposition 2.3) hence $u \in C^1([\tau,T];E)$.

Writing

$$B_n u_n - Bu = -nR(n,B)u'_n + B_n R(n,A(\tau))x_1 + nR(n,B)u' + (B_n - B)u \tag{4.14}$$

we get

$$\lim_{n\to\infty} \|Bu - B_n u_n\|_{L^1(\tau,T;E)} = 0 \tag{4.15}$$

and so from (4.3)

$$\lim_{n\to\infty} \|\Lambda u_n + Bu + f\|_{L^1(\tau,T;E)} = 0. \tag{4.16}$$

As $\Lambda$ is a closed operator we deduce $u \in D(\Lambda)$ and

$$\Lambda u = -Bu + f, \tag{4.17}$$

i.e. $u$ is a strict solution of (1.1) . □

## References

[1] DALECKII J.L., KREIN M.G., *Stability of solutions of differential equations in Banach space*, Am. Math. Assoc. (1974).

[2] DA PRATO G., IANNELLI M., *On a method for studying abstract evolution equations in the hyperbolic case*, Comm. in Partial Differential Equations **1** (1976), 585–608.

[3] DA PRATO G., SINESTRARI E., *On the Phillips and Tanabe regularity theorems*, Semesterbericht Funktionalanalysis, Tubingen, Sommersemester, 117–124.

[4] DA PRATO G., SINESTRARI E., *Differential operators with Non Dense Domain*, Ann.Sc. Norm. Sup. Pisa **14** (1987), 285–344.

[5] KATO T., *Linear evolution equations of "hyperbolic type"*, J.Fac.Sc.Univ.of Tokyo **17** (1970), 241–258.

[6] KATO T., *Linear evolution equations of "hyperbolic type" II*, J.Math.Soc.Japan **25** (1973), 648–666.

[7] KATO T., *Abstract Differential Equations and Nonlinear Mixed Problems*, Lezioni Fermiane, Scuola Normale Superiore di Pisa.

[8] PAZY A., *Semigroups of Linear Operators and Applications to Partial Differential Equations*, Springer-Verlag, 1983.

Giuseppe Da Prato
Scuola Normale Superiore di Pisa
I-56126 Pisa, Italy

Eugenio Sinestrari
Dipartimento di Matematica
Università di Roma "La Sapienza"
Piazzale Aldo Moro 2
I-00185 Roma, Italy

# Towards a Numerical Analysis of the Escalator Boxcar Train

A.M. DE ROOS
J.A.J. METZ

Department of Pure and Applied Ecology, University of Amsterdam
Institute of Theoretical Biology, Leiden University

## 1. Introduction

Within a biological population individuals usually exhibit differences in their population dynamical behaviour. Here we use the term behaviour to refer to those processes that have an influence upon the dynamics of the total population, as, for example, reproduction, mortality or the feeding on a limiting food resource. The differences can first of all arise from the fact that individuals occupy different positions in the space in which the population lives, and hence experience a different local environment. Another source of observed variation in behaviour is the fact that individuals are physiologically different. Finally, differences in behaviour are also observed between individuals that seem physiologically identical. In recent years developments in the field of modelling the dynamics of biological populations have been largely aimed at accounting for this variation in behaviour.

In this paper we will only consider the general class of models that are developed to account for variation in behaviour which stems from differences in the physiological characteristics of an individual. These models are usually referred to as physiologically structured population models. The term structure refers to the subdivision of the population on the basis of traits that characterize the individuals. The physiological traits of an individual often have a strong influence upon its behaviour. Body size is one of the most prominent example in this respect [1]. Physiologically structured population models are suitable tools to investigate the population dynamical impact of, for example, variations in individual body size within the population and the processes that are related with it, such as growth. These models hence establish links between phenomena that are observed at the level of the population and behavioural processes at the level of the individual. An example of the type of insight gained from a detailed investigation of a structured population model can be found in [2].

The theoretical framework for physiologically structured population models has recently been described in some detail [3]. Physiologically structured population models always give rise to a first-order hyperbolic partial differential equation (PDE) for the distribution of the population over its "space"-domain (an exact description of this domain follows in section 2). The equations are usually non-linear as the coefficients may depend on, for example, the total population size. Moreover, the boundary condition contains a linear functional of the distribution itself, reflecting the reproduction process as a function of the individuals present. Together these characteristics constitute the quite specific nature of the equations.

Following the development of the theoretical framework, an efficient numerical method, the *Escalator Boxcar Train*, has been developed to integrate numerically this type of PDE [4,5,6]. This method fully exploits the biological nature of the equations and is hence rather unusual from a numerical point of view. It can be shown, however, that the method is biologically very relevant, since it links the class of continuously structured population models, as described by Metz & Diekmann [3] with the class of Leslie matrix models, which describe the dynamics of populations in a discrete fashion. Hence, the *Escalator Boxcar Train* can be seen as a generalization of the (in biology more popular) Leslie matrix models [5,6].

Despite its biological relevance, the numerical properties of the *Escalator Boxcar Train* are poorly understood. In this paper we will investigate to some extent the numerical properties of the method. After a short introduction to the class of physiologically structured population models and to the *Escalator Boxcar Train* itself, we will show that the method consistently approximates the original model equations in a weak sense. Subsequently, we will discuss some results on the convergence of the method for a specific physiologically structured model, for which an analytical solution can be obtained.

## 2. The model structure

### 2.1. The model on the individual level.

In physiologically structured population models individuals are characterized by a set of variables, the *i-state variables*, together unambiguously defining the *i-state* (here and below *i*- refers to *individual* properties). These variables contain, by assumption, the information about the past of the individual, necessary and sufficient to describe completely its future, as far as this is relevant for the dynamics of the population. Examples of frequently used *i-state variables* are: age, size, energy reserves, amount of foliage, *etc.* The *i-state* $x = (x^1, \ldots, x^q)^T$ of an individual takes its value in the reachable *i-state-space* $\Omega \subset \mathbb{R}^q$, the set of all states that an individual might attain (the superscript $^T$ indicates the transpose of a vector). In a similar way the environment, in which the population lives, is characterized by a set of variables, $E(t) = (E^0(t), \ldots, E^s(t))^T \in \mathbb{R}^s$, the environmental state variables. These variables can represent quantities like food availability or the density of predators in the environment.

On the individual level a model has to be specified which mathematically describes those processes that change the *i-state* and those which lead to creation or destruction of individuals. The processes that change the *i-state* smoothly are assumed to be completely deterministic. These smooth changes are described by means of an ODE:

$$\frac{dx}{dt} = v(E, x), \tag{1}$$

where $v = (v^1, \ldots, v^q)^T : \mathbb{R}^s \times \mathbf{\Omega} \to \mathbb{R}^q$ describes the direction of and the velocity along the trajectory followed by an individual. Examples of such smooth changes are aging and growing. The processes which lead to the creation and destruction of individuals are usually stochastic, any individual having a certain chance to die or give birth. However, it will be assumed that the population is large enough to describe the consequences of these *chance* processes as deterministic rates. Examples of such stochastic processes are fission of unicellular organisms and death of individuals due to predation.

## 2.2. The model on the population level.

To describe the dynamics of the population, a density-function $\eta(t, \cdot)$ on $\mathbf{\Omega}$ is introduced. This density-function can be interpreted as the mathematical analogue of the biological population. More precisely:

$$\int_{\Omega_i} \eta(t, \xi) \, d\xi,$$

equals the number of individuals at time $t$ with an *i-state* in $\Omega_i$, where $\Omega_i \subset \mathbf{\Omega}$. Following the lines set out by Metz & Diekmann [3] it is possible to derive the following hyperbolic partial differential equation (PDE) for $\eta(t, x)$ that summarizes the impact of the processes on the individual level on the dynamics of the population:

$$\frac{\partial \eta(t, x)}{\partial t} + \nabla \cdot \big(v(E, x)\eta(t, x)\big) = -d(E, x)\eta(t, x), \tag{2a}$$

$$\psi \cdot \big(v(E, x_0)\eta(t, x_0)\big) = B\big(E, x_0, \eta(t, \cdot)\big), \qquad x_0 \in \partial^+ \mathbf{\Omega}. \tag{2b}$$

Here $\partial^+ \mathbf{\Omega}$ denotes that part of $\partial \mathbf{\Omega}$, i.e., the boundary of the reachable *i-state-space* $\mathbf{\Omega}$, where the vector $v(E, x_0)$ points inwards, i.e., the inner product of $v$ and the inward normal $\psi$ is positive. In this formulation $\nabla \cdot (v\eta)$, the divergence of the flux $v\eta$ at $x$, describes the convection of individuals through $\mathbf{\Omega}$ due to the smooth changes in their *i-state*. $B(E, x_0, \cdot)$ is a linear functional acting on the density-function $\eta$ and describes the entrance (creation) of individuals at the boundary of $\Omega$. We have assumed that the age of an individual is one of the *i-state variables*, which ensures that all individuals enter $\Omega$ across its boundary. (Application of the *Escalator Boxcar Train* requires that all individuals enter $\Omega$ across its boundary; the latter is, however, not required in general.) The product $d(E,x)\eta(t,x)$ describes the removal (death) of individuals from $\Omega$. The function $d(E,x): \mathbb{R}^s \times \Omega \to \mathbb{R}$ denotes the death rate of individuals with *i-state* $x$.

The linear operator $B$ describing the birth rate on the boundary of the *i-state-space* $\partial^+\Omega$, is generally expressed as:

$$B(E, y, \eta(t, \cdot)) = \int_\Omega s(\xi, y) b(E, \xi) \eta(t, \xi) \, d\xi$$

$$\text{with } y \in \partial^+\Omega \text{ and } \oint_{\partial^+\Omega} s(x, \sigma) \, d\sigma = 1, \qquad \forall x \in \Omega.$$

(3)

In this formulation the function $b(E,x): \mathbb{R}^s \times \Omega \to \mathbb{R}$ denotes the reproduction rate of individuals with *i-state* $x$ and $s(x,y): \Omega \times \partial^+\Omega \to \mathbb{R}$ is a function describing the distribution of the offspring of parents with *i-state* $x$ over the *i-states* $y \in \partial^+\Omega$.

The system of equations (2) is a general form of the PDE and boundary condition occurring in structured population models. Note that (2a) represents the "convection" of individuals through $\Omega$ along their trajectories, together with the deaths occurring during this convection. Equation (2b) represents the inflow of new individuals across $\partial^+\Omega$ due to birth processes. Note also that the trajectories of the individuals coincide with the characteristics of the hyperbolic PDE.

Non-linearities are always incorporated in (2) by means of the environmental variables $E(t)$. The non-linearities occurring in structured population models belong generally to one of the two following classes. The first class may be referred to as direct density-dependence, in which case some population statistic(s) directly influence(s) the coefficient functions. An example of a biological, non-linear phenomenon that belongs to this category is cannibalism or intraspecific predation. In this case the death rate depends, for example, on the total population size. The second class of non-linearities may be referred to as environmental feed-back loops. In this case the population itself induces changes in its environment and hence indirectly in the behaviour of the individuals. An example is the feeding of a population on an external food source, the abundance of which in turn influences individual growth and reproduction.

In the case of direct density-dependence one or more of the environmental variables $E(t)$ can be conceived as a weighted integral of $\eta(t,x)$ over the *i-state-space* $\Omega$, representing the density-dependent effects. In case of the environmental feed-back loops the dynamics of the environmental variables $E(t)$ are described by a separate system of ODEs, containing again one or more weighted integral of $\eta(t,x)$. These integrals now represent the influence of the population on its environment.

## 3. The escalator boxcar train: A short derivation

To integrate the type of PDE and boundary condition (2), a special numerical method was developed, the *Escalator Boxcar Train* or EBT-method [4]. This method relies on the biological interpretation of the PDE, rather than on its mathematical properties. The EBT-method only takes care of the discretization of the "space"-domain $\Omega$ and hence gives rise to a system of ODEs which subsequently have to be solved by a standard numerical method for ODEs. Instead of approximating the density-function $\eta(t,x)$ at some nodal points in its domain, the EBT-method deals with moments of this density-function over small subdomains in $\Omega$. These moments have a biological interpretation as the total number of individuals with an *i-state* in the subdomain, the mean *i-state* of the individuals in the subdomain, the variance around this mean, *etc*. Another important aspect of the method is that it does not assume a certain smoothness of the density-function itself, but only assumes that the coefficient functions in the model ($v(E,x)$, $d(E,x)$, $b(E,x)$ and $s(x,y)$) are continuously differentiable up to a desired order.

In the following we will first give a short derivation of the method. In the derivation we will implicitly make assumptions about the differentiability of the coefficient functions. These assumptions will be explicitly stated when we prove the consistency of the EBT-method in the next section. The EBT-method was derived in a second and a third order form [4], in which the neglected terms in the approximation contained, respectively, only terms that scaled with the square or the cube of the mesh width in $\Omega$. In the rest of this paper we will exclusively focus upon the second order version of the EBT-method. Since the EBT-method deals with the discretization of the *i-state-space* $\Omega$, we will study this part of the numerical solution of the system (2) only. Solving the resulting systems of ODEs can be done, using any ODE time integration method. The numerical properties of this time integration method will not be considered.

### 3.1. The internal subdomains.

Assume that equation (1), describing the smooth changes in the *i-state* generates a unique solution, such that $x(t, t_0, x_0)$ represents the *i-state* of an individual at time $t \geq t_0$, whose *i-state* at time $t_0$ was equal to $x_0$. Assume that the interior of $\Omega$ is subdivided into arbitrary domains $\Omega_i(t_0)$ at the initial time $t_0$ and define $\Omega_i(t)$ as:

$$\Omega_i(t) = \{x(t, t_0, x_0) | x_0 \in \Omega_i(t_0)\}. \tag{4}$$

The definition of $\Omega_i(t)$ shows clearly that the domain is transported along the characteristics of the PDE in time and is thus effectively closed to transport processes across its boundary. In the following we will focus for the time being on one subdomain $\Omega_i(t)$ in the interior of $\Omega$.

To describe the evolution of $\eta(t,x)$ within $\Omega_i(t)$ the following quantities are defined:

(a) $0^{\text{th}}$ moment of $\eta(t,\cdot)$ within $\Omega_i(t)$:

$$\lambda_i(t) = \int_{\Omega_i(t)} \eta(t,\xi)\, d\xi. \tag{5}$$

This moment equals the total number of individuals having an *i*-state within $\Omega_i(t)$.

(b) $1^{\text{st}}$ moment of the conditional distribution derived from $\eta(t,\cdot)$ within $\Omega_i(t)$:

$$\mu_i(t) = \frac{\int_{\Omega_i(t)} \xi\, \eta(t,\xi)\, d\xi}{\lambda_i(t)}. \tag{6}$$

The quantity $\mu_i(t) = (\mu_i^1(t), \ldots, \mu_i^q(t))^T$ equals the mean *i*-state of the individuals within $\Omega_i(t)$.

These quantities together can be used to approximate integrals of $\eta(t,x)$ with arbitrary, sufficiently smooth weight-functions $\psi$ over the *i*-state-space $\Omega$, which are the kind of output usually required from this type of models.

The dynamics of the quantities $\lambda_i(t)$ and $\mu_i(t)$ within $\Omega_i(t)$ can approximately be described by the following system of ODEs [4]:

## I. Second order EBT scheme: Internal subdomains

$$\begin{cases} \dfrac{d\lambda_i}{dt} = -d(E,\mu_i)\lambda_i, \\ \dfrac{d\mu_i}{dt} = v(E,\mu_i). \end{cases}$$

Every subdomain $\Omega_i(t)$ in $\Omega$ is characterized by a set of variables as defined in equations (5) and (6). The dynamics of these set of variables $(\lambda_i(t), \mu_i(t))$ characterizing every subdomain is hence described by a set of ODEs of the above form.

The system of ODEs I represents a second-order approximation of the original PDE (2a), when integrated for all the subdomains $\Omega_i(t)$ within the interior of $\Omega$. It is easily seen that scheme I exactly describes the dynamics of a group of identical individuals (usually called a cohort of individuals) in a population which is not continuously distributed over an *i*-state-space, but concentrated into discrete classes (*e.g.* fish-populations with definite year-classes). In this sense the discretization method, derived above, seems to be a very natural one.

## 3.2. The boundary subdomains.

The system of ODEs I presented in the previous subsection only describes the dynamics of the quantities $\lambda_i$ and $\mu_i$ in an internal subdomain in $\Omega$, i.e., a subdomain that is transported along the characteristics of the PDE (2a) and hence closed to transport processes across its boundary. These equations can not account for the boundary condition (2b). At the instream boundary of $\Omega$ we have to carry out a different approximation procedure.

At the instream boundary of $\Omega$, $\partial^+\Omega$, a collection of subdomains $\omega_j(t)$ is defined such that each $\omega_j$ contains some portion of $\partial^+\Omega$, and $\partial^+\Omega \subset \bigcup_j \omega_j$ (throughout this section a subdomain bordering $\partial^+\Omega$ will be denoted with $\omega_j(t)$ and a subdomain in the interior of $\Omega$ with $\Omega_i(t)$). These domains are treated as the domains in the interior of $\Omega$ in the previous section except that the part of the boundary of $\omega_j(t)$ that constitutes $\partial^+\Omega$, does not move along with the convection defined by the PDE (2a), but stays put, until $\omega_j(t)$ has reached a certain threshold size. Upon reaching this threshold the subdomain $\omega_j(t)$ takes off from the boundary $\partial^+\Omega$ and becomes a subdomain $\Omega_i(t)$ in the interior. At the same moment a new subdomain $\omega_j(t)$ is created, taking the place along $\partial^+\Omega$ which is left open by the leaver.

Within the subdomains $\omega_j(t)$ again a set of moments of the original density-function $\eta(t, x)$ is defined, though in a slightly different way. The reason for this modification is that at the empty (i.e., $\lambda = 0$) start of a boundary subdomain the (now non-linear) system of ODEs needs to be carefully analyzed for the correct initial condition and time-derivative. This problem is circumvented by slightly modifying the definition of the moments. The analogues of equations (5) and (6) now are:

$$\lambda_j(t) = \int_{\omega_j(t)} \eta(t, \xi) \, d\xi, \tag{7}$$

$$\pi_j(t) = \int_{\omega_j(t)} (\xi - a_j) \eta(t, \xi) \, d\xi, \tag{8}$$

where $a_j$ is some fixed point of $\partial^+\omega_j$, the instream boundary of $\omega_j(t)$.

Upon reaching the threshold the subdomain $\omega_j(t)$ is closed and the following transformation of moments is necessary due to the difference in definition of the moments of an interior subdomain $\Omega_i(t)$ and of a boundary subdomain $\omega_j(t)$:

$$\lambda_\Omega = \lambda_\omega, \tag{9}$$

$$\mu_\Omega = \frac{\pi_\omega}{\lambda_\omega} + a_j. \tag{10}$$

The dynamics of the variables $\lambda_j(t)$ and $\pi_j(t)$ characterizing a boundary subdomain $\omega_j(t)$ can now approximately be described by the following set of ODEs [4]:

## II. Second order EBT-method: Boundary subdomains

$$\begin{cases} \dfrac{d\lambda_j}{dt} = -d(E,a_j)\lambda_j - D_x d(E,a_j)\pi_j + \sum_p b(E,\mu_p)\lambda_p S_j^\lambda(\mu_p), \\ \dfrac{d\pi_j}{dt} = v(E,a_j)\lambda_j + D_x v(E,a_j)\pi_j - d(E,a_j)\pi_j + \sum_p b(E,\mu_p)\lambda_p S_j^\mu(\mu_p) \end{cases}$$

in which $D_x$ is the total derivative with respect to $x$, i.e., if $f(x) : \mathbb{R}^m \to \mathbb{R}^n$ $D_x f$ is defined as:

$$D_x f(a) = \begin{pmatrix} \dfrac{\partial f^1}{\partial x^1} & \dfrac{\partial f^1}{\partial x^2} & \cdots & \dfrac{\partial f^1}{\partial x^m} \\ \dfrac{\partial f^2}{\partial x^1} & \dfrac{\partial f^2}{\partial x^2} & \cdots & \dfrac{\partial f^2}{\partial x^m} \\ \vdots & \vdots & \ddots & \vdots \\ \dfrac{\partial f^n}{\partial x^1} & \dfrac{\partial f^n}{\partial x^2} & \cdots & \dfrac{\partial f^n}{\partial x^m} \end{pmatrix}_{x=a}.$$

The summation over $p$ in the ODEs above refers to a summation over all internal and boundary subdomains that produce offspring. To determine the contribution of a boundary subdomain to this sum, $\pi_j(t)$, which characterizes the subdomain, should first be transformed using equation (10) into its equivalence $\mu_j(t)$, characterizing an internal subdomain. (Obviously, if $\lambda_j(t) = 0$ the reproductive contribution of the boundary subdomain is 0 and the subdomain is not included in the sum.) Evaluation of the function $b(E,x)$ at $x = \mu_p(t)$ for both the internal and the boundary subdomains is necessary for the EBT-method to be a consistent approximation to the system (2), as can be inferred from the calculations in the next section.

The quantities $S_j^\lambda$ and $S_j^\mu$ in these sums are defined as follows:

$$S_j^\lambda(\mu_p) = \oint_{\partial^+\omega_j} s(\mu_p,\sigma)\,d\sigma,$$

$$S_j^\mu(\mu_p) = \oint_{\partial^+\omega_j} (\sigma - a_j) s(\mu_p,\sigma)\,d\sigma.$$

As opposed to the other coefficients appearing in the approximation equations which are all evaluations of known functions and derivatives of known functions at some point $\mu_i$ (cf. $b(E,\mu_i)$, $d(E,\mu_i)$), the coefficients $S$ represent moments of the (known) density-function $s(\mu_p,y)$ over the surface $\partial^+\omega_j$. The method to determine these coefficients heavily depends upon the exact nature of the density-function $s(x,y)$. In some cases the moments can be determined analytically (see [4] for an example), but usually this has to be done by means of some numerical integration method.

The system of ODEs II is analogous with the system of ODEs I and describes the time evolution of the moments (7) and (8) in a boundary subdomain $\omega_j(t)$. The differences between the systems I and II are due to the slightly different definition of the moments within the subdomain, and to the fact that the system II also contains a second-order approximation of the boundary condition (2b).

Together the systems of ODEs I and II constitute a numerical approximation of the original set of equations (2). If we assume, as is done hereafter, that the initial distribution $\eta(0, x)$ has bounded support, the *i-state-space* $\Omega$ can initially be subdivided into a finite set of internal and boundary subdomains. The dynamics of the local measures, characterizing the density-function $\eta(t, x)$ in these subdomains, are now given by the systems of ODEs I and II for internal and boundary subdomains, respectively. These ODEs are subsequently solved on a time interval $[0, \tau]$, where $\tau$ is that time at which one of the boundary subdomains $\omega_j(t)$ has reached a certain threshold size $\omega_{max}$. The boundary subdomains are subsequently closed off and transformed into internal subdomains using the equations (9) and (10). Thereupon a new set of boundary subdomains is created and the described integration cycle starts anew.

The choice of the threshold size $\omega_{max}$ (or equivalently the threshold time $\tau$) is arbitrary, but ultimately determines the mesh width on $\Omega$. Smaller values of the threshold can be chosen to yield a better approximation of the original system of equations at the expense of increasing the number of subdomains and hence the number of ODEs to solve. (See [4] for an illustration of how the threshold affects the performance of the EBT-method.) At present we do not have a method to select an optimal threshold size (time) $\omega_{max}$ ($\tau$), comparable to the procedures for adapting the step size, which are used in some time integration methods.

It should be noted that the number of subdomains continuously increases during the integration. Given a finite set of initial subdomains, however, the total number of subdomains remains bounded over a finite time interval $[0, T]$. We will exploit this feature in the analysis in the next sections. For practical applications it is often desirable to restrict the total number of subdomains that make up the population. If the smooth changes in the *i*-state, described by equation (1), are such that the characteristics of the PDE (2a) converge, the width of a subdomain ultimately decreases. This can result in two internal subdomains having almost identical values of $\mu_i(t)$. These subdomains can then be lumped into one new subdomain, with $\lambda_i(t)$ of the new subdomain equal to the sum of the old $\lambda_i(t)$ values and $\mu_i(t)$ of the new subdomain equal to the weighted average of the old $\mu_i(t)$ values (weighted with the old $\lambda_i(t)$). Alternatively, the total number of subdomains can be kept within bounds by deleting subdomains with negligible values of $\lambda_i(t)$. Based on our experience both methods seem to yield satisfactory results (see, for instance, [2]).

## 4. The consistency of the escalator boxcar train

As stated in the introduction, the EBT-method is rather unusual from a numerical point of view and its properties are ill-understood. To prove that the

EBT-method yields a consistent approximation to the set of equations (2) is also not a standard procedure. First of all, the mathematical theory to provide a rigorous justification and interpretation of the general framework of physiologically structured population models is still in its infancy (see [7] for a survey). The existence and uniqueness of solutions to (2) cannot be proven in general and the precise interpretation of the equations (2) is still beset with problems. Furthermore, the density-function $\eta(t,x)$ cannot, even approximately, be retrieved from the collection of its moments, without making further assumptions. Hence, conventional methods to prove the consistency of a numerical method for PDEs, which study the difference between the exact dynamics of the complete density-function $\eta(t,x)$ and the discrete, approximate dynamics of $\eta(t,x)$ at a set of nodal points, are not applicable. We will therefore study the consistency of the EBT-method, using the concept of approximation in a "weak" sense, *i.e.*, we will show that the exact dynamics of <u>measures</u> of the density-function $\eta(t,x)$, as specified by equations (2), are consistently approximated, when we apply the EBT-method to these equations. (As an aside it should be noted that this weak concept of approximation fits in well with the functional analytic underpinning of the general framework [7].)

Our weak concept of approximation is inspired by the fact that the biologically relevant output quantities from a structured population model are (weighted) integrals of $\eta(t,x)$ over the *i-state-space* $\Omega$. This output is thus generally of the form:

$$\Psi(t) = \int_\Omega \psi(x)\eta(t,x)\,dx \qquad (11)$$

where $\psi(x): \Omega \to \mathbb{R}$ is some arbitrary weighing function over the *i-state-space* $\Omega$. These integrals usually have a clear biological interpretation, *e.g.* number of individuals, total biomass, *etc.*

In the following it will be shown that (A) the output quantity $\Psi(t)$ can be approximated consistently to a certain order of accuracy by a quantity $\widehat{\Psi}(t)$, which is defined in terms of the local moments $\lambda$ and $\mu$ of $\eta(t,x)$ within the internal subdomains and the local moments $\lambda$ and $\pi$ of $\eta(t,x)$ within the boundary subdomains and (B) the dynamics of $\Psi(t)$ as determined by the equations (2) are approximated to the same order of accuracy by the dynamics of $\widehat{\Psi}(t)$, as governed by the numerical schemes I and II, which characterize the EBT-method. From these two observations we conclude that the EBT-method yields a consistent approximation of the set of equations (2) in the weak sense that we introduced in this section.

Let the total population size be denoted by

$$N = \int_\Omega \eta(t,x)\,dx$$

and assume that $N$ is bounded. Let $\eta(t,x)$ at time $t$ have bounded support in $\Omega$ and let $\Omega$ be subdivided into a finite set of $m$ internal subdomains $\Omega_p(t)$, $p =$

$1, \cdots, m$, with $\forall_p \Omega_p(t) \cap \partial^+\Omega = \emptyset$, and $n$ boundary subdomains $\omega_r(t)$, $r = 1, \cdots, n$, with $\forall_r \omega_r(t) \cap \partial^+\Omega = \partial^+\omega_r$, in which $\partial^+\Omega$ denotes the instream boundary of $\Omega$. Given the bounded support of $\eta(t,x)$ we may assume that $\eta(t,x) = 0$ for $x \notin \bigcup_{p=1}^m \Omega_p(t) \cup \bigcup_{r=1}^n \omega_r(t)$. (Here and in the following the index $p$ always refers to the subdomains $\Omega_p(t)$ in the interior of the $\Omega$, while the index $r$ always refers to subdomains $\omega_r(t)$ along the boundary $\partial^+\Omega$ of $\Omega$.)

Let every internal subdomain $\Omega_p(t)$ now be characterized by quantities $\lambda_p$ and $\mu_p$, as specified by (5) and (6), and each boundary subdomain $\omega_r(t)$ by quantities $\lambda_r$ and $\pi_r$, given by (7) and (8). Moreover, let $\mu_r$ denote the transformed moment in a boundary subdomain $\omega_r(t)$, i.e., after applying the transformation (10) to $\pi_r$. Obviously,

$$\frac{d\mu_r}{dt} = \frac{1}{\lambda_r}\left\{\frac{d\pi_r}{dt} - (\mu_r - a_r)\frac{d\lambda_r}{dt}\right\}. \tag{12}$$

The quantity $\Psi(t)$ can be approximated by the quantity $\widehat{\Psi}(t)$, defined as (see [4]):

$$\widehat{\Psi}(t) = \sum_{p=1}^m \psi(\mu_p)\lambda_p + \sum_{r=1}^n \psi(\mu_r)\lambda_r. \tag{13}$$

Let $h$ denote the maximum width of a subdomain in the $i$-state-space $\Omega$, defined as:

$$h := \max\bigl(\max_p \|\Omega_p(t)\|, \max_r \|\omega_r(t)\|\bigr)$$
$$= \max\bigl(\max_p \sup_{x,y \in \Omega_p(t)} \|x-y\|_\infty, \max_r \sup_{x,y \in \omega_r(t)} \|x-y\|_\infty\bigr),$$

in which $\|x-y\|_\infty := \max_i |x^i - y^i|$ is the maximum norm on $\Omega$.

If $\psi \in \mathbb{C}2$ the following inequality can be derived for the difference between (11) and (13) using the mean value theorem:

$$|\Psi(t) - \widehat{\Psi}(t)| = \frac{1}{2}\left|\sum_{p=1}^m \int_{\Omega_p(t)} (\sigma - \mu_p)^T D_{xx}\psi(\xi(\sigma))(\sigma - \mu_p)\eta(t,\sigma)\,d\sigma\right.$$
$$\left. + \sum_{r=1}^n \int_{\omega_r(t)} (\sigma - \mu_r)^T D_{xx}\psi(\tau(\sigma))(\sigma - \mu_r)\eta(t,\sigma)\,d\sigma\right| \tag{14}$$
$$\leq \frac{1}{2}\|D_{xx}\psi\|Nh^2.$$

Here, $D_{xx}\psi$ is the (symmetric) $q \times q$ matrix with second order partial derivatives of $\psi$ with respect to $x$ and $\|D_{xx}\psi\| := \sup_{x \in \Omega} \|D_{xx}\psi(x)\|_\infty$ is the matrix norm of $D_{xx}\psi(x)$ corresponding to the earlier defined maximum norm on $\Omega$. In the integrals above, $\xi(\sigma)$ and $\tau(\sigma)$ are arbitrary points in $\Omega_p(t)$ and $\omega_r(t)$, respectively. These integrals represent the rest terms in the local expansion of

$\psi(x)$ around the points $\mu_p$ and $\mu_r$ in every $\Omega_p(t)$ and $\omega_r(t)$, respectively, using Taylor's formula.

The difference between $\Psi(t)$ and $\widehat{\Psi}(t)$ is hence $O(h^2)$, if the following assumption holds:

ASSUMPTION 1: The width of the subdomains in the *i-state-space* $\Omega$ and the total population size stay bounded away from infinity, *i.e.*, $h < \infty$ and $N < \infty$.

Because the internal subdomains are transported along the characteristics during the integration, the assumed boundedness of all $\omega_r(t)$ and all $\Omega_p(t)$ requires that the characteristics of the PDE (2a) do not diverge unlimited.

In the following an ordinary differential equation (ODE) for the quantity $\Psi(t)$ will be derived using the equations (2) for $\eta(t,x)$. Moreover, on the basis of the numerical schemes I and II an approximate ODE will be derived for the quantity $\widehat{\Psi}(t)$. It will be shown that the difference term between these two ODEs is $O(h^2)$. From this observation it is concluded that the numerical schemes I and II constitute a consistent approximation of the PDE and boundary condition (2) to the same order of approximation as the difference between the two quantities $\Psi(t)$ and $\widehat{\Psi}(t)$. During the calculations also the basic assumptions underlying the *Escalator Boxcar Train* will come into play and will be stated explicitly.

Differentiation of expression (11) for the quantity $\Psi(t)$ and using the equations (2) leads to:

$$\begin{aligned}
\frac{d\Psi}{dt} &= \int_\Omega \bigl(v(E,x)\eta(t,x)\bigr)\cdot\nabla\psi(x)\,dx - \int_\Omega \psi(x)d(E,x)\eta(t,x)\,dx \\
&\quad + \oint_{\partial^+\Omega} \psi(y)B\bigl(E,y,\eta(t,\cdot)\bigr)\,dy \\
&= \int_\Omega \bigl(v(E,x)\eta(t,x)\bigr)\cdot\nabla\psi(x)\,dx - \int_\Omega \psi(x)d(E,x)\eta(t,x)\,dx \\
&\quad + \oint_{\partial^+\Omega} \psi(y) \int_\Omega s(x,y)b(E,x)\eta(t,x)\,dx\,dy.
\end{aligned} \qquad (15)$$

In this derivation is explicitly used:

ASSUMPTION 2: The *i-state-space* $\Omega$ contains no exit boundary $\partial^-\Omega$.

Differentiation of the quantity $\widehat{\Psi}(t)$ (expression (13)) with respect to time results in:

$$\begin{aligned}
\frac{d\widehat{\Psi}}{dt} &= \sum_{p=1}^m \psi(\mu_p)\frac{d\lambda_p}{dt} + \sum_{p=1}^m D_x\psi(\mu_p)\frac{d\mu_p}{dt}\lambda_p \\
&\quad + \sum_{r=1}^n \psi(\mu_r)\frac{d\lambda_r}{dt} + \sum_{r=1}^n D_x\psi(\mu_r)\frac{d\mu_r}{dt}\lambda_r.
\end{aligned}$$

If we substitute the expression (12) for $d\mu_r/dt$ and the numerical schemes I and II that represent the EBT-method into the equation above, we obtain an ODE for $\widehat{\Psi}$ that represents the EBT-approximation of the ODE (15) for $\Psi$. Here and in the following the simplifying assumption will be made that $b(E,x)$ equals 0 in all boundary subdomains $\omega_r(t)$. (This assumption can easily be relaxed at the cost of introducing a few additional, though messy terms in the resulting formulas below.) The substitution yields:

$$\frac{d\widehat{\Psi}}{dt} = -\sum_{p=1}^{m}\psi(\mu_p)d(E,\mu_p)\lambda_p - \sum_{r=1}^{n}\psi(\mu_r)d(E,a_r)\lambda_r$$
$$-\sum_{r=1}^{n}\psi(\mu_r)D_x d(E,a_r)(\mu_r - a_r)\lambda_r + \sum_{p=1}^{m}D_x\psi(\mu_p)v(E,\mu_p)\lambda_p$$
$$+\sum_{r=1}^{n}D_x\psi(\mu_r)v(E,a_r)\lambda_r + \sum_{r=1}^{n}D_x\psi(\mu_r)D_x v(E,a_r)(\mu_r - a_r)\lambda_r$$
$$+\sum_{p=1}^{m}\sum_{r=1}^{n}\psi(\mu_r)S_r^\lambda(\mu_p)b(E,\mu_p)\lambda_p + \sum_{p=1}^{m}\sum_{r=1}^{n}D_x\psi(\mu_r)S_r^\mu(\mu_p)b(E,\mu_p)\lambda_p$$
$$-\sum_{p=1}^{m}\sum_{r=1}^{n}D_x\psi(\mu_r)(\mu_r - a_r)S_r^\lambda(\mu_p)b(E,\mu_p)\lambda_p + R_\psi$$

(16)

in which $R_\psi$ is defined as:

$$R_\psi = \sum_{r=1}^{n}D_x\psi(\mu_r)(\mu_r - a_r)D_x d(E,a_r)(\mu_r - a_r)\lambda_r$$
$$\Rightarrow \quad |R_\psi| \leq \|D_x\psi\|\|D_x d\|Nh^2.$$

(17)

To show that the ODE (16) for $\widehat{\Psi}$ is a consistent approximation of the ODE (15) for $\Psi$, we will now expand the integrals in (15) around specific points and determine the difference between the resulting expressions and the appropriate sums in equation (16). Each integral in the ODE (15) will be investigated separately.

Given the discretization of the *i-state-space* $\Omega$ the first integral in the ODE (15) can be expressed as:

$$\int_\Omega (v(E,x)\eta(t,x))\cdot\nabla\psi(x)\,dx = \int_\Omega D_x\psi(x)v(E,x)\eta(t,x)\,dx$$
$$= \sum_{p=1}^{m}\int_{\Omega_p(t)} D_x\psi(x)v(E,x)\eta(t,x)\,dx$$
$$+ \sum_{r=1}^{n}\int_{\omega_r(t)} D_x\psi(x)v(E,x)\eta(t,x)\,dx.$$

(18)

In the expression above, the product of the functions $D_x\psi(x)$ and $v(E,x)$ in the integral for every internal subdomain is subsequently substituted by Taylor's formula around the mean $i$-state $\mu_p$ within that subdomain, while in the integral for every boundary subdomain the function $D_x\psi(x)$ is replaced by Taylor's formula around the mean $i$-state $\mu_r$ within the subdomain and the function $v(E,x)$ by Taylor's formula around the fixed point $a_r$ on the boundary $\partial^+\omega_r$ of the subdomain. These substitutions yield:

$$\int_\Omega (v(E,x)\eta(t,x)) \cdot \nabla\psi(x)\,dx = \sum_{p=1}^m D_x\psi(\mu_p)v(E,\mu_p)\lambda_p$$

$$+ \sum_{r=1}^n D_x\psi(\mu_r)v(E,a_r)\lambda_r \qquad (19)$$

$$+ \sum_{r=1}^n D_x\psi(\mu_r)D_xv(E,a_r)(\mu_r - a_r)\lambda_r + R_v$$

in which $R_v$ is defined as:

$$R_v = \sum_{p=1}^m \frac{1}{2} \int_{\Omega_p(t)} (\sigma - \mu_p)^T D_{xx}\big[D_x\psi(\xi(\sigma))v(E,\xi(\sigma))\big](\sigma - \mu_p)\eta(t,\sigma)\,d\sigma$$

$$+ \sum_{r=1}^n \frac{1}{2} \int_{\omega_r(t)} D_x\psi(\mu_r)(\sigma - a_r)^T D_{xx}v(E,\tau(\sigma))(\sigma - a_r)\eta(t,\sigma)\,d\sigma$$

$$+ \sum_{r=1}^n \int_{\omega_r(t)} \Big(D_{xx}\psi(\mu_r)(\sigma - \mu_r)\Big) \cdot \Big(D_xv(E,a_r)(\sigma - a_r)\Big)\eta(t,\sigma)\,d\sigma \qquad (20)$$

$$+ \sum_{r=1}^n \frac{1}{2} \int_{\omega_r(t)} \Big(D_{xx}\psi(\mu_r)(\sigma - \mu_r)\Big) \cdot \Big((\sigma - a_r)^T D_{xx}v(E,\tau(\sigma))(\sigma - a_r)\Big)\eta(t,\sigma)\,d\sigma$$

$$+ \sum_{r=1}^n \frac{1}{2} \int_{\omega_r(t)} \Big((\sigma - \mu_r)^T D_{xxx}\psi(\tau(\sigma))(\sigma - \mu_r)\Big) \cdot v(E,a_r)\eta(t,\sigma)\,d\sigma$$

$$+ \sum_{r=1}^n \frac{1}{2} \int_{\omega_r(t)} \Big((\sigma - \mu_r)^T D_{xxx}\psi(\tau(\sigma))(\sigma - \mu_r)\Big) \cdot \Big(D_xv(E,a_r)(\sigma - a_r)\Big)\eta(t,\sigma)\,d\sigma$$

$$+ \sum_{r=1}^n \frac{1}{4} \int_{\omega_r(t)} \Big((\sigma - \mu_r)^T D_{xxx}\psi(\tau(\sigma))(\sigma - \mu_r)\Big)$$

$$\times \Big((\sigma - a_r)^T D_{xx}v(E,\tau(\sigma))(\sigma - a_r)\Big)\eta(t,\sigma)\,d\sigma.$$

In the expression above $\xi(\sigma)$ and $\tau(\sigma)$ are again points within $\Omega_p(t)$ and $\omega_r(t)$, respectively, that stem from the rest terms in Taylor's formula. Note that the summation terms that occur in expression (19) are exactly the same terms incorporating the function $v$, which occur in the ODE (16) for $\widehat{\Psi}$. In the expression above $D_{xxx}\psi$ denotes the $q \times q \times q$ matrix with third order partial derivatives of $\psi$ with respect to $x$. $D_{xx}v$ is also a $q \times q \times q$ matrix containing, however, the second order partial derivatives of $v$ with respect to $x$, i.e., the element with index (i,j,k) in this matrix is $\frac{\partial^2 v^j}{\partial x^i \partial x^k}$. $D_{xx}[D_x\psi(x)v(E,x)]$ is the $q \times q$ matrix with second order partial derivatives of the product function $D_x\psi v: \mathbb{R}^s \times \Omega \to \mathbb{R}$ and hence comparable with $D_{xx}\psi$, defined above.

If the following assumptions are made on the functions $\psi$ and $v$:

ASSUMPTION 3: $\psi \in \mathbb{C}3$

and

ASSUMPTION 4: $v \in \mathbb{C}2$

the quantity $R_v$ satisfies the inequality:

$$|R_v| \leq \left\{\frac{1}{2}\|D_x\psi\|\|D_{xx}v\| + \|D_{xx}\psi\|\|D_x v\| + \frac{1}{2}\|D_{xxx}\psi\|\|v\|\right\}Nh^2 + O(h^3). \quad (21)$$

The assumptions 3 and 4 hence ensure that $R_v$ is bounded and $O(h^2)$. Moreover, these assumptions also imply that all previous conditions on the differentiability of $\psi$ and $v$ are fulfilled.

In a similar way as was done above for the first integral in the ODE (15) the second integral can be expressed as:

$$-\int_\Omega \psi(x)d(E,x)\eta(t,x)\,dx = -\sum_{p=1}^m \int_{\Omega_p(t)} \psi(x)d(E,x)\eta(t,x)\,dx$$
$$-\sum_{r=1}^n \int_{\omega_r(t)} \psi(x)d(E,x)\eta(t,x)\,dx. \quad (22)$$

Subsequently the product of the functions $\psi(x)$ and $d(E,x)$ in the integral for every internal subdomain is again substituted by Taylor's formula around the mean i-state $\mu_p$ within that subdomain, while in the integral for every boundary subdomain the function $\psi(x)$ is replaced by Taylor's formula around the mean i-state $\mu_r$ within the subdomain and the function $d(E,x)$ by Taylor's formula around the fixed point $a_r$ on the boundary $\partial^+\omega_r$ of the subdomain. These substitution yield:

$$-\int_\Omega \psi(x)d(E,x)\eta(t,x)\,dx = -\sum_{p=1}^m \psi(\mu_p)d(E,\mu_p)\lambda_p$$
$$-\sum_{r=1}^n \psi(\mu_r)d(E,a_r)\lambda_r \quad (23)$$
$$-\sum_{r=1}^n \psi(\mu_r)D_x d(E,a_r)(\mu_r - a_r)\lambda_r - R_d$$

in which $R_d$ is defined as:

$$R_d = \sum_{p=1}^{m} \frac{1}{2} \int_{\Omega_p(t)} (\sigma - \mu_p)^T D_{xx}[\psi(\xi(\sigma))d(E,\xi(\sigma))](\sigma - \mu_p)\eta(t,\sigma)\,d\sigma$$

$$+ \sum_{r=1}^{n} \frac{1}{2} \int_{\omega_r(t)} \psi(\mu_r)(\sigma - a_r)^T D_{xx}d(E,\tau(\sigma))(\sigma - a_r)\eta(t,\sigma)\,d\sigma$$

$$+ \sum_{r=1}^{n} \int_{\omega_r(t)} D_x\psi(\mu_r)(\sigma - \mu_r) D_x d(E,a_r)(\sigma - a_r)\eta(t,\sigma)\,d\sigma$$

$$+ \sum_{r=1}^{n} \frac{1}{2} \int_{\omega_r(t)} D_x\psi(\mu_r)(\sigma - \mu_r)(\sigma - a_r)^T D_{xx}d(E,\tau(\sigma))(\sigma - a_r)\eta(t,\sigma)\,d\sigma$$

$$+ \sum_{r=1}^{n} \frac{1}{2} \int_{\omega_r(t)} (\sigma - \mu_r)^T D_{xx}\psi(\tau(\sigma))(\sigma - \mu_r)d(E,a_r)\eta(t,\sigma)\,d\sigma \qquad (24)$$

$$+ \sum_{r=1}^{n} \frac{1}{2} \int_{\omega_r(t)} (\sigma - \mu_r)^T D_{xx}\psi(\tau(\sigma))(\sigma - \mu_r) D_x d(E,a_r)(\sigma - a_r)\eta(t,\sigma)\,d\sigma$$

$$+ \sum_{r=1}^{n} \frac{1}{4} \int_{\omega_r(t)} (\sigma - \mu_r)^T D_{xx}\psi(\tau(\sigma))(\sigma - \mu_r)(\sigma - a_r)^T$$
$$\times D_{xx}d(E,\tau(\sigma))(\sigma - a_r)\eta(t,\sigma)\,d\sigma.$$

In this expression the definition of $D_{xx}d$ is equivalent with the definition of $D_{xx}\psi$ given before, i.e., $D_{xx}d$ is the $q \times q$ matrix of second order partial derivatives of $d$ with respect to $x$. Making the additional assumption

ASSUMPTION 5: $d \in \mathbb{C}2$

the quantity $R_d$ satisfies the inequality:

$$|R_d| \leq \left\{ \frac{1}{2}\|\psi\|\|D_{xx}d\| + \|D_x\psi\|\|D_x d\| + \frac{1}{2}\|D_{xx}\psi\|\|d\| \right\} Nh^2 + O(h^3). \qquad (25)$$

Hence assumption 5 implies that $R_d$ is bounded and $O(h^2)$. As before, the summation terms in expression (23) are equivalent with the summation terms found in the ODE (16), that incorporate the function $d$.

Given the (simplifying) assumption made above that $b(E,x)$ equals 0 in all the boundary subdomains, the last integral in (15) can be rewritten as:

$$\oint_{\partial+\Omega} \psi(y) \int_{\Omega} s(x,y)b(E,x)\eta(t,x)\,dx\,dy =$$

$$\sum_{p=1}^{m} \int_{\Omega_p(t)} b(E,x) \oint_{\partial+\Omega} \psi(y)s(x,y)\,dy\,\eta(t,x)\,dx. \qquad (26)$$

If we define the function $F(x): \Omega \to \mathbb{R}$ as:

$$F(x) = \oint_{\partial^+\Omega} \psi(y) s(x,y)\, dy$$

and if the product of the functions $b(E,x)$ and $F(x)$ in the integral for every internal subdomain in expression (26) is substituted by Taylor's formula around the mean $i$-state $\mu_p$ within that subdomain, the following equality is obtained:

$$\oint_{\partial^+\Omega} \psi(y) \int_\Omega s(x,y) b(E,x) \eta(t,x)\, dx\, dy = \sum_{p=1}^m b(E,\mu_p) F(\mu_p) \lambda_p + R'_b \qquad (27)$$

with

$$R'_b = \frac{1}{2} \sum_{p=1}^m \int_{\Omega_p(t)} (\sigma-\mu_p)^T D_{xx}[b(E,\xi(\sigma))F(\xi(\sigma))](\sigma-\mu_p)\eta(t,\sigma)\, d\sigma. \qquad (28)$$

When the function $\psi(y)$ in the definition of $F(x)$ is replaced by Taylor's formula around the mean $i$-state $\mu_r$ $F(\mu_p)$ can be rewritten as:

$$F(\mu_p) = \sum_{r=1}^n \oint_{\partial^+\omega_r} \psi(\sigma) s(\mu_p,\sigma)\, d\sigma$$

$$= \sum_{r=1}^n \psi(\mu_r) \oint_{\partial^+\omega_r} s(\mu_p,\sigma)\, d\sigma + \sum_{r=1}^n D_x \psi(\mu_r) \oint_{\partial^+\omega_r} (\sigma-\mu_r) s(\mu_p,\sigma)\, d\sigma \qquad (29)$$

$$+ \frac{1}{2} \sum_{r=1}^n \oint_{\partial^+\omega_r} (\sigma-\mu_r)^T D_{xx}\psi(\tau(\sigma))(\sigma-\mu_r) s(\mu_p,\sigma)\, d\sigma.$$

Hence, the integral (26) can also be expressed as:

$$\oint_{\partial^+\Omega} \psi(y) \int_\Omega s(x,y) b(E,x)\eta(t,x)\, dx\, dy$$

$$= \sum_{p=1}^m \sum_{r=1}^n \psi(\mu_r) S_r^\lambda(\mu_p) b(E,\mu_p) \lambda_p$$

$$+ \sum_{p=1}^m \sum_{r=1}^n D_x\psi(\mu_r) S_r^\mu(\mu_p) b(E,\mu_p) \lambda_p \qquad (30)$$

$$- \sum_{p=1}^m \sum_{r=1}^n D_x\psi(\mu_r)(\mu_r - a_r) S_r^\lambda(\mu_p) b(E,\mu_p) \lambda_p + R_b$$

which

$$R_b = R_b' + \frac{1}{2}\sum_{p=1}^{m}\sum_{r=1}^{n} \oint_{\partial+\omega_r} (\sigma-\mu_r)^T D_{xx}\psi(\tau(\sigma))(\sigma-\mu_r) s(\mu_p,\sigma)\,d\sigma\, b(E,\mu_p)\lambda_p. \quad (31)$$

Obviously, if

ASSUMPTION 6: $\oint_{\partial+\Omega} b(E,x)s(x,y)\psi(y)\,dy \in \mathbb{C}2$

and

ASSUMPTION 7: $b \in \mathbb{C}1$

hold, $R_b$ fulfills the following inequality:

$$|R_b| \leq \left\{\frac{1}{2}\|D_{xx}[bF]\| + \frac{1}{2}\|D_{xx}\psi\|\|b\|\right\} Nh^2 \quad (32)$$

and is hence $O(h^2)$ and bounded.

Given the equivalencies derived above for the three integrals, occurring in the ODE (15) for $\Psi(t)$ we can conclude that

$$\left|\frac{d\Psi}{dt} - \frac{d\widehat{\Psi}}{dt}\right| \leq |R_\psi| + |R_v| + |R_d| + |R_b| \quad (33)$$

and that the difference between the exact ODE (15) for $\Psi(t)$ and the EBT-approximation (16) for $\widehat{\Psi}(t)$ is $O(h^2)$. This difference is therefore of the same order as the difference between the quantities $\Psi(t)$ and $\widehat{\Psi}(t)$ themselves. From this observation we conclude that applying the *Escalator Boxcar Train* to the PDE and boundary condition (2) results in a consistent numerical approximation of these equations. The local approximation error is bounded and vanishes if the discretization of the *i-state-space* $\Omega$ is made finer and finer, *i.e.*, if $h \to 0$.

## 5. The convergence of the escalator boxcar train

In this section we will focus upon the convergence of the EBT-method. As in the previous section, the fact that a rigorous interpretation of the equations (2) is still lacking and that the density-function $\eta(t,x)$ cannot be retrieved from its local moments without further assumptions, hampers the study of the convergence of the EBT-method and bars the application of more traditional methods to prove convergence. Therefore, the same approach will be taken as was used in the consistency analysis: The convergence of the EBT-method will be studied in a weak sense, *i.e.*, by studying the dynamics and approximation of a measure of the density-function $\eta(t,x)$, as defined by expression (11). More specifically, to study the convergence of the EBT-method we assume that $\widehat{\Psi}(0) = \Psi(0)$ and compare the exact value of $\Psi(T)$ at some $t = T$ with its approximation $\widehat{\Psi}(T)$, which results from applying the EBT-method during the time interval $[0;T]$.

In the previous section it was proven that the difference between the exact ODE (15) for $\Psi(t)$ and the EBT-approximation (16) for $\widehat{\Psi}(t)$ is $O(h^2)$. Given that $\widehat{\Psi}(0) = \Psi(0)$, this implies that the difference between $\Psi(t)$ and $\widehat{\Psi}(t)$ would remain $O(h^2)$ on the time interval $[0;\tau]$, if all the ODEs involved could be solved explicitly. Here $\tau$ indicates the threshold time at which the first boundary subdomain reaches the threshold size $\omega_{max}$ (see section 3). At $t = \tau$ the boundary subdomains are transformed into internal subdomains, applying equations (9) and (10), and replaced by a new set of boundary subdomains. Hence, if all ODEs could be solved explicitly and if the (non-linear) transformation of the boundary subdomains would leave the order of approximation unchanged, the difference between $\Psi(T)$ and $\widehat{\Psi}(T)$ would still be $O(h^2)$. This would imply that also the global discretization error at $t = T$ would vanish if $h \to 0$, proving the convergence of the EBT-method.

We have not been able to prove the convergence of the EBT-method in such a general sense. First of all, we cannot show that the transformation of the boundary subdomains leaves the order of approximation unchanged. This is all the more important, since, if $h \to 0$, necessarily $\tau \to 0$ and the number of transformations on a finite interval $[0;T]$ approaches $\infty$. Moreover, the sets of ODEs I and II can usually not be solved explicitly and an additional time integration method should be used to solve these equations numerically. Solving the ODEs numerically could in principle interact with the errors, introduced by the discretization of $\Omega$, or with the transformation procedure of the boundary subdomains and cause the global discretization error to diverge.

As an alternative, we could focus on a model for which both $\Psi(T)$ and $\widehat{\Psi}(T)$ can be obtained and determine the global discretization error directly for such a particular case. In general it is impossible to find the analytical solution of (2) and hence the exact value of $\Psi(T)$, especially if the model incorporates non-linear interactions with the environment. For this reason, we will, as an example, focus on a simple, linear model, in which the *i-state-space* $\Omega$ is one-dimensional: $\Omega := [1, \infty)$. This implies that all individuals are born with an *i-state* $x = 1$. At the same time this value of $x$ is the point at the boundary of $\Omega$ that characterizes the boundary subdomain, *i.e.*, $a_1 := 1$. Because of the linearity, we will denote the coefficient functions describing the growth, death and reproduction rate of the individuals by $v(x)$, $d(x)$ and $b(x)$, respectively.

The model can be described by the following equations:

$$\frac{\partial \eta(t,x)}{\partial t} + \frac{\partial v(x)\eta(t,x)}{\partial x} = -d(x)\eta(t,x), \tag{34a}$$

$$v(1)\eta(t,1) = \int_1^\infty b(x)\eta(t,x)\,dx. \tag{34b}$$

A possible choice for the coefficient functions that constitute a model, for which we can obtain an analytical solution, can be derived from example 2 in the publication by Metz & Diekmann in this volume. The example is based upon the following assumptions about the individual behaviour: (1) Individuals acquire food at a rate $\alpha x$, where $x$ is the individual size. (2) The acquired food is partitioned into a fraction $\kappa(x)$, which is spent on reproduction and a fraction $1 - \kappa(x)$ spent on basal metabolism and growth. (3) The costs of producing offspring biomass equals the costs of producing parent biomass, which means that growth is as expensive as reproduction. Here, these costs are arbitrarily set to unity without loss of generality. (4) Finally, metabolism is proportional to size with proportionality constant $m$ and the death rate $\gamma$ is size-independent. The total biomass of the population will be the required output quantity of the model and hence $\psi(x) := x$.

The assumptions above lead to the following choice for the coefficient functions in the model equations (34):

$$v(x) = \alpha(1 - m - \kappa(x))x$$
$$b(x) = \alpha\kappa(x)x$$
$$d(x) = \gamma$$
$$\psi(x) = x$$

and hence

$$\Psi(t) = \int_\Omega x\eta(t,x)\,dx.$$

Differentiating the expression for $\Psi(t)$ and using the PDE (34a) to replace $\partial \eta/\partial t$ it turns out that the dynamics of $\Psi$ can be described by the following ODE:

$$\frac{d\Psi}{dt} = (\alpha - m - \gamma)\Psi \quad \Rightarrow \quad \Psi(T) = \Psi_0 e^{(\alpha - m - \gamma)T}.$$

The quantity $\Psi$ grows or decays exponentially, depending upon the sign of $(\alpha - m - \gamma)$.

Considering the discretization errors given by the expressions in the previous section we conclude that for the current model these discretization errors are:

$$R_\psi = 0$$
$$R_d = 0$$
$$R_v = -\sum_{p=1}^{m} \int_{\Omega_p(t)} [\alpha x \kappa(x)]''_{x=\xi(x)} (x - \mu_p)^2 \eta(t, x)\, dx$$
$$- \int_{\omega_1(t)} [\alpha x \kappa(x)]''_{x=\tau(x)} (x - 1)^2 \eta(t, x)\, dx$$
$$R_b = R'_b.$$

With respect to the error $R'_b$ we have to remark that in the previous section it was assumed that the reproduction rate $b(E, x) = 0$ within the boundary subdomains $\omega_r(t)$. In the current model this is not the case and we have to include this contribution to the reproduction from the boundary subdomain. This reproduction from the boundary subdomain is given by:

$$\int_{\omega_1(t)} \alpha x \kappa(x) \eta(t, x)\, dx. \tag{35}$$

This integral can be expanded around the boundary point $x = 1$. The system of ODEs which describe the dynamics of the quantities $\lambda_0$ and $\pi_0$ in the boundary subdomain then becomes:

$$\begin{aligned}\frac{d\lambda_0}{dt} &= -\gamma\lambda_0 + \alpha\kappa(1)\lambda_0 + \alpha[\kappa(1) + \kappa'(1)]\pi_0 + \sum_{p=1}^{m} \alpha\kappa(\mu_p)\mu_p \lambda_p, \\ \frac{d\pi_0}{dt} &= (\alpha(1 - \kappa(1)) - m)\lambda_0 + (\alpha(1 - \kappa(1)) - \kappa'(1)) - m)\pi_0 - \gamma\pi_0 \end{aligned} \tag{36}$$

and the discretization error $R'_b$ equals:

$$R'_b = \sum_{p=1}^{m} \int_{\Omega_p(t)} [\alpha x \kappa(x)]''_{x=\xi(x)} (x - \mu_p)^2 \eta(t, x)\, dx$$
$$+ \int_{\omega_1(t)} [\alpha x \kappa(x)]''_{x=\tau(x)} (x - 1)^2 \eta(t, x)\, dx = -R_v.$$

The total discretization error $R$ hence turns out to be 0. Obviously all the local truncation errors cancel. This is not too surprising since the dynamics of the output quantity $\Psi(t)$ was independent of the choice of $\kappa(x)$ and all the discretization errors were due to the occurrence of $\kappa(x)$ in $v(x)$ and $b(x)$.

Hence, in the context of this simple model we are also unable to study the convergence of the EBT-method. Unfortunately, this represents the state of the art: We have not found a possible route to study the convergence of the EBT-method in either a linear, example model or in a general sense. Since the method is qualitatively rather different from other numerical methods for PDEs, conventional methods to study convergence do not seem appropriate. The convergence of the method is therefore still an open question.

Nonetheless, the current example does show something interesting: In the previous section it was shown that an approximation of the desired output quantity could be obtained by expanding the weighing function $\psi(x)$ in a boundary subdomain around $x = \mu_r$, in which $\mu_r$ was defined by the transformation (10). In principle, the total reproduction of the population is nothing different than such an output quantity and in general the integral (35) describing the reproduction from the boundary subdomain should indeed be expanded around $x = \mu_1$. However, the specific choice of the coefficient functions in the current example lead to a better approximation of the dynamics of the output quantity if the integral (35) is expanded around $x = 1$. This means that the numerical schemes I and II, although constituting a consistent approximation of the general set of equations (2), can sometimes be changed slightly to obtain a more optimal numerical approximation of a specific model.

## 6. Discussion

In this paper we have focused on the numerical properties of the *Escalator Boxcar Train*, which was especially designed to integrate numerically the type of PDE that occurs in physiologically structured population models. For reasons related to the exact interpretation of the equations and the characteristics of the numerical method, we first introduced a weak concept of consistency and convergence in terms of measures of the original density-function. Most importantly, this weak concept fits in with the way in which the functional analytic underpinning of the general framework of physiological structured population models is developed. Subsequently, we studied the consistency and the convergence of the second order variant of the EBT-method, although all results obtained can be shown to hold for the third order variant as well.

From the analysis in the previous sections we have to conclude that the numerical properties of the *Escalator Boxcar Train* are not yet understood completely. We have shown that in a weak sense the method yields a consistent approximation of the exact dynamics. However, the convergence of the method could not be proven. More specifically, there is no guarantee that after applying the EBT-method over some time interval $[0, T]$ the EBT-solution still approximates the exact solution of the equations. Up to now only indirect evidence exists that the contrary is true. For example, in [2] it is shown for a specific model which incorporates non-linear interactions, that the numerical results obtained by applying the EBT-method match very well with analytical results from the stability analysis of the model. Based on our own experience, the EBT-method works satisfactorily for a wide variety of problems and appears to yield results which are consistent with results obtained along alternative routes (*cf.* [2]). More importantly, however, it is especially the biological interpretation of the EBT-method that yields the most reassuring support for its validity.

Apart from the convergence of the method, there still remain a few other properties of the method to be investigated. As noted in section 3, the threshold size $\omega_{max}$, or equivalently the threshold time $\tau$, which represents the maximum size that a boundary subdomain can attain before being closed off, is up to now arbitrarily chosen. It would be desirable to base the choice of this threshold upon some measure of the error, caused by approximating the density-function with a set of its moments. Furthermore, the influence of the procedures to restrict the total number of subdomains in $\Omega$, as described in section 3, should be investigated. These open problems we like to leave for the mathematically more adept.

## References

[1] J.R. Sauer and N.A. Slade, *Size-based demography of vertebrates*, Ann. Rev. Ecol. Syst. **18** (1987), 71–90.

[2] A.M. De Roos, J.A.J. Metz, E. Evers and A. Leipoldt, *A size dependent predator-prey interaction: Who pursues whom?*, J. Math. Biol (1990), (in press).

[3] J.A.J. Metz and O. Diekmann, *The dynamics of physiologically structured populations*, Springer Lect. Notes in Biomath. **68**, Springer-Verlag, Heidelberg, 1986.

[4] A.M. De Roos, *Numerical methods for structured population models: The Escalator Boxcar Train*, Num. Meth. Part. Diff. Eqs. **4** (1988.), 173–195.

[5] A.M. De Roos, J.A.J. Metz, and O. Diekmann, *The Escalator Boxcar Train: Basic theory and an application to Daphnia population dynamics*, CWI report AM-R8814, Center for Mathematics and Computer Science, Amsterdam (1988).

[6] A.M. De Roos, J.A.J. Metz, and O. Diekmann, *Studying the dynamics of structured population models: A versatile technique and its application to Daphnia*, Amer. Nat (1990), (in press).

[7] O. Diekmann, *On semigroups and populations*, Advanced topics in the theory of dynamical systems, G. Fusco, M. Ianelli and L. Salvadori, eds., Academic Press, pg. 125–135, 1989.

A.M. De Roos
Department of Pure and Applied Ecology
University of Amsterdam
Kruislaan 302
1098 SM Amsterdam, The Netherlands

J.A.J. Metz
Institute of Theoretical Biology
Leiden University
Kaiserstraat 63
2311 GP Leiden, The Netherlands

# An Application of Polynomial Operator Matrices to a Second Order Cauchy Problem

Klaus-J. Engel

Mathematisches Institut, Universität Tübingen

In this note we will apply the abstract results on polynomial operator matrices (see [E1]-[E3], [EN]) to a concrete example. Thereby we will derive some interesting new results on the well posedness of a second order Cauchy problem, reestablish some results given by Gutowski [Gu] and improve an estimate for the growth bound for the solution of this problem. More precisely we consider the second order Cauchy problem

$$u_{tt}(t,x) + 2w\, u_{tx}(t,x) + 2b\, u_t(t,x) + c\, u(t,x) = a\, u_{xx}(t,x),$$
$$t \geq 0,\ x \in [0,1],$$
$$u(0,.) = u_0,\quad u_t(0,.) = v_0,$$
$$u(t,0) = 0,\quad u(t,1) = 0,\quad t \geq 0,$$
$$(CP_2)$$

which describes the linear transversal vibration of a string taking into account the linear exterior damping and the flow of a fluid in the interior of the string. Here $u(t,x)$ denotes the transversal displacement of the string and

$$w, b, c, a > 0$$

are constants determining the mass, tension, damping of the string and the pressure, velocity of the fluid (see [Gu] for more details). Moreover for physical reasons it is assumed that

$$b^2 - c < 0.$$

Denoting by $A$ the first derivative $\frac{d}{dx}$ on the Banach space $E := L^2[0,1]$ with domain $D(A) := \{f \in E : f \text{ absolutely continuous}, f' \in E, f(0) = f(1)\}$ the usual reduction $v := u_t$ of (CP$_2$) yields an "equivalent" first order system (see e.g. [F]). This system can be written in matrix form

$$\dot{w}(t) = \mathcal{A}w(t), \quad t \geq 0,$$
$$w(0) = w_0 \qquad \qquad \text{(ACP)}$$

where $w(t) = \binom{u(t)}{v(t)} \in E^2$, $w_0 = \binom{u_0}{v_0}$ and

$$\mathcal{A} = \begin{pmatrix} 0 & Id \\ aA^2 - cId & -2(wA + bId) \end{pmatrix}.$$

In the sequel we will study the operator matrix $\mathcal{A}$. In particular we will show that a restriction $\mathcal{A}_G$ of $\mathcal{A}$ to an "abstract energy space" $\mathcal{E}_G$ is the generator of a strongly continuous semigroup, i.e. (ACP) is well posed on $\mathcal{E}_G$. Once the well posedness is proved we compute the spectrum $\sigma(\mathcal{A}_G)$ and the spectral bound s($\mathcal{A}_G$) of $\mathcal{A}_G$. Finally we show that the growth bound $\omega(\mathcal{A}_G)$ of $\mathcal{A}_G$ coincides with the spectral bound s($\mathcal{A}_G$). Hence the location of the spectrum of $\mathcal{A}_G$ determines the asymptotic behavior of the solutions of (ACP) and (CP$_2$).

First we have to specify an appropriate domain $D(\mathcal{A})$ of $\mathcal{A}$ in order to make the operator matrix a closed, densely defined operator. The following result is a special case of [E1], Proposition 1.3.

**Proposition 1.** *The operator matrix*

$$\mathcal{A} := \begin{pmatrix} 0 & Id \\ aA^2 - cId & -2(wA + bId) \end{pmatrix}$$

*with domain*

$$D(\mathcal{A}) := \left\{ \begin{pmatrix} x \\ y \end{pmatrix} \in E \times E : x \in D(A), aAx - 2wy \in D(A) \right\}$$

*is a densely defined and closed operator on* $\mathcal{E} := E \times E$.

Having defined $\mathcal{A}$ as a closed operator on $\mathcal{E}$ we now turn to the question whether $\mathcal{A}$ is the generator of a strongly continuous semigroup. Since $A$ is unbounded it follows from [E2], Theorem 2.6(iv) that this is not true on the Banach space $\mathcal{E}$, hence $\mathcal{E}$ is not an appropriate space for this problem. Using the theory of "abstract energy spaces" for polynomial operator matrices (see [EN]) we now prove that the restriction $\mathcal{A}_G$ of $\mathcal{A}$ to the Banach space $\mathcal{E}_G := [D(A)] \times E$ (here [D(A)] denotes the vector space $D(A)$ equipped with the graph norm) is always a generator. For this define an operator matrix

$$G(A) := \begin{pmatrix} \sqrt{a}\left(\sqrt{\frac{c}{a}} - A\right) & \frac{w+\sqrt{w^2+a}}{\sqrt{a}} Id \\ -(w + \sqrt{w^2 + a})\left(\sqrt{\frac{c}{a}} - A\right) & Id \end{pmatrix}.$$

Since $\sigma(A) = 2\pi i\mathbb{Z}$ the operator $G(A)$ is invertible by [E1], Theorem 2.3 with bounded inverse $G^{-1}(A) \in \mathcal{L}(\mathcal{E})$. Thus by [EN], Proposition 2.1 the restriction $\mathcal{A}_G$ of $\mathcal{A}$ to $\mathcal{E}_G := [D(G(A))] = [D(A)] \times E$ is similar to

$$\mathcal{A}_0 := G(A)\mathcal{A}G^{-1}(A)$$
$$= \begin{pmatrix} \lambda A + \alpha Id & \beta Id \\ \gamma Id & \mu A + \delta Id \end{pmatrix} \quad (1)$$

with domain $D(\mathcal{A}_0) := D(A) \times D(A)$ on $\mathcal{E}$ where

$$\lambda := -\left(w + \sqrt{w^2 + a}\right),$$

$$\mu := -\left(w - \sqrt{w^2 + a}\right),$$

$$\beta := \frac{-2b\sqrt{a}\left(w + \sqrt{w^2 + a}\right) + \sqrt{c}\left(a + \left(w + \sqrt{w^2 + a}\right)^2\right)}{a + \left(w + \sqrt{w^2 + a}\right)^2},$$

$$\alpha := -\frac{2b\left(w + \sqrt{w^2 + a}\right)^2}{a + \left(w + \sqrt{w^2 + a}\right)^2},$$

$$\gamma := \frac{-2b\sqrt{a}\left(w + \sqrt{w^2 + a}\right) - \sqrt{c}\left(a + \left(w + \sqrt{w^2 + a}\right)^2\right)}{a + \left(w + \sqrt{w^2 + a}\right)^2},$$

$$\delta := -\frac{2ba}{a + \left(w + \sqrt{w^2 + a}\right)^2}.$$

Since $\lambda, \mu \in \mathbb{R}$ and $A$ generates a group on $E$ the matrix $\mathcal{A}_0$ also generates a group on $\mathcal{E}$ by the bounded perturbation theorem (see [Go], Chapter 1, 6.4). Using that $\mathcal{A}_0$ and $\mathcal{A}_G$ are similar we obtain the following result.

**Theorem 2.** *The restriction $\mathcal{A}_G$ of $\mathcal{A}$ to the Banach space $\mathcal{E}_G := [D(A)] \times E$ with domain*

$$D(\mathcal{A}_G) := \{x \in \mathcal{E}_G : G(A)x \in D(\mathcal{A}_0)\}$$
$$= D(A^2) \times D(A)$$

*is the generator of a strongly continuous group.*

Once we have well posedness of (ACP) on $\mathcal{E}_G$ we are interested in the asymptotic behavior of the solutions of this problem. As known from general semigroup theory spectral theory yields the appropriate tools for this investigation (see e.g. [N2], A-III, A-IV). Thus we have to calculate the spectrum $\sigma(\mathcal{A}_G)$ of $\mathcal{A}_G$ which coincides with $\sigma(\mathcal{A}_0)$ since $\mathcal{A}_G$ and $\mathcal{A}_0$ are similar. From the spectral mapping theorem for polynomial operator matrices applied to $\mathcal{A}_0$ (see [E1], Theorem 2.8) we obtain the following result.

**Proposition 3.** *The spectrum of the operator matrix $\mathcal{A}_G$ is given by*
$$\sigma(\mathcal{A}_G) = \sigma(\mathcal{A}_0)$$
$$= \left\{ -(w\mu + b) \pm \sqrt{(w^2 + a)\mu^2 + b^2 - c + 2wb\mu} : \mu \in 2\pi i \mathbb{Z} \right\}.$$

Now a straightforward calculation using the fact that $b^2 - c < 0$ shows that the function
$$\mathbb{R} \ni s \mapsto \left| \operatorname{Re} \sqrt{(w^2 + a)(is)^2 + b^2 - c + 2wbis} \right| =: g(s)$$
is even and strictly increasing for $s \geq 0$. Hence a short calculation yields
$$\sup_{s \in \mathbb{Z}} g(s) = \lim_{s \to \infty} g(s) = \frac{bw}{\sqrt{w^2 + a}}$$
and we obtain the following corollary.

**Corollary 4.** *The spectral bound of the operator matrix $\mathcal{A}_G$ is given by*
$$s(\mathcal{A}_G) = -b \left( 1 - \frac{w}{\sqrt{w^2 + a}} \right).$$

In a final step we show that the growth bound $\omega(\mathcal{A}_G)$ coincides with the spectral bound $s(\mathcal{A}_G)$. Since none of the usual results implying "$s(A) = \omega(A)$" (see [N2], A-III, 6 and 7) is applicable in this situation we use the explicit formula for the semigroup generated by $\mathcal{A}_0$ given in [N1], 5.1.

**Lemma 5.** *The semigroup $(\mathcal{T}_0(t))$ generated by the operator matrix*
$$\mathcal{A}_0 = \begin{pmatrix} \lambda A + \alpha Id & \beta Id \\ \gamma Id & \mu A + \delta Id \end{pmatrix}$$
*with domain $D(\mathcal{A}_0) = D(A) \times D(A)$ is given by $\mathcal{T}_0(t) := (T_{ij}(t))_{2 \times 2}$ where*

$$T_{11}(t) := S(t) - \int_0^t J_1\left(2\sqrt{bcs(s-t)}\right) \sqrt{\frac{bc(s-t)}{s}} S(t-s)T(s)ds,$$

$$T_{12}(t) := \beta \int_0^t J_0\left(2\sqrt{bcs(s-t)}\right) S(s)T(t-s)ds,$$

$$T_{21}(t) := \gamma \int_0^t J_0\left(2\sqrt{bcs(s-t)}\right) S(s)T(t-s)ds,$$

$$T_{22}(t) := T(t) - \int_0^t J_1\left(2\sqrt{bcs(s-t)}\right) \sqrt{\frac{bc(s-t)}{s}} S(s)T(t-s)ds.$$
(2)

Here $J_\nu$ denotes the $\nu$-th Bessel function ($\nu = 0, 1$) and $S(t) := e^{t(\lambda A + \alpha Id)}$, $T(t) := e^{t(\mu A + \delta Id)}$.

**Remark.** From the identity $\mathcal{A}_G = G^{-1}(A)\mathcal{A}_0 G(A)$ it follows that the semigroup $(\mathcal{T}_G(t))$ generated by $\mathcal{A}_G$ is given by $\mathcal{T}_G(t) = G^{-1}(A)\mathcal{T}_0(t)G(A)$. Using this and (2) yields an explicit representation of the semigroup $(\mathcal{T}_G(t))$. In particular an explicit formula for the solution of (CP$_2$) is given by the first coordinate of the function $t \mapsto \mathcal{T}_G(t)\binom{u_0}{v_0}$.

Observe now that, since $A$ generates a group of isometries, we have $\omega(\lambda A) = \omega(\mu A) = 0$. Moreover from the definition of $\beta$ and $\gamma$ it follows $\beta\gamma \leq b^2 - c < 0$, hence the arguments of the Bessel functions in (2) are real. Using this and the fact that $|J_\nu(s)| \leq 1$ for $s \in \mathbb{R}$, $\nu = 0, 1$ (see [W], III.3.31) one obtains the following estimate for the growth bound of $\mathcal{A}_0$ (see [E4] for a systematic treatment of a more general situation).

**Corollary 6.** *For the growth bound $\omega(\mathcal{A}_0)$ one has*
$$\omega(\mathcal{A}_0) \leq \max\{\alpha, \delta\}.$$

From this estimate and the similarity of $\mathcal{A}_G$ and $\mathcal{A}_0$ we conclude

$$\omega(\mathcal{A}_G) = \omega(\mathcal{A}_0)$$
$$\leq \max\left\{\frac{-2b\left(w + \sqrt{w^2 + a}\right)^2}{a + \left(w + \sqrt{w^2 + a}\right)^2}, \frac{-2ba}{a + \left(w + \sqrt{w^2 + a}\right)^2}\right\}$$
$$= \frac{-2ba}{a + \left(w + \sqrt{w^2 + a}\right)^2}$$
$$= -b\left(1 - \frac{w}{\sqrt{w^2 + a}}\right)$$
$$= s(\mathcal{A}_G).$$

Since the growth bound always dominates the spectral bound we have proved the following result.

**Theorem 7.** *The spectral bound and the growth bound of $\mathcal{A}_G$ coincide and are given by*
$$\omega(\mathcal{A}_G) = s(\mathcal{A}_G) = -b\left(1 - \frac{w}{\sqrt{w^2 + a}}\right).$$

This improves a result of Gutowski ([Gu], (3.62)) stating
$$\omega(\mathcal{A}_G) \leq -b\left(1 - \frac{w}{\sqrt{a}}\right). \tag{3}$$

Since the growth bound of $\mathcal{A}_G$ is less than zero we arrive at the final conclusion.

**Corollary 8.** *The semigroup generated by $\mathcal{A}_G$ on the Banach space $\mathcal{E}_G$ is uniformly exponentially stable.*

Note that in order to derive a similar result from (3) one needs the additional assumption $1 - \frac{w}{\sqrt{a}} > 0$.

## References

[E1] Engel, K.-J., *A spectral mapping theorem for polynomial operator matrices*, Diff. Integral Eq. **2** (1989), 203–215.

[E2] Engel, K.-J., *Polynomial operator matrices as semigroup generators: the $2 \times 2$ case*, Math. Ann. **284** (1989), 563–576.

[E3] Engel, K.-J., *Polynomial operator matrices as semigroup generators: the general case*, Int. Eq. Operator Th. **13** (1990), 175–192.

[E4] Engel, K.-J., *An explicit formula for semigroups generated by $2 \times 2$ operator matrices*, Preprint 1990.

[EN] Engel, K.-J.; Nagel, R., *Cauchy problems for polynomial operator matrices on abstract energy spaces*, Forum Math. **2** (1990), 89–102.

[F] Fattorini, H.O., *Second Order Linear Differential Equations in Banach Spaces*, North-Holland Mathematics Studies 108, North-Holland 1985.

[Go] Goldstein, J.A., *Semigroups of Linear Operators and Applications*, Oxford University Press 1985.

[Gu] Gutowski, R., *On the stability of some material systems II*, Conf. Sem. Mat. Bari **212** (1986).

[N1] Nagel, R., *Well-posedness and positivity for systems of linear evolution equations*, Conf. Sem. Mat. Bari **203** (1985).

[N2] Nagel, R. (Ed.), *One-parameter Semigroups of Positive Operators*, Lect. Notes Math. bf 1184, Springer-Verlag, Berlin-Heidelberg-New York-Tokyo, 1986.

[W] Watson, G.N., *Theory of Bessel Functions*, Cambridge University Press, 1966.


Klaus-J. Engel
Mathematisches Institut
Universität Tübingen
Auf der Morgenstelle 10
D-7400 Tübingen, FRG
Electronic address:
mina001@dtuzdv5a.bitnet

# Asymptotic Convergence for a Class of Autocatalytic Chemical Systems

W.W. Farr*
W.E. Fitzgibbon* **
J.J. Morgan***
S.J. Waggoner*

Department of Mathematics, Worcester Polytechnic Institute
Department of Mathematics, University of Houston
Department of Mathematics, Texas A & M University
Department of Mathematics, Furman University

## 1. Introduction

In this note we are concerned with the asymptotic behavior of a system of ordinary differential equations which describe a class of chemical reactions. We consider a model chemical system consisting of $m$ species $A_1, \ldots, A_m$ and $m-1$ independent reactions. The reaction mechanism for this system is assumed to be

$$
\begin{aligned}
A_1 + a_1 A_2 &\; \underset{k^1_{-1}}{\overset{k^1_1}{\rightleftarrows}} \; (a_1+1)A_2 \\
A_2 + a_2 A_3 &\; \underset{k^2_{-2}}{\overset{k^2_2}{\rightleftarrows}} \; (a_2+1)A_3 \\
&\vdots \\
A_{m-1} + a_{m-1} A_m &\; \underset{k^{m-1}_{-(m-1)}}{\overset{k^{m-1}_{m-1}}{\rightleftarrows}} \; (a_{m-1}+1)A_m
\end{aligned}
\qquad (1.1)
$$

where $k_{\pm i}$, $i = 1$ to $m$ are positive rate constants and $a_i \in \mathbb{Z}^+$, $i = 1$ to $m$. Here we generalize schemes of Gray and Scott [11], [12] who considered (1.1) with

---

The authors gratefully acknowledge the following support:
* TARP Grant #1100; ** NSF Grant DMS 8803151 and ONR Grant N0014-89-J-1011;
*** NSF Grant DMS 8813071.

$m = 3$, $a_2 = 0$, $a_i = 1,2$ and $k_{-i} = 0$, $i = 1,2$ and D'Anna, Lignola and Scott [5] who allowed $a_i \in \mathbb{Z}$. Typically, $a_i = 0$, 1 or 2 but as in Gray-Scott [11] we do not require that the reactions be elementary and therefore values of $a_i \geq 2$ do not require physically unrealistic encounters between two or more species. Thus while (1.1) does not correspond to any real system it is consistent with chemical principles.

The system (1.1) can be conceptualized as a set of $m - 1$ sequential reactions

$$A_1 \underset{\leftarrow}{\rightarrow} A_2 \underset{\leftarrow}{\rightarrow} \ldots \underset{\leftarrow}{\rightarrow} A_m \qquad (1.2)$$

with the rate of reaction being a nonlinear function (if $a_i > 0$) of the concentrations of $A_i$ and $A_{i+1}$. By common abuse of notation will shall use $A_i$ interchangeably to refer to a species $A_i$ and its concentration.

Using the principles of mass action kinetics, cf. [14], the rate (forward rate minus backward rate) of the $i^{th}$ reaction is

$$r_i = k_i A_i A_{i+1}^{a_i} - k_{-i} A_i^{a_i+1}. \qquad (1.3)$$

An important feature of (1.3) is that formation of the product $A_{i+1}$ increases the rate of the forward reaction whenever $a_i > 0$. This acceleration is known as autocatalysis and it is a common feature of nearly all of the chemical systems which exhibit oscillatory or other exotic behavior [11] in experiments, for example the Belousov-Zhabotinski (B–Z) reaction [18]. These real chemical systems are complicated and detailed kinetic models do not as yet exist. Thus Gray and Scott [12] proposed their simple model as prototype of autocatalytic behavior. Surprisingly, the model yields results which are in qualitative agreement with many features of the experiments in stirred reactors.

More recently pattern formation in the B–Z reaction for unstirred reactors has aroused considerable interest [17], especially waves and rotating waves. The Gray-Scott model or a generalization such as (1.1) are likely candidates for modeling and trying to understand the experiments.

In the work at hand we analyze the asymptotic behavior of the kinetic equations associated with (1.1) and indicate results for the well posedness and asymptotic behavior of the system of partial differential equations obtained by adding a linear diffusion mechanism to each component of the kinetic equations. The details of the analysis of the reaction diffusion system will appear in a subsequent paper.

## 2. Kinetic equations

Utilizing standard principles we are led to associate with (1.1) the following system of ordinary differential equations

$$\begin{aligned}
\dot{A}_1 &= -k_1^1 A_1 A_2^{a_1} + k_{-1}^1 A_2^{a_1+1}, \\
\dot{A}_2 &= k_1^1 A_1 A_2^{a_1} - k_{-1}^1 A_2^{a_1+1} - k_1^2 A_2 A_3^{a_2} + k_{-1}^2 A_3^{a_2+1}, \\
&\vdots \qquad \vdots \\
\dot{A}_{m-1} &= k_1^{m-2} A_{m-2} A_{m-1}^{a_{m-2}} - k_{-1}^{m-2} A_{m-1}^{a_{m-2}+1} \\
&\quad - k_1^{m-1} A_{m-1} A_m^{a_{m-1}} - k_{-1}^{m-1} A_m^{a_{m-1}+1}, \\
\dot{A}_m &= k_1^{m-1} A_{m-1} A_{m-1}^{a_{m-1}} - k_{-1}^{m-1} A_m^{a_{m-1}+1}
\end{aligned} \quad (2.1a)$$

with initial conditions

$$A_i(0) = A_{0_i} \geq 0. \quad (2.1b)$$

We simplify the notation by denoting the ith component of (2.1a) by $f_i(A_1, \ldots, A_m) = f_i(A)$. In this way we obtain a vector field $f(A) = (f_i(A))_{i=1}^m$. Under this notational convention, system (2.1a – b) assumes the form

$$\dot{A}_i = f_i(A), \qquad i = 1 \text{ to } m, \quad (2.2a)$$
$$A_i(0) = A_{0_i} \geq 0, \quad i = 1 \text{ to } m. \quad (2.2b)$$

We notice that for $A \in \mathbb{R}_+^m$ (the positive orthant) and $A_i = 0$ we have $f_i(A) \geq 0$. Moreover, this condition implies that if the initial values $A_{0_i} \geq 0$ then the solution $A(\cdot) = (A_i(\cdot))_{i=1}^m$ remains confined to the positive orthant $\mathbb{R}_+^m$. Further observe that our vector field balances, i.e.,

$$\sum_{i=1}^m f_i(A) = 0 \quad \text{for } A \in \mathbb{R}^m. \quad (2.3)$$

We define constants $c_1, \ldots, c_m$ as follows

$$\begin{aligned}
c_1 &= 1, \\
c_i &= \prod_{j=1}^i (k_{-1}^j / k_1^j) \quad \text{for } i = 2, \ldots, m.
\end{aligned} \quad (2.4)$$

Direct algebraic computation produces two continua of steady states for (2.1 a – b). We have:

**Proposition 2.5.** *The zeros of $f(A) = (f_i(A))_{i=1}^m$ in $\mathbb{R}_+^m$ are given by*

$$\mathcal{L}_1(\Theta) = \{(\Theta, 0, \ldots, 0) | \Theta \geq 0\} \quad (2.6a)$$

and

$$\mathcal{L}_2(\Theta) = \{(\Theta/c_1, \Theta/c_2, \ldots, \Theta/c_m) | \Theta \geq 0\}. \quad (2.6b)$$

Thus we have two families of steady states for (2.2a – b), one of which lies along the $A_1$ axis and the other which is a line penetrating the interior of the positive orthant $\mathbb{R}_+^m$. Consequently we will have no global attractor. However as the next result indicates we will have asymptotic convergence of solutions to steady states.

**Theorem 2.7.** If $A_{0_i} \geq 0$ and $1 \leq i \leq m$ and $\sum_{i=1}^m A_{0_i} = k$ then for all $t \geq 0$, $\sum_{i=1}^m A_i(t) = k$. Moreover if $\Theta^*$ is such that

$$\sum_{i=1}^m \Theta^*/2c_i = k \tag{2.7a}$$

and

$$\sum_{i=1}^m c_i(A_{0_i} - \Theta^*/2c_i)^2 < \left(k - \frac{\Theta^*}{2}\right)^2 + \sum_{i=2}^m (\Theta^*)^2/4c_i \tag{2.7b}$$

then for $i = 1$ to $m$

$$\lim_{t \to \infty} A_i(t) = \Theta^*/2c_i. \tag{2.8}$$

**Proof.** The first assertion is a direct consequence of the fact that the vector field is balanced, i.e., $\sum_{i=1}^m f_i(A) = 0$. For the second assertion we construct the function

$$V(A) = \sum_{i=1}^m c_i(A_i - \Theta^*/2c_i)^2$$

and argue that along solutions to (2.1a – b)

$$\frac{dV(A)}{dt} \leq 0$$

with strict inequality holding if $(A_{0_1}, \ldots, A_{0_m}) \neq (k, 0, \ldots, 0)$ and $(A_{0_1}, \ldots, A_{0_m}) \neq (\Theta^*/2c_1, \ldots, \Theta^*/2c_m)$.

We point out that condition (2.7b) includes all points of the hyperplane $\sum_{i=1}^m A_i = k$ which lie in the positive orthant except the point lying along the $A_1$ axis. Thus the point $(\Theta^*/2c_1, \ldots, \Theta^*/2c_m)$ is an attractor for all solutions whose initial data does not lie on the $A_1$ axis and satisfies $\sum_{i=1}^m A_{0_i} = k$.

## 3. Reaction diffusion systems

We first observe that there exist higher order Lyapunov functions for our vector field. For each positive integer $p$ we introduce a separable convex functional $H_p(A)$ defined by the formula

$$H_p(A) = \sum_{i=1}^m h_i(p, A_i) = \sum_{i=1}^m (c_i)^{p-1} A_i^p \tag{3.1}$$

where the $c_i$'s are specified via (2.4). Direct computation yields the following result.

**Lemma 3.2.** If $A = (A_1, \ldots, A_m) \in \mathbb{R}_+^m$ then $\nabla H(p, A) f(A) = \sum_{i=1}^m h_i'(p, A_i) f_i(A) = \sum_{i=1}^m p c_i^{p-1} A_i^{p-1} f_i(A) \leq 0$.

We add diffusion to each component of our system of kinetic equations and consider the spatially dependent system:

$$\partial_t A_i = d_i \Delta A_i + f_i(A), \quad i = 1 \text{ to } m, x \in \Omega, \, t \geq 0, \tag{3.3a}$$
$$\partial A_i / \partial n = 0, \quad i = 1 \text{ to } m, x \in \partial\Omega, \, t \geq 0, \tag{3.3b}$$
$$A_i(x, 0) = A_{0_i}(x), \quad i = 1 \text{ to } m, x \in \Omega. \tag{3.3c}$$

We remark that $H_p(A)$ will provide a generalized Lyapunov structure for the reaction system. The utility of such a structure was recognized in [16] and it has subsequently been exploited for elliptic systems in [6], [7], [8] and for parabolic systems in [15], [9], [10]. Geometrically Lemma 3.2 implies that reaction vector field $f(A) = (f_i(A))_{i=1}^m$ does not point out of the bounded convex region determined by level curves $H_p(A) =$ constant and the coordinate hyperplanes of the positive orthant. Here we assume that $\Omega$ is a bounded Lipschitz domain with $C^{2+\epsilon}$ boundary. The diffusivities $d_i$ are distinct and positive. The initial data $A_0(\cdot) = (A_{0_i}(\cdot))_{i=1}^m$ is assumed to belong to $C(\bar{\Omega}, \mathbb{R}_i^m)$. We have the following theorem.

**Theorem 3.4.** If $A_0(\cdot) \in C(\bar{\Omega}, \mathbb{R}_+^m)$, then there exists a unique nonnegative classical solution to (3.3a - c) for all $t > 0$. For each $i$, $x \in \Omega$ and $t \geq 0$, $A_i(x, t) \geq 0$. Moreover, solutions to (3.3a - c) can be uniformly bounded in terms of the initial data.

**Proof.** Local well posedness follows from classical theory and global results therefore will be predicated upon the establishment of a priori bounds. The fact $f_i(A) \geq 0$ when $A_i = 0$ and $A \in \mathbb{R}_+^m$ implies that the positive orthant is an invariant region; hence solutions with nonnegative initial data remain nonnegative on their interval of existence.

In order to obtain a priori bounds we multiply the $i^{th}$ component of (3.3a) by $h_i'(p_1, A_i)$ and integrate on $\Omega$ to produce

$$d/dt \int_\Omega h_i'(p, A) dx + \int_\Omega d_i p(p-1)(c_i)^{p-1}(A_i(x,t))^{p-2} |\nabla A_i(x,t)|^2 dx$$
$$\leq h_i'(p, A) f_i(A). \tag{3.5}$$

Summing from $1 = 1$ to $m$ and invoking Lemma 3.2 we have

$$d/dt \int_\Omega H(p, A) dx \leq 0. \tag{3.6}$$

The above inequality can be integrated to produce a uniform bound for $\|H(p,A)\|_{1,\Omega}$ and because the terms on the right hand side are polynomials in the $A_i$'s we can thereby produce $L_q$ bounds for any integer power of the $f_i(A)$'s. Straightforward adoption of Lemma 3.6, [15] yields the existence of $N > 0$ so that for each $i$,

$$\|A_i(\cdot, t)\|_{\infty,\Omega} \leq N. \tag{3.7}$$

We further add that the constant $N$ may be explicitly computed in terms of the initial data and the parameters $c_i$.

We are able to produce analogous convergence results for the spatially dependent system (3.3a – c). If $A = (A_i)$ is a solution we denote the spatial average of the $i^{th}$ component by $\bar{A}_i(t)$, namely

$$\bar{A}_i(t) = \frac{1}{|\Omega|} \int_\Omega A_i(x,t)dx \tag{3.8}$$

where $|\Omega|$ denotes the $m$-dimensional volume of $\Omega$. We state the following theorem.

**Theorem 3.9.** For $1 \leq i \leq m$, $\lim_{t \to \infty} \|A_i(\cdot, t) - \bar{A}_i(t)\|_{\infty,\Omega} = 0$. If $\sum \bar{A}_i(0) = k > 0$ and $\Theta^* > 0$ is such that $\sum_{i=1}^m \Theta^*/2c_i = k$ and $\sum_{i=1}^m c_i(\bar{A}_i(0) - \Theta^*(2c_i))^2 < (k - \Theta^*/2)^2 + \sum_{i=2}^m (\Theta^*)^2/4c_i$, then for $1 \leq i \leq m$ we have

$$\lim_{t \to \infty} \|A_i(\cdot, t) - \Theta^*/2c_i\|_{\infty,\Omega} = 0.$$

We point out that this result is consistent with thermodynamic reasoning which indicated that a balanced reaction diffusion system with no flux boundary conditions should converge to a constant solution. Moreover periodic solutions or more exotic behavior of reaction diffusion systems obtained from the kinetic mechanism (1.1) are artifacts of inhomogeneities arising from the boundary or acting as source terms.

The proof of Theorem 3.9 proceeds by a lengthy argument consisting of lemmas and estimates and will not be given here. It will appear in a subsequent paper by the authors which will deal with (3.3a – c) in greater generality.

## References

[1] H. Amann, *Quasilinear Evolution Equations and Parabolic Systems*, Trans. Amer. Math. Soc. **293** (1986), 191–227.

[2] P. Bates, *Containment for Weakly Coupled Parabolic Systems*, Houston J. Math. **11** (1985), 151–158.

[3] N. Cheuh, C. Conley and J. Smoller, *Positively Invariant Regions for Systems of Nonlinear Diffusion Equations*, Indiana U. Math. J. **26** (1977), 373–392.

[4] E. Conway, D. Hoff and S. Smoller, *Large Time Behavior of Solutions of Systems of Nonlinear Reaction Diffusion Equations*, SIAM J. Appl. Math. **35** (1978), 1–16.

[5] A. D'Anna, P.G. Lignola and S.K. Scott, *The Application of Singularity Theory to Isothermal Autocatylic Systems: the Elementary Scheme $A + mB = (m + 1)B$*, Proc. Royal London Soc. A **403** (1986), 341–363.

[6] W. Fitzgibbon and J. Morgan, *Existence of Solutions for Weakly Coupled Semilinear Elliptic Systems*, J. Diff. Equations **77** (1989), 351–368.

[7] W. Fitzgibbon and J. Morgan, *Steady State Solutions for a Class of Reaction Diffusion Systems*, Nonlinear Anal. (to appear).

[8] W. Fitzgibbon, J. Morgan and S. Waggoner, *A Priori Bounds for a Class of Stationary Diffusion Systems*, Comm. Partial Diff. Equations (to appear).

[9] W. Fitzgibbon, J. Morgan and S. Waggoner, *Forward Containment for Semilinear Parabolic Systems*, preprint.

[10] W. Fitzgibbon, J. Morgan and S. Waggoner, *Generalized Lyapunov Structure for a Class of Semilinear Parabolic Systems*, J. Math. Anal. Appl. (to appear).

[11] P. Gray and S. Scott, *Autocatylic Reactions in the Isothermal Continuous Stirred Tank Reactor*, Chem. Eng. Sci. **38** (1983), 29–43.

[12] P. Gray and S. Scott, *Sustained Oscillations and Other Exotic Patterns of Behavior in Isothermal Reactions*, J. Phys. Chem. **89** (1985), 22–32.

[13] S. Hollis, R. Martin and M. Pierre, *Global Existence and Boundedness in Reaction Diffusion Systems*, SIAM J. Math. Anal. **18** (1987), 744–760.

[14] F. Horn and R. Jackson, *General Mass Action Kinetics*, Arch. Rat. Mech. Anal. **47** (1972), 81–116.

[15] J. Morgan, *Boundedness and Decay Results for Parabolic Systems*, SIAM Math. Anal. (to appear).

[16] J. Morgan, *Global Existence for Solutions to Semilinear Parabolic Systems*, SIAM J. Math. Anal. **20** (1989), 1128–1144.

[17] S. Müller, T. Plesser and B. Hess, *The Structure of the Core of the Spiral Wave in the Belousov–Zhabotinski Reaction*, Science **230** (1985), 661–663.

[18] R. Noyes, R. Field and E. Koros, *Oscillations in Chemical Systems*, J. Aus. Chem. Soc. **94** (1972), 1394–1401.

[19] Z. Nosticzius, W. Horsthemke, W.D. McCormick, H.L. Swinney and W.Y. Tam, *Sustained Chemical Waves in an Annular Gel Reaction: A Chemical Pinwheel*, Nature (1987).

[20] J. Smoller, *Shock Waves and Reaction Diffusion Equations*, Springer–Verlag, New York, 1983.


W.W. Farr
Department of Mathematics
Worcester Polytechnic Institute
Worcester, Massachusetts 01806, USA


W.E. Fitzgibbon
Department of Mathematics
University of Houston
Houston, Texas 77204, USA


J.J. Morgan
Department of Mathematics
Texas A & M University
College Station, Texas 77483, USA

# Asymptotic convergence for a class of autocatalytic chemical systems

S.J. Waggoner
Department of Mathematics
Furman University
Greenville, South Carolina 29613, USA

# Second Order Parabolic Equations in Banach Space

ANGELO FAVINI

Dipartimento di Matematica, Università di Bologna

## 1. Introduction and Preliminaries

Of concern is the local solvability in time of the nonlinear Cauchy problem

$$u'' = f(t, u(t), u'(t)), \quad 0 \le t \le T^*,$$
$$u(0) = u_0 = u_0, \quad u'(0) = u_1, \tag{1}$$

where $f$ is a given function from $[0, T] \times Y_0 \times Y_1$ into $X$. $X$, $Y_0$, $Y_1$ are (real or complex) Banach spaces, with continuous embeddings $Y_0$, $Y_1 \hookrightarrow X$, and $u_0$, $u_1$ are elements of $Y_0$, $Y_1$, respectively.

More precisely, we are interested in giving conditions on $f$ in such a way that (1) has "regular" solutions with second derivative $u''$ Hölder–continuous from $[0, T^*]$ in $X$.

Our aim is, substantially, to extend the treatment of [FB 1,2,3], relative to first order equations, by linearization of the equation in (1).

To this end, the nonlinear term $f$ is supposed to be sufficiently smooth to apply the Banach fixed point Theorem in a suitable space of continuous functions.

A particularly interesting example of (1), also widely studied in the literature, is given by

$$u''(t) + A_1 u'(t) + A_0 u(t) = F(t, u(t), u'(t)), \quad 0 \le t \le T^*,$$
$$u(0) = u_0, \ u'(0) = u_1, \tag{2}$$

---

Research partially supported by Ministero della Pubblica Istruzione, Italy (Fondi 40%), Università di Bologna (Fondi 60%), and the GNAFA of CNR.

where $A_0$, $A_1$ are closed linear operators in $X$, such that the corresponding *linear* equation

$$u''(t) + A_1 u'(t) + A_0 u(t) = f(t), \quad 0 \leq t \leq T^*, \tag{3}$$

is parabolic; of course, it has to be assumed that the term $F$ in (2) acts from $[0,T] \times Y_0 \times Y$, into $X$, with continuous embeddings $D(A_0) \hookrightarrow Y_0$, $D(A_1) \hookrightarrow Y_1$.

We recall what we mean for parabolicity of (3). See [FO]:

**Definition 1.** *Equation (3) or, equivalently, the operator pencil*

$$P(z) = z^2 + zA_1 + A_0, \quad z \in \mathbb{C},$$

*given by $D(P) = D(A_0) \cap D(A_1)$, $P(z)u = (z^2 + zA_1 + A_0)u$, $u \in D(P)$, is parabolic, if*

a) *there are $K \geq 0$, $\theta_0 \in (\pi/2, \pi)$ such that $\Sigma \subset \rho(B)$, where $\Sigma = \{z \in \mathbb{C} \ / \ |z| \geq K, \ |\arg z| \leq \theta_0\}$, $\rho(P) = \{z \in X / \exists P(z)^{-1} \in \mathcal{L}(X)\}$ is the resolvent set of $P$.*

*Here $\mathcal{L}(X,Y)$ denotes the space of bounded linear operators from $X$ into $Y$, $X$ and $Y$ two Banach spaces, and $\mathcal{L}(X) = \mathcal{L}(X,X)$.*

b) *there exists $M \geq 0$ such that*

$$\|P(z)^{-1}; \mathcal{L}(X)\| \leq M|z|^{-1},$$
$$\|A_1 P(z)^{-1}; \mathcal{L}(X)\| \leq M|z|^{-1},$$
$$\|A_0 P(z)^{-1}; \mathcal{L}(X)\| \leq M, \forall z \in \Sigma.$$

We refer to [FO] for various conditions on the operators $A_0$, $A_1$ entailing that $P(z)$ is parabolic. We also remark that if $A_0$ $A_1$ are perturbed by closed linear operators $B_0$, $B_1$ with $A_0$-bound ($A_1$-bound) zero, according to [K], then the new pencil $z^2 + z(A_1 + B_1) + A_0 + B_0$ is parabolic, as it is easily verified.

By means of a slight change of the proof given in [FO], we obtain

**Proposition 1.** *Let $-A_1$ be the infinitesimal generator of an analytic semigroup in $X$ (possibly with a non dense domain) and $A_0 = CA_1$, where $-C$ is another generator of an analytic semigroup in $X$. If $C$ is $A_1$-bounded, with $A_1$-bound equal to 0, then the pencil $P(z)$ is parabolic.*

**Remark.** If there are $K > 0$ and $\theta \in (0,1)$ such that

$$\|Cu; X\| \leq K \|A_1 u; X\|^{1-\theta} \|u; X\|^{\theta}, \ u \in D(A_1) \hookrightarrow D(C), \tag{4}$$

then $C$ is $A_1$-bounded and has $A_1$-bound equal to 0.

Moreover, (4) holds iff the real interpolation space $(D(A_1), X)_{\theta,1} \hookrightarrow D(C)$. Now, the complex interpolation space $[D(A_1), X]_\theta$ contains $(D(A_1), X)_{\theta,1}$ and, if $A_1$ is positive and has bounded imaginary powers, $[D(A_1), X]_\theta = D(A_1^{1-\theta})$. See, for example [T, p.103].

We conclude that under the preceding hypotheses on $A_1$, (4) holds if $D(A_1^\omega) \hookrightarrow D(C)$ for a certain $\omega \in (0,1)$.

Let us describe a concrete case.

If $\Omega$ is a bounded domain in $\mathbb{R}^n$, $n \geq 1$, with a smooth boundary $\partial\Omega$, it is well known that the operator $A_1 u = -(-\Delta)^m u$, $u \in C_0^{2m}(\Omega)$, $m \in \mathbb{N}$, has a self-adjoint extension to $X = L^2(\Omega)$, which is positive and has $H^{2m}(\Omega) \cap H_0^m(\Omega)$ as its domain.

Let $0 < \alpha < 1$, $m\alpha$ a positive integer. Then [H, p.29]

$$D((-\Delta)^{m\alpha}) \subset H^{2m\alpha}(\Omega) \cap H_0^{m\alpha}(\Omega).$$

More generally, one finds in [LM, p.117] that if $B_j$ is a suitable differential operator on the boundary $\partial\Omega$, of order $m_j < 2m$, $0 \leq j \leq \nu$, and

$$H_B^{2m}(\Omega) = \{u \in H^{2m}(\Omega)/B_j u = 0 \text{ on } \partial\Omega, 0 \leq j \leq \nu\},$$

with $0 < \theta < 1$, $2(1-\theta)m \neq$ integer $+1/2$, then

$$[H_B^{2m}(\Omega), L^2(\Omega)]_\theta =$$
$$\{u \in H^{2(1-\theta)/m}(\Omega)/B_j u = 0 \text{ on } \partial\Omega, m_j < 2(1-\theta)m - 1/2\}.$$

It follows from Proposition 1 that we could take as $C$ any generator of an analytic semigroup on $L^2(\Omega)$ for which $\{u \in H^{2s}(\Omega)/B_j u = 0$ on $\partial\Omega$, $m_j < 2s - 2/2\} \subseteq D(C)$, $1 \leq s < m$.

Analogously, one can generalize these considerations to the case $L^p(\Omega) p \neq 2$, by means of the spaces $H^{2m}_{p,\{B_j\}}$, $B^{2m}_{p,q,\{B_j\}}$, [T. pp.320–321].

Let us go back to analyse the nonlinear equation in (1), seeking to make precise regularity assumptions of $f$.

If the concerned Banach spaces are real, the hypothesis that $f = f(t,u,v)$ has Fréchet-derivatives with respect to $u$ and $v$ doesn't appear to be too onerous, as it may become if the spaces are complex.

We recall that, with regard to the first order problem analogous to (1), K. Masuda has studied its solvability, with an analytic $f$; see [M].

In order to avoid some complications, we shall suppose that our Banach spaces are real, we shall linearize the equation and study the corresponding linear pencil obtained by complexification.

Applying the definition itself of complexification of an operator, we shall return to an $X$-valued solution corresponding to an $X$-valued function $f = f(\cdot)$.

If our function $f$ in (1) has the partial derivatives $\frac{\partial f}{\partial u}(t,u,v)$ and $\frac{\partial f}{\partial v}(t,u,v)$ for all $0 \leq t \leq T$, $u \in Y_0$, $v \in Y_1$, that are bounded linear operators from $Y_0$ into $X$ and from $Y_1$ into $X$, respectively, let us

$$\frac{\partial f}{\partial u}(0, u_0, u_1) = -A_0,$$
$$\frac{\partial f}{\partial v}(0, u_0, u_1) = -A_1,$$
$$f(t, x, y) + A_0 x + A_1 y = F(t, x, y).$$

Then problem (1) takes the form

$$u''(t) + A_1 u'(t) + A_0 u(t) = F(t, u(t), u'(t)), \quad 0 \leq t \leq T^*,$$
$$u(0) = u_0, \; u'(0) = u_1. \tag{5}$$

Assume for a moment that there exists a $\varphi \in C^2[0, T; X]$ such that $\varphi \in C[0, T; D(A_0)] = C[0, T; Y_0]$, $\varphi' \in C[0, T; D(A_1)] = C[0, T; Y_1]$, $\varphi(0) = u_0$, $\varphi'(0) = u_1$, $\varphi''(0) = F(0, u_0, u_1) - A_0 u_0 - A_1 u_1 = f(0, u_0, u_1)$.
Putting $u(t) - \varphi(t) = v(t)$, (5) becomes

$$v''(t) + A_1 v'(t) + A_0 v(t) = G(t, v(t), v'(t)), \quad 0 \leq t \leq T^*,$$
$$v(0) = v'(0) = 0, \tag{6}$$

where $G(t, v(t), v'(t)) = f(t, v(t) + \varphi(t), v'(t) + \varphi'(t)) + A_0 v(t) + A_1 v' - \phi''(t)$.

Clearly it is then important to have an existence – uniqueness – regularity theorem for the solution of the *linear* problem.

$$B^2 u + A_1 B u + A_0 u = h, \tag{7}$$

where $B$ is a suitable linear operator (in our case, $B = \frac{d}{dt}$ in a certain function space). This is achieved by the following

**Theorem 1.** *Let $A_0$, $A_1$, $B$ be closed linear operators in the complex Banach space $X$ such that $\overline{D(B)} = X$ and*
  (A) $B^{-1} P(z)^{-1} = P(z)^{-1} B^{-1}$, $z \in \Sigma : \operatorname{Re} z \geq a_0 - b_0 |\operatorname{Im} z|$, $b_0 > 0$.
  (B) $\|(B - z)^{-1}, \mathcal{L}(X)\| \leq C(1 + |z|)^{-1}$, $\operatorname{Re} z \geq a_1 - b_1 |\operatorname{Im} z|$, $0 < a_0 < a_1$, $b_0 > \frac{a_0}{a_1} b_1$.
  (C) $\|P(z)^{-1}; \mathcal{L}(E)\| \leq C(1 + |z|)^{-2}$, $z \in \Sigma$, $P(z) = z^2 + z A_1 + A_0$.
  (D) $\|A_0 P(z)^{-1}, \mathcal{L}(E)\| \leq C$, $z \in \Sigma$.

Then for any $\theta \in (0, 1)$ and all $h \in (E_1 D(B))_{\theta, \infty} = V_\theta$ there is a unique $u \in E$ such that (7) holds and $A_0 u$, $A_1 B u$, $B^2 u \in V_\theta$.
Moreover, $u = Sh$, $Bu = S_1 h$, $h \in V_\theta$, where

$$S = (2\pi i)^{-1} \int_\gamma P(z)^{-1} (B - z)^{-1} dz,$$

$$S_2 = (2\pi i)^{-1} \int_\gamma z P(z)^{-1} (B - z)^{-1} dz,$$

and $\gamma$ is the contour in the complex plane given by $\operatorname{Re} z = a_2 - b_0 |\operatorname{Im} z|$, $a_0 < a_2 < a_1$, oriented upwards.

For a proof of Theorem 1, we refer to [D,G]. We also note that $A_0 S$ and $A_1 S_1 \in \mathcal{L}(V_\theta)$.

To treat problem (6) we take, with usual notation,

$$E = C_0[0, T_1; X], \text{ for any } 0 < T_1 \leq T,$$
$$D(B) = \{u \in C_0^1[0, T_1; X] : u'(0) = 0\}, (Bu)(t) = u'(t),$$

so that [G]:
$V_\theta = C_0^\theta[0, T_1; X]$, $0 < \theta < 1$, is the space of all $X$-valued Hölder-continuous functions $[0, T_1]$, vanishing at $t = 0$.

In fact, this operator $B$ satisfies (A), (B), in Theorem 1. Concerning (C), (D), these are the parabolicity hypotheses on the pencil $z^2 - z\frac{\partial f}{\partial v}(0, u_0, u_1) - \frac{\partial f}{\partial u}(0, u_0, u_1)$.

We observe that it is not restrictive to suppose the constant $C$ in (B), (C), (D) to be *independent* of $T_1 \in ]0, T]$, so that

$$\|A_0 S; \mathcal{L}(V_\theta)\| \leq m_0 \text{ (independent of } T_1 \in ]0, T]),$$
$$\|A_1 S_1; \mathcal{L}(V_\theta)\| \leq m_1 \text{ (independent of } T_1 \in ]0, T]).$$

Our aim is to seek a solution for (6) of the form $v(t) = (Sh)(t)$, with $h \in C_0^\theta[0, T^*, X]$, $0 < T^*$ a suitably small number.

In view of Theorem 1, the problem is reduced to find $h \in C_0^\theta[0, T^*, X]$ such that

$$h(t) = G(t, (Sh)(t), (S_1 h)(t)), t \in [0, T^*].$$

It is then natural to ask Hölder-regularity for our $\varphi$.

To this purpose, it is well known by [G] that if $A$ is a positive operator in a Banach space $X$, then for any $u_0 \in (X, D(A^3))_{\frac{2+\theta}{3}, \theta}(\Leftrightarrow A^2 u_0 \in (X, D(A))_{\theta, \infty})$, $u_1 \in (X, D(A^3))_{\frac{1+\theta}{3}, \infty}(\Leftrightarrow A u_1 \in (X, D(A))_{\theta, \infty})$, $u_2 \in (X, D(A^3))_{\frac{\theta}{3}, \infty}(\Leftrightarrow u_2 \in (X, D(A))_{\theta, \infty})$, there is a function $\varphi$ such that $t^{1-\theta}\varphi^{(j)} \in L^\infty(0, \infty; D(A^{3-j}))$, $j = 0, 1, 2, 3$, and $\varphi^{(j)}(0) = u_j$, $j = 0, 1, 2$. This easily implies that

$$\|A^{2-j}[\varphi^{(j)}(t) - \varphi^{(j)}[; X\| \leq C|t - s|^\theta, \ 0 \leq t \leq T, \ j = 0, 1, 2.$$

Let us suppose

$$A = A_1, D(A_1^2) \subset^1 D(A_0),$$
$$A_1^2 u_0 \in (X, D(A_1))_{\theta, \infty} A_1 u_1 \in (X, D(A_1))_{\theta; \infty}, \quad \text{(H)}$$
$$u_2 = f(0, u_0, u_1) \in (X, D(A_1))_{\theta, \infty}.$$

What we have recalled above allows to affirm that under (H) there is a $\varphi$ from $[0, T]$ into $X$ such that $\varphi(0) = u_0$, $\varphi'(0) = u_1$, $\varphi''(0) = f(0, u_0, u_1)$ and $A_0\varphi$, $A_1\varphi'$, $\varphi'' \in C^\theta[0, T; X]$, $0 < \theta < 1$.

In the sequel we shall show that under certain Hölder regularity assumptions on $f$ and $F$, problem (1) or (2) has solutions with corresponding time regularity.

We shall describe some applications of these abstract results to some concrete initial–boundary–value problems in another subsequent paper.

As for comparision with different approaches, we quote [Y] for problem (2), where it is studied by reducing it to a first order equation.

On the other hand, our methods are close to the ones by A. Lunardi, relative to fully nonlinear parabolic equation of first order in Banach space [L]. See also [DL].

## 2. Solvability of (1) and of (2)

To begin with, we introduce the concept of a strict solution of (1) (respectively (2)).

**Definition 1.** *An $X$-valued function $u$ on $[0, T^*]$, $0 < T^* \leq T$, is a strict solution of problem (1) (resp. (2)) if $u \in C[0, T^*, Y_0]\,(C[0, T^*; D(A_0)])$, there is $u' \in C[0, T^*; Y_1]\,(C[0, T^*; D(A_1)])$, the second derivative $u'' \in D[0, T^*, X]$ and (1) (resp. (2)) is satisfied.*

We shall show that under certain assumptions of $f$, (resp. $F$) $u_0$, $u_1$, there is a strict solution of (1), (resp. (2)), with higher regularity in time.

To this end, we need the following assumptions:

(K) For any $t \in [0, T]$, $u \in Y_0$, $v \in Y_1$, there exist the partials $\frac{\partial f}{\partial u}(t, u, v)$, $\frac{\partial f}{\partial v}(t, u, v)$, satisfying

$$\max\{\|\frac{\partial f}{\partial u}(t_1, u_1, v_1) - \frac{\partial f}{\partial u}(t_2, u_2, v_2); \mathcal{L}(Y_0; X)\|,$$

$$\|\frac{\partial f}{\partial v}(t_1, u_1, v_1)\frac{\partial f}{\partial v}(t_2, u_2, v_2); \mathcal{L}(Y_1; X)\|\}$$

$$\leq C(\|u_i; Y_0\|, \|v_i; Y_1\|)\{|t_1 - t_2|^\omega + \|u_1 - u_2; Y_0\| + \|v_1 - v_2; Y_1\|\},$$

$0 < \omega \leq 1$, $0 \leq t_2$, $t_2 \leq T$, $C$ being a monotone function of its arguments. There are $m > 0$, $k > 0$ such that

$$\|f(t_1, u, v) - f(t_2, u; v); X\| \leq k|t_1 - t_2|^\omega,$$

for all $t_1, t_2 \in [0, T]$, $u \in Y_0$, $v \in Y_1$, $\|u - u_0; Y_0\| \leq m$, $\|v - u_1; Y_1\| \leq m$.

Then one has

**Theorem 2.** *Assume (H) and (K), with $0 < \theta \leq \omega$, ($0 < \theta < 1$, if $\omega = 1$).*

*Then problem (1) has a strict solution $u$ on a suitably small interval $[0, T^*]$, such that $u'' \in C^\theta[0, T^*, X]$.*

**Proof.** We extend the technique used in [FP2].

Put $W = \{h \in V_\theta; \|h; V_\theta| \leq r\}$, where $r$ is a positive constant to be specified afterwards.

Making use of the notation we have introduced in §1, we need an estimate of the Hölder–norm of $G(h)$, $h \in W$.

Let $h_1, h_2 \in W$. Then
$$G(h_1)(t) - G(h_2)(t)$$
$$= \int_0^1 \{[\frac{\partial f}{\partial u}(t, \varphi(t) + (Sh_2)(t) + \sigma[(Sh_1)(t) - (Sh_2)(t)], \varphi'(t) + (S_1h_2)(t)$$
$$+ \sigma[(S_1h_1)(t) - (S_1h_2)(t)] - \frac{\partial f}{\partial u}(0, u_0, u_1)][(Sh_1)(t) - (Sh_2)(t)]$$
$$+ [\frac{\partial f}{\partial v}(t, \varphi(t) + (Sh_2)(t) + \sigma[(Sh_1)(t) - (Sh_2)(t)], \varphi'(t) + (S_1h_2)(t)$$
$$+ \sigma[(S_1h_1)(t) - (S_1h_2(t)]) - \frac{\partial f}{\partial v}(0, u_0, u_1)][(S_1h_1)(t) - (S_1h_2)(t)]\} ds.$$

Since $\|A_0[\varphi(t) - \varphi(0)]; X\| \leq Ct^\theta$, $\|A_1[\varphi'(t) - \varphi'(0)]; X\| \leq Ct^\theta$, and $\|A_0 S; \mathcal{L}(V_\theta)\| \leq m_0$, $\|A_1 S_1; \mathcal{L}(V_\theta)\| \leq m_1$,
$$\sup_{0 \leq t \leq T_1} \|A_0(Sh)(t); X\| \leq m_0 r T_1^\theta, \sup_{0 \leq t \leq T_1} \|A_1(S_1h)(t); X\| \leq m_1 r T_1^\theta, 0 < T_1 \leq T,$$
one deduces that
$$\sup_{0 \leq t \leq T_1} \|G(h_1)(t) - G(h_2)(t); X\| \leq C(T_1, r) T_1^\theta \{\|A_0 S(h_1 - h_2); C_0^\theta[0, T_1; X]\|\}$$
$$+ \|A_1 S_1(h_1 - h_2); C_0^\theta[0, T_1; X]\|\|, 0 < T_1 \leq T.$$

In a similar way we estimate
$$\|G(h_1)(t) - G(h_2)(t) - G(h_1)(s) + G(h_2)(s); X\|$$
$$\leq C_1(T_1, r) T_1^\theta \|h_1 - h_2; C_0^\theta[0, T_1; X]\|\}, 0 < T_1 \leq T.$$

Clearly one has $C(T_1, r), C_1(T_1, r) \leq C_2(r)$ for $0 \leq T_1 \leq T$.

Hence, if we take $T_1^\theta < [4C_2(r)]^{-1}$, $G$ is a contractive map from $W$ into $C_0^\theta[0, T_1; X] = V_\theta$.

Let us fix this $T_1$. If $h \in W$ and we show $G(0) \in V_\theta$, then $\|G(h); V_\theta\| \leq \|G(h) - G(0); V_\theta\| + \|G(0), V_\theta\| \leq r/2 + \|G(0); V_\theta\| \leq r$, provided that $r \geq 2\|G(0); V_\theta\|$.

Now, $G(0)(t) = f(t, \varphi(t), \varphi'(t)) - \varphi''(t)$ and thus
$$\|G(0)(t_1) - G(0)(t_2); X\|$$
$$\leq \|f(t_1, \varphi(t_1), \varphi'(t_1)) - f(t_2, \varphi(t_1), \varphi'(t_1)); X\|$$
$$+ \|f(t_2, \varphi(t_1), \varphi'(t_1)) - f(t_2, \varphi(t_2), \varphi'(t_2)); X\| + \|\varphi''(t_1) - \varphi''(t_2); X\|$$
$$\leq k|t_1 - t_2|^\omega + \|\varphi''(t_1) - \varphi''(t_2); X\|$$
$$+ \|\int_0^1 \{\frac{\partial f}{\partial u}(t_2, \varphi(t_2, \varphi(t_2)$$
$$+ \sigma[\varphi(t_1) - \varphi(t_2)], \varphi'(t_2) + \sigma[\varphi'(t_1) - \varphi'(t_2)])(\varphi(t_1) - \varphi(t_2))$$
$$+ \frac{\partial f}{\partial v}(t_2, \varphi(t_2) + \sigma[\varphi(t_1) - \varphi(t_2)], \varphi'(t_2)$$
$$+ \sigma[\varphi'(t_1) - \varphi'(t_2)])(\varphi'(t_1) - \varphi'(t_2))\}\|d\sigma.$$

We then use again (H) and (K) to deduce that there is $K > 0$ such that $\|G(0)(t_1) - G(0)(t_2), X\| \leq K|t_1 - t_2|^\theta$. This ends the proof.

Now we turn to problem (2).

Let $F$ be defined from $[0,T] \times Y_0 \times Y_1$ into $X$ and satisfy the same assumptions (K) as $f$. This time the operators in $X$, $A_0$, $A_1$ are given independently on $F$, and on these operators we assume

$$\begin{cases} D(A_0) \hookrightarrow Y_0, \|\frac{\partial F}{\partial u}(0,u_0,u_1)x; X\| \leq C\|A_0 x; X\|^{1-\theta_0}\|x; X\|^{\theta_0}, x \in D(A_0) \\ D(A_1) \hookrightarrow Y_1, \|\frac{\partial F}{\partial v}(0,u_0,u_1)y; X\| \leq C\|A_1 y_1; X\|^{1-\eta_0}\|y; X\|^{\eta_0}, y \in D(A_0) \\ 0 < \theta_0, \eta_0 < 1. \end{cases}$$

(L) could be replaced by the more general hypothesis that $\frac{\partial F}{\partial u}(0, u_0, u_1)$ (respectively, $\frac{\partial F}{\partial v}(0, u_0, u_1)$ be $A_0$ (resp. $A_1$) – bounded with $A_0$ (resp. $A_1$)-bound equal to 0.

Then the equation in (2) is written accordingly

$$u''(t) = F(t,u), u'(t) - A_1 u'(t) - A_0 u(t) = f(t, u(t), u'(t));$$

that is, $f(t,u,v) = F(t,u,v) - A_1 v - A_0 u$, $0 \leq t \leq T$ $u \in D(A_0)$, $v \in D(A_1)$.

In order to apply Theorem 2, we regard $f$ as a map $[0,T] \times D(A_0) \times D(A_1)$ into $X$. The embeddings $D(A_1) \hookrightarrow Y_1$ $i = 0, 1$ ensure that (K) holds for $f$ too, with the above mentioned choice of domains.

On the other hand, in view of (L), if the pencil $z^2 + zA_1 + A_0$ is parabolic, then also $z^2 + z(A_1 - \frac{\partial F}{\partial v}(0, u_0, u_1)) + A_0 - \frac{\partial F}{\partial u}(0, u_0, u_1))$ is parabolic, as it is shown in [FO]. Put $A_2 = \frac{\partial F}{\partial v}(0, u_0, u_1)$. Using Theorem 2, we obtain

**Theorem 3.** *Suppose that $F$ satisfies (K), with the obvious substitution of $F$ for $f$. Further, assume (L) and the parabolicity of the pencil $z^2 + zA_1 + A_0$, with $D((A_1 - A_2)^2) \subseteq D(A_0)$.*

*If $u_0 \in D(A_0)$, $u_1 \in D(A_1)$ satisfy $(A_1 - A_2)^2 u_0 \in (X, D(A_1))_{\theta,\infty} = Z$, $(A_1 - A_2)u_1 \in Z$, $F(0, u_0, u_1) - A_1 u_1 - A_0 u_0 \in Z$, then problem (2) has one local strict solution $u$ such that $u''(\cdot) \in C^\theta[0, T^*, X]$, $0 < \theta \leq \omega (0 < \theta < 1$, if $\omega = 1)$, where $0 < T^* \leq T$ is suitably small.*

We remark that when applied to some partial differential equations, the condition $D((A_1 - A_2)^2) \subseteq D(A_0)$ implies that the partial derivative $\frac{\partial F}{\partial v}(0, u_0, u_1)$ must satisfy assumptions analogous to the ones in [DP].

For examples of differentiable non linearities with the required regularity, we refer to [P, pp. 179 and following].

## References

[DG] G. Da Prato and P. Grisvard, *Equations d'évolution abtraites non linéaires de type parabolique*, Ann. Mat. Pura Appl. **120** (1979), 329–396.

[DL] G. Da Prato and A. Lunardi, *Stability, instability and center manifold theroem for fully nonlinear autonomous parabolic equations in Banach space*, Archive Rat. Mech. Anal. **101** (1988), 115–141.

[D]   J.U.A. Dubinskii, *On some differential–operator equations of arbitrary order*, Mat. Sbornik **90**(132) (1973), Engl. Trans.: Math. USSR Sbornik **19** (1973), 1–21.

[FO]  A. Favini and E. Obrecht, *Conditions for parabolicity of second order abstract differential equations*, Preprint.

[FP1] A. Favini and P. Plazzi, *On some abstract degenerate problems of parabolic type 1: The linear case*, Nonlinear Anal. **12** (1988), 1017–1027.

[FP2] A. Favini and P. Plazzi, *On some abstract degenerate problems of parabolic type 2: The nonlinear case*, Nonlinear Anal. **13** (1989), 23–31.

[FP3] A. Favini and P. Plazzi, *On some abstract degenerate problems of parabolic type 3: Applications to linear and nonlinear problems*, to appear in Osaka Math. J.

[G]   P. Grisvard, *Spazi di tracce e applicazioni*, Rend. Mat (6)**5** (1972), 657–729.

[H]   D. Henry, *Geometric theory of semilinear parabolic equations*, Lecture Notes in Mathematics **840**, ed. Springer, 1966.

[K]   T. Kapo, *Perturbation theory for linear operators*, ed. Springer, 1966.

[LM]  J.L. Lions and E. Magenes, *Problèmes aux limites non homogènes et applications Vol.1*, ed. Dunot, 1968.

[L]   A. Lunardi, *Interpolation spaces between domains of elliptic operators and spaces of continuous functions with applications to nonlinear parabolic equations*, Math. Nachr. **121** (1985), 295–318.

[M]   K. Masuda, *Analytic solutions to some nonlinear diffusion equations*, Math. Z. **187** (1984), 61–73.

[P]   B. Pini, *Lezioni di analisi matematica di secondo livello, parte seconda*, ed. CLUEB, 1986.

[T]   H. Triebel, *Interpolation theory, function spaces, differential operators*, ed. North-Holland, 1978.

[Y]   S.Yu. Yakubov, *A nonlocal boundary value problem for a class of Petrovskii well posed equations*, Mat. Sb (N.S.) **118**(69) (1982), Engl. Trans.: Math. USSR-Sb **46**(1983), 255–265.


Angelo Favini
Dipartimento di Matematica
Università di Bologna
Piazza di Porta S. Donato 5
I-40126 Bologna, Italy

# On the Modified Korteweg-deVries Equation

F. GESZTESY

Department of Mathematics, University of Missouri

## 1. Introduction

In this paper we review our earlier results on the modified Korteweg-deVries equation in [Ges1], [Ge-Si], [GSS] and continue these investigations into several directions. Specifically, in Section 2, we review the connections between the Korteweg-deVries (KdV) and the modified Korteweg-deVries (mKdV) equations based on Miura's transformation [Miu], and commutation methods. Appendix A summarizes the necessary commutation formulas needed in Section 2. In Section 3 we study soliton-like solutions of the mKdV-equation (i.e., solutions that tend to (time-independent) finite asymptotic values as $x \to \pm\infty$ sufficiently fast). In particular, due to our more general Hypothesis (H.3.1), Theorem 3.2 considerably extends our earlier findings in [GSS]. Section 4 reviews our derivation of pure soliton solutions in [GSS]. Both, Sections 3 and 4 are supported by Appendix B which summarizes spectral and scattering properties of one-dimensional Schrödinger and Dirac operators with nontrivial spatial asymptotics in the corresponding potential terms. Section 5 is devoted to spatially periodic solutions of the mKdV-equation. While Theorem 5.3 summarizes our results on periodic solutions in [GSS], the rest of this section presents new material.

In order to keep Section 5 within a reasonable length we decided to put all necessary background material on periodic Schrödinger and Dirac operators into Appendices C-G: Appendix C summarizes basic Floquet theory, Appendix D treats spectra associated with various boundary conditions and the isospectral manifold of periodic potentials for Schrödinger operators, Appendix E recalls trace relations, Appendix F utilizes the connection between the Schrödinger and an associated Riccati-type equation, and Appendix G collects various formulas in the special case where the Schrödinger operator has only finitely many simple periodic and antiperiodic eigenvalues.

Finally, we would like to point out that our methods are by no means confined to the mKdV-equation. In fact, they also have been successfully applied to the modified Kadomtsev-Petviashvili equation [GHSS], [Ge-Sc] as well as to the Kac-van Moerbeke system [GHSZ].

## 2. Connections between the KdV and mKdV-equation

In this section we review the main results of [Ges 1], [Ge-Si], [GSS] concerning links between the Korteweg-deVries (KdV) and the modified Korteweg-deVries (mKdV) equation based on Miura's transformation [Miu].

We start by introducing the hypothesis

**(H.2.1).** $f \in C^\infty(\mathbb{R}^2)$ real-valued, $\partial_x^n f \in L^\infty(\mathbb{R}^2), n = 0, 1$.

Then assuming that $V, \phi$ satisfy (H.2.1), the KdV and mKdV-equation are defined by

$$\text{KdV}(V) := V_t - 6VV_x + V_{xxx} = 0, \quad (t,x) \in \mathbb{R}^2, \tag{2.1}$$

$$\text{mKdV}(\phi) := \phi_t - 6\phi^2 \phi_x + \phi_{xxx} = 0, \quad (t,x) \in \mathbb{R}^2, \tag{2.2}$$

and Miura's transformation and identity read [Miu]

$$V_j(t,x) := \phi(t,x)^2 + (-1)^j \phi_x(t,x), \ j = 1, 2, \ (t,x) \in \mathbb{R}^2, \tag{2.3}$$

$$\text{KdV}(V_j) = \left[2\phi + (-1)^j \partial_x\right] \text{mKdV}(\phi), \ j = 1, 2. \tag{2.4}$$

Evidently (2.4) implies that any solution $\phi$ of the mKdV-equation (2.2) yields by (2.3) two solutions $V_1, V_2$ of the KdV-equation (2.1). In order to reverse that process, i.e., starting with a solution, say $V_2$, of (2.1) and then construct a solution $\phi$ (resp. $V_1$) of (2.2) (resp. (2.1)) such that (2.3) holds, we need some preparations.

First we recall

**Theorem 2.2.** *[Lax 1], [Tan]*

(i) *Suppose $V$ satisfies (H.2.1), $V_t(t,.) \in L^\infty(\mathbb{R}), t \in \mathbb{R}$ and KdV(V)=0. Define in $L^2(\mathbb{R})$ the one-dimensional Schrödinger operator $H(t)$*

$$H(t) := -\partial_x^2 + V(t,.) \text{ on } H^2(\mathbb{R}), \ t \in \mathbb{R}. \tag{2.5}$$

*Then there exists a family of unitary operators $U(t)$, $t \in \mathbb{R}$, $U(0) = 1$, in $L^2(\mathbb{R})$ such that*

$$U(t)^{-1} H(t) U(t) = H(0), \ t \in \mathbb{R}. \tag{2.6}$$

(ii) *Suppose $\phi$ satisfies (H.2.1), $\phi_{xx} \in L^\infty(\mathbb{R}^2), \phi_t(t,.), \phi_{tx}(t,.) \in L^\infty(\mathbb{R}), t \in \mathbb{R}$ and $mKdV(\phi) = 0$. Define in $L^2(\mathbb{R}) \otimes \mathbb{C}^2$ the one-dimensional Dirac operator*

$$Q(t) := \begin{pmatrix} 0 & -\partial_x + \phi(t,.) \\ \partial_x + \phi(t,.) & 0 \end{pmatrix} \text{ on } H^1(\mathbb{R}) \otimes \mathbb{C}^2, \ t \in \mathbb{R}. \qquad (2.7)$$

*Then there exists a family of unitary operators $W(t), t \in \mathbb{R}, W(0) = 1$ in $L^2(\mathbb{R}) \otimes \mathbb{C}^2$ such that*

$$W(t)^{-1}Q(t)W(t) = Q(0), \ t \in \mathbb{R}. \qquad (2.8)$$

The main ingredient for the proof of part (i) is the Lax pair $(H(t), B_V(t))$, where

$$B_V(t) := -4\partial_x^3 + 6V(t,.)\partial_x + 3V_x(t,.) \text{ on } H^3(\mathbb{R}), \ t \in \mathbb{R}, \qquad (2.9)$$

in $L^2(\mathbb{R})$, and the commutation relation (on $H^5(\mathbb{R})$)

$$\dot{H} - [B_V, H] = \text{KdV}(V). \qquad (2.10)$$

For part (ii) one studies the Lax pair $(Q(t), B(t))$, where

$$B(t) := B_{\phi^2 - \phi_x}(t) \oplus B_{\phi^2 + \phi_x}(t), \ t \in \mathbb{R}, \qquad (2.11)$$

in $L^2(\mathbb{R}) \otimes \mathbb{C}^2$, and the corresponding commutation relation ( on $H^4(\mathbb{R}) \otimes \mathbb{C}^2$)

$$\dot{Q} - [B, Q] = \text{mKdV}(\phi) \begin{pmatrix} 0 & 1 \\ 1 & 0 \end{pmatrix}. \qquad (2.12)$$

(Here "$\cdot$" denotes $\dfrac{d}{dt}$.)

While Theorem 2.2, in particular, contains the result that the spectra of $H(t)$ and $Q(t)$ are $t$-independent as long as $V$ resp. $\phi$ satisfy (2.1) resp. (2.2), it is operator theoretic in nature and therefore requires somewhat stronger assumptions on $V, \phi$ than necessary for that result. Because of this fact and our need of distributional solutions $\psi$ of $H(t)\psi(t) = \lambda\psi(t), \lambda \in \mathbb{R}$ later on, we also give an ODE-version of Theorem 2.2. (By distributional solutions we always mean locally absolutely continuous solutions w.r.t. $x$ with an appropriate number of $x$-derivatives that are locally absolutely continuous as well.)

**Lemma 2.3.** *[Ge-Si] Assume $V \in C^\infty(\mathbb{R}^2) \cap L^\infty(\mathbb{R}^2)$ is real-valued and $KdV(V) = 0$. Let $\psi_0(k,.) \in C^\infty(\mathbb{R})$ be a real-valued, distributional solution of $H(0)\psi_0(k) = k^2\psi_0(k), \ k^2 \in \mathbb{R}, \text{Im} k \geq 0$. Then*

$$H(t)\psi(t,k) = k^2\psi(t,k), \ (t, k^2) \in \mathbb{R}^2, \qquad (2.13)$$

has a unique, real-valued, distributional solution $\psi(.,k,.) \in C^\infty(\mathbb{R}^2)$ satisfying

$$\psi_t(t,k,x) = (B_V(t)\psi)(t,k,x), \quad (t,x) \in \mathbb{R}^2, \tag{2.14}$$

or equivalently, using (2.13),

$$\psi_t(t,k,x) = 2\left[V(t,x) + 2k^2\right]\psi_x(t,k,x) - V_x(t,x)\psi(t,k,x), \quad (t,x) \in \mathbb{R}^2, \tag{2.15}$$

with the initial condition

$$\psi(0,k,x) = \psi_0(k,x), \quad k^2, x \in \mathbb{R}. \tag{2.16}$$

(Clearly $\psi(t,k,x)$ is smooth w.r.t. $k$ if $\psi_0(k,x)$ is.) Moreover, if $\psi(t,k,x)$, $\tilde{\psi}(t,k,x)$, $(t,k^2,x) \in \mathbb{R}^3$ are two such solutions of (2.13) with initial functions $\psi_0(k,x)$, $\tilde{\psi}_0(k,x)$ respectively, then their Wronskian

$$W(\psi(t,k), \tilde{\psi}(t,k)) = W(\psi_0(k), \tilde{\psi}_0(k)) \tag{2.17}$$

is independent of $(t,x) \in \mathbb{R}^2$. (Here $W(f,g) := fg' - f'g$ denotes the Wronskian of $f$ and $g$.)

This lemma follows by investigating the Volterra integral equation

$$\psi(t,k,x) = c(t,k)\cos(kx) + d(t,k)k^{-1}\sin(kx)$$
$$+ \int_0^x dx' k^{-1} \sin[k(x-x')] V(t,x') \psi(t,k,x'), \quad (t,k^2,x) \in \mathbb{R}^3, \tag{2.18}$$

with real-valued $C^\infty$-coefficients $c$ and $d$. It evidently implies the $t$-independence of the spectrum $\sigma(H(t))$ of $H(t)$.

Next, following [Ad-Mo], [Dei], we look into Miura's transformation (2.3) in more detail. Assuming (H.2.1) we define in $L^2(\mathbb{R})$

$$A(t) := \partial_x + \phi(t,.) \text{ on } H^1(\mathbb{R}), \ t \in \mathbb{R}, \tag{2.19}$$

and

$$H_j(t) := -\partial_x^2 + V_j(t,.) \text{ on } H^2(\mathbb{R}), \ j = 1,2, \ t \in \mathbb{R}. \tag{2.20}$$

Then

$$H_1(t) = A(t)^* A(t), \ H_2(t) = A(t)A(t)^*, \ Q(t) = \begin{pmatrix} 0 & A(t)^* \\ A(t) & 0 \end{pmatrix}, \ t \in \mathbb{R}, \tag{2.21}$$

and

$$Q(t)^2 = \begin{pmatrix} A(t)^* A(t) & 0 \\ 0 & A(t)A(t)^* \end{pmatrix} = H_1(t) \oplus H_2(t), \ t \in \mathbb{R},$$
$$B(t) = B_{V_1}(t) \oplus B_{V_2}(t), \ t \in \mathbb{R}, \tag{2.22}$$

shows the close connection between the Lax pairs

$$(H_j, B_{V_j}) \xleftarrow{V_j=\phi^2+(-1)^j\phi_x} (Q,B), \; j=1,2, \qquad (2.23)$$

of the KdV and mKdV-eq. effected by (2.3), i.e.,

$$\text{KdV}(V_j) = 0 \xleftarrow{V_j=\phi^2+(-1)^j\phi_x} \text{mKdV}(\phi) = 0, \; j=1,2. \qquad (2.24)$$

Since
$$H_j(t) \geq 0, \; j=1,2, \; t \in \mathbb{R}, \qquad (2.25)$$

by (2.21), we modify (H.2.1) as follows.

**(H.2.4).**
   (i) $V$ satisfies (H.2.1) and $\text{KdV}(V) = 0$.
   (ii) $H(t) \geq 0$ for some (and hence for all) $t \in \mathbb{R}$.

In order to reverse the arrow in (2.24) we need a few more facts.

**Lemma 2.5.** *[Har] Suppose $V$ satisfies (H.2.1) and assume the existence of a $0 < \psi \in C^\infty(\mathbb{R}^2)$ with $H(t)\psi(t) = 0$, $t \in \mathbb{R}$ in the distributional sense. Define*

$$\phi(t,x) := \psi_x(t,x)/\psi(t,x), \; (t,x) \in \mathbb{R}^2. \qquad (2.26)$$

*Then $\phi$ and $\tilde{V} := \phi^2 - \phi_x$ satisfy (H.2.1).*

The proof of Lemma 2.5 is based on a comparison result, Corollary XI.6.5 of [Har].

**Lemma 2.6.** *[Lax 2] Suppose $V$ satisfies (H.2.4)(i) and assume $\psi \in C^\infty(\mathbb{R}^2)$ to satisfy $H(t)\psi(t) = 0$, $t \in \mathbb{R}$ and (2.15) with $k=0$ in the distributional sense. If $\psi(t, x(t)) = 0$, $t \in \mathbb{R}$ then $x$ solves*

$$\dot{x}(t) = -2V(t,x), \; t \in \mathbb{R}. \qquad (2.27)$$

*Conversely, if $\psi(t_0, x_0) = 0$ for some $(t_0, x_0) \in \mathbb{R}^2$, solve (2.27) with $x(t_0) = x_0$ to get $\psi(t, x(t)) = 0$, $t \in \mathbb{R}$. In particular, if $\psi(t_0, x) > 0$ for all $x \in \mathbb{R}$ then $\psi(t,x) > 0$ for all $(t,x) \in \mathbb{R}^2$.*

Now we can state the main result of [Ge-Si].

**Theorem 2.7.** *[Ge-Si] (see also [Ges 1], [GSS]) Suppose $V_2$ satisfies (H.2.4). Let $0 < \psi_{2,\pm} \in C^\infty(\mathbb{R}^2)$ be distributional solutions of $H_2(t)\psi_{2,\pm}(t) = 0$, $t \in \mathbb{R}$ and of (2.15) with $k=0$. Define*

$$\psi_{2,\sigma}(t,x) := 2^{-1}[1-\sigma(t)]\psi_{2,-}(t,x) + 2^{-1}[1+\sigma(t)]\psi_{2,+}(t,x), \qquad (2.28)$$

$$\phi_\sigma(t,x) := \psi_{2,\sigma,x}(t,x)/\psi_{2,\sigma}(t,x), \qquad (2.29)$$

$$V_{1,\sigma}(t,x) := \phi_\sigma(t,x)^2 - \phi_{\sigma,x}(t,x); \ (t,x) \in \mathbb{R}^2, \tag{2.30}$$

where $\sigma : \mathbb{R} \to [-1,1]$, $\sigma \in C^\infty(\mathbb{R})$. Then $\phi_\sigma$, $V_{1,\sigma}$ satisfy (H.2.1). In addition,

$$\text{mKdV}(\phi_\sigma) = 0, \ \text{KdV}(V_{1,\sigma}) = 0 \text{ iff } \dot\sigma = 0 \text{ or } W(\psi_{2,-}, \psi_{2,+}) = 0. \tag{2.31}$$

For the proof one computes

$$\text{mKdV}(\phi_\sigma) = -\psi_{2,\sigma}^{-2} \dot\sigma \, W(\psi_{2,-}, \psi_{2,+})/2, \tag{2.32}$$

$$\text{KdV}(V_{1,\sigma}) = 2\psi_{2,\sigma}^{-3} \psi_{2,\sigma,x} \dot\sigma \, W(\psi_{2,-}, \psi_{2,+}). \tag{2.33}$$

In order to interpret this theorem it is appropriate to recall some results from the oscillation theory for Schrödinger operators (for a treatment of general Sturm-Liouville operators on an arbitrary interval with most general conditions on the coefficients see [Ge-Zh] and the references therein): According to [Sim] and [Mur], the nonnegative Schrödinger operator $H(0) \geq 0$ is called critical iff $H(0)\psi = 0$ has a unique, positive, distributional solution $\psi_0 > 0$ (up to multiples of constants). $H(0) \geq 0$ is called subcritical iff there are two linearly independent, positive distributional solutions $\psi_\pm > 0$ of $H(0)\psi = 0$. Since $H(0)$ is assumed to be nonnegative this case distinction exhausts all possibilities, i.e., either $H(0) \geq 0$ is critical or subcritical. Combining Lemmas 2.3 and 2.6 we infer

**Lemma 2.8.** *Assume (H.2.4). Then $H(t)$ is (sub)critical for all $t \in \mathbb{R}$ iff $H(t)$ is (sub)critical for some $t \in \mathbb{R}$.*

**Remark 2.9.** Theorem 2.7 yields a unique solution, say $\phi_0$, of the mKdV-equation (2.2) and of Miura's transformation (2.3) iff $H_2(0)$ is critical. In this case we can dispense with the condition that $\psi_0$ satisfies (2.15) with $k = 0$ since by Lemma 2.3 a certain time-dependent multiple of $\psi_0$ certainly will satisfy it. (The time-dependent factor, however, drops out in the definition of $\phi_0$ and hence is irrelevant.) Otherwise, i.e., iff $H_2(0)$ is subcritical, we get a one-parameter family of solutions $\phi_\sigma$ of (2.2) and (2.3) indexed by $\sigma \in [-1,1]$. Moreover, since $\phi_\sigma = \psi_{2,\sigma,x}/\psi_{2,\sigma}, \sigma \in [-1,1]$ is the general solution of the Riccati-type eq. $\phi_x + \phi^2 = V_2$ on $\mathbb{R}$, the explicit construction (2.29) yields all smooth solutions of (2.2) and (2.3).

**Remark 2.10.** The "if part" of Theorem 2.7 has been known for quite a while and follows e.g. from prolongation methods developed in [Wa-Es] (see also [De-Sp], [Sa-Zu], [Zh-Ch]). In general, these techniques do not distinguish between singular and nonsingular solutions of the mKdV-equation. It is the "only if part" in Theorem 2.7 that yields a classification into singular and nonsingular solutions $\phi$ (depending on whether $H_2(0)$ is nonnegative or not) and a uniqueness result for $\phi$ in Miura's transformation (2.3) depending on whether $H_2(0)$ is critical or subcritical. A different approach to this problem in the context of rapidly decreasing solutions of the KdV-equation can be found in Section 38 of [BDT].

**Remark 2.11.** Given $\phi_\sigma$ in (2.29), $H_2(t)$ is recovered from $\phi_\sigma$ via

$$H_2(t) = -\partial_x^2 + V_2(t,.) = A_\sigma(t)A_\sigma(t)^* \text{ on } H^2(\mathbb{R}), \ t \in \mathbb{R}, \quad (2.34)$$

$$A_\sigma(t) := \partial_x + \phi_\sigma(t,.) \text{ on } H^1(\mathbb{R}), \ t \in \mathbb{R}, \quad (2.35)$$

$$V_2(t,x) = \phi_\sigma(t,x)^2 + \phi_{\sigma,x}(t,x), \ (t,x) \in \mathbb{R}^2; \ \sigma \in [-1,1], \quad (2.36)$$

and similarly $H_1(t) := -\partial_x^2 + V_{1,\sigma}(t,.)$ on $H^2(\mathbb{R})$, $t \in \mathbb{R}$ is recovered from $\phi_\sigma$ via

$$H_{1,\sigma}(t) = A_\sigma(t)^* A_\sigma(t), \ t \in \mathbb{R}, \quad (2.37)$$

$$V_{1,\sigma}(t,x) = \phi_\sigma(t,x)^2 - \phi_{\sigma,x}(t,x), \ (t,x) \in \mathbb{R}^2; \ \sigma \in [-1,1]. \quad (2.38)$$

We also state the elementary

**Lemma 2.12.** Assume that $V_j$, $j = 1,2$ satisfy (H.2.1) and Miura's transformation (2.3) for some $\phi \in C^\infty(\mathbb{R}^2)$. Then $\phi$ satisfies (H.2.1) and

$$\phi(t,x) = 2^{-1}[V_{1,x}(t,x) + V_{2,x}(t,x)][V_2(t,x) - V_1(t,x)]^{-1}, \ (t,x) \in \mathbb{R}^2. \quad (2.39)$$

**Remark 2.13.** The fact that $\phi \to -\phi$ is a symmetry of the mKdV-equation (2.2), i.e., if $\phi(t,x)$ is a solution of (2.2) then so is $-\phi(t,x)$, evidently corresponds to the interchange $H_2(t) \longrightarrow H_1(t)$.

Finally, let us mention that every result in this section generalizes to the entire (m)KdV-hierarchy (see e.g. [Ad-Mo], [Ch-Pe], [Lax 1]) as shown in [Ges 1].

## 3. Soliton-like solutions of the mKdV-equation

Here we concentrate on soliton-like solutions of the (m)KdV-equation (i.e., solutions that tend to certain finite asymptotic values sufficiently fast as $x \to \pm\infty$).

We introduce the hypothesis

**(H.3.1).** Let $0 < w_\pm^{-1} \in C^0([0,\pm\infty)) \cap L^1((0,\pm\infty))$, $w_+$ (resp. $w_-$) monotonically increasing (resp. decreasing) in $(0,\infty)$ (resp. $(-\infty,0)$).

(i) $V \in C^\infty(\mathbb{R}^2)$ is real-valued, $\partial_x^n V \in L^\infty(\mathbb{R}^2)$, $n = 0,1$,
$$\lim_{x \to \pm\infty} V(t,x) = V_\pm \in \mathbb{R}, \ 0 < V_- \leq V_+,$$
$$\pm \int_0^{\pm\infty} dx \, w_\pm(x) |V(t,x) - V_\pm| < \infty, \ t \in \mathbb{R}.$$

(ii) $\phi \in C^\infty(\mathbb{R}^2)$ is real-valued, $\partial_x^n \phi \in L^\infty(\mathbb{R}^2)$, $n = 0,1$,
$$\lim_{x \to \pm\infty} \phi(t,x) = \phi_\pm \in \mathbb{R}, \ 0 < \phi_-^2 \leq \phi_+^2,$$
$$\pm \int_0^{\pm\infty} dx \, w_\pm(t,x) [|\phi(t,x) - \phi_\pm| + |\phi_x(t,x)|] < \infty, \ t \in \mathbb{R}.$$

(Here $V_\pm, \phi_\pm$ are $t$-independent constants.)

Next, supposing $V_2$ satisfies (H.3.1)(i) and $\text{KdV}(V_2) = 0$, we define

$$H_2(t) := -\partial_x^2 + V_2(t,.) \text{ on } H^2(\mathbb{R}), \ t \in \mathbb{R}, \tag{3.1}$$

and assume

$$\mathcal{E}_2; = \inf[\sigma(H_2(0))] \geq 0. \tag{3.2}$$

Then, according to Theorem 2.7, we have to distinguish the cases $\mathcal{E}_2 = 0$ and $\mathcal{E}_2 > 0$. We start with the case

$\mathcal{E}_2 = 0$ (**i.e.,** $H_2(t)$ **is critical,** $t \in \mathbb{R}$):
Since $\sigma_{\text{ess}}(H_2(t)) = [V_-, \infty), V_- > 0, \mathcal{E}_2 = 0$ implies

$$0 \in \sigma_d(H_2(t)), \ t \in \mathbb{R}, \tag{3.3}$$

and thus $H_2(t)$ is critical for all $t \in \mathbb{R}$ [Ge-Zh]. This yields the existence of a unique (up to possibly $t$-dependent multiples of constants), positive distributional solution $0 < \psi_{2,0} \in C^\infty(\mathbb{R}^2)$ of

$$H_2(t)\psi_{2,0}(t) = 0, \ 0 < \psi_{2,0}(t,.) \in H^2(\mathbb{R}), \ t \in \mathbb{R}. \tag{3.4}$$

Without loss of generality we may assume that $\psi_{2,0}$ satisfies the time evolution (2.15) with $k = 0$, i.e.,

$$\psi_{2,0,t}(t,x) = 2V_2(t,x)\psi_{2,0,x}(t,x) - V_{2,x}(t,x)\psi_{2,0}(t,x), \ (t,x) \in \mathbb{R}^2, \tag{3.5}$$

(otherwise multiply $\psi_{2,0}$ with an appropriate time-dependent factor). From (3.5) and (B.7) one infers that

$$\psi_{2,0}(t,x) \underset{x \to \pm\infty}{=} e^{\mp V_\pm^{\frac{1}{2}}(x+2V_\pm t)} + o\left(e^{\mp V_\pm^{\frac{1}{2}} x}\right), \ t \in \mathbb{R}, \tag{3.6}$$

since

$$k_\pm(0) = iV_\pm^{\frac{1}{2}}, \ W(f_{2,-}(t,0), f_{2,+}(t,0)) = 0,$$
$$\psi_{2,0}(t,x) = e^{\mp 2V_\pm^{\frac{3}{2}} t} f_{2,\pm}(t,0,x), \ (t,x) \in \mathbb{R}^2. \tag{3.7}$$

According to (2.29) we then define

$$\phi_0(t,x) := \psi_{2,0,x}(t,x)/\psi_{2,0}(t,x), \ (t,x) \in \mathbb{R}^2. \tag{3.8}$$

Next we consider the case

$\mathcal{E}_2 > 0$ (**i.e.,** $H_2(t)$ **is subcritical,** $t \in \mathbb{R}$):
Then $0 \notin \sigma(H_2(t))$ and hence $H_2(t)$ is subcritical for all $t \in \mathbb{R}$ [Ge-Zh]. This guarantees the existence of two linearly independent, positive distributional solutions $0 < \psi_{2,\pm} \in C^\infty(\mathbb{R}^2)$ of

$$H_2(t)\psi_{2,\pm}(t) = 0, \ 0 < \psi_{2,\pm}(t,.) \in L^\infty_{\text{loc}}(\mathbb{R}), \ t \in \mathbb{R}, \tag{3.9}$$

and again we may without loss of generality (cf. Lemma 2.3) assume that $\psi_{2,\pm}$ satisfy (2.15) with $k = 0$, i.e.,

$$\psi_{2,\pm,t}(t,x) = 2V_2(t,x)\psi_{2,\pm,x}(t,x) - V_{2,x}(t,x)\psi_{2,\pm}(t,x), \ (t,x) \in \mathbb{R}^2. \quad (3.10)$$

From (3.10 and B.7) one infers

$$\psi_{2,\pm}(t,x) = e^{\mp 2V_\pm^{\frac{3}{2}}t} f_{2,\pm}(t,x), \ (t,x) \in \mathbb{R}^2, \quad (3.11)$$

and

$$\psi_{2,+}(t,x) \underset{x\to\pm\infty}{=} e^{-V_\pm^{\frac{1}{2}}(x+2V_\pm t)} + o\left(e^{-V_\pm^{\frac{1}{2}}x}\right),$$
$$\psi_{2,-}(t,x) \underset{x\to\pm\infty}{=} e^{V_\pm^{\frac{1}{2}}(x+2V_\pm t)} + o(e^{V_\pm^{\frac{1}{2}}x}). \quad (3.12)$$

In accordance with (2.29) we now define

$$\psi_{2,\sigma}(t,x) := 2^{-1}(1-\sigma)\psi_{2,-}(t,x) + 2^{-1}(1+\sigma)\psi_{2,+}(t,x),$$
$$\sigma \in [-1,1], \ (t,x) \in \mathbb{R}^2, \quad (3.13)$$

(where $\sigma$ is $t$-independent) and

$$\phi_\sigma(t,x) := \psi_{2,\sigma,x}(t,x)/\psi_{2,\sigma}(t,x), \ \sigma \in [-1,1], (t,x) \in \mathbb{R}^2. \quad (3.14)$$

Then we obtain the following characterization of soliton-like solutions of the mKdV-equation.

**Theorem 3.2.** Assume $V_2$ satisfies (H.3.1) (i) and $\text{KdV}(V_2) = 0$. Moreover, suppose $\mathcal{E}_2 = \inf[\sigma(H_2(0))] \geq 0$. Then $\phi_\sigma$, defined in (3.8) resp. (3.14), satisfies (H.3.1) (ii) and $V_{1,\sigma}$ defined by

$$V_{1,\sigma}(t,x) = \phi_\sigma(t,x)^2 - \phi_{\sigma,x}(t,x), \ \sigma = 0 \text{ resp. } \sigma \in [-1,1], (t,x) \in \mathbb{R}^2, \quad (3.15)$$

satisfies (H.3.1) (i). Moreover,

$$\text{mKdV}(\phi_\sigma) = 0, \ \text{KdV}(V_{1,\sigma}) = 0, \ \sigma = 0 \text{ resp. } \sigma \in [-1,1]. \quad (3.16)$$

In addition,

$$\phi_{\sigma,\pm} := \lim_{x\to\pm\infty} \phi_\sigma(t,x) = \begin{cases} \mp V_\pm^{\frac{1}{2}}, & \sigma = 0, \quad \mathcal{E}_2 = 0, \\ -V_\pm^{\frac{1}{2}}, & \sigma = 1, \quad \mathcal{E}_2 > 0, \\ +V_\pm^{\frac{1}{2}}, & \sigma = -1, \quad \mathcal{E}_2 > 0, \\ \pm V_\pm^{\frac{1}{2}}, & \sigma \in (-1,1), \quad \mathcal{E}_2 > 0, \end{cases} \quad (3.17)$$

and

$$0 \in \sigma_d(H_2(t)), \ 0 \notin \sigma_d(H_{1,\sigma}(t)), \ \phi_{\sigma,+} < 0 < \phi_{\sigma,-},$$
$$0 \notin \sigma_d(H_2(t)), \ 0 \in \sigma_d(H_{1,\sigma}(t)), \ \phi_{\sigma,-} < 0 < \phi_{\sigma,+},$$
$$0 \notin \{\sigma_d(H_2(t)) \cup \sigma_d(H_{1,\sigma}(t))\}, \ \mathrm{sgn}(\phi_{\sigma,-}) = \mathrm{sgn}(\phi_{\sigma,+}),$$
$$\sigma = 0 \ \text{resp.} \ \sigma \in [-1,1], \ t \in \mathbb{R}, \quad (3.18)$$

where

$$H_{1,\sigma}(t) := -\partial_x^2 + V_{1,\sigma}(t,.) \ \text{on} \ H^2(\mathbb{R}), \ t \in \mathbb{R}. \quad (3.19)$$

In the special case, where $w_\pm(x) = 1 + x^2$, Theorem 3.2 first appeared in [GSS]. Here, concentrating on that part in its proof that deviates from the one given in [GSS], we formulate

**Lemma 3.3.** *Under the hypotheses of Theorem 3.2 we get*

$$\phi_\sigma(t,x) \underset{x \to \pm\infty}{=} \phi_{\sigma,\pm} + O(w_\pm(x)^{-1}) h_{\sigma,\pm}(t,x),$$
$$\sigma = 0 \ \text{resp.} \ \sigma \in [-1,1], \ t \in \mathbb{R}, \quad (3.20)$$

*where*

$$h_{\sigma,\pm}(t,.) \in L^1((0,\pm\infty)), \ \sigma = 0 \ \text{resp.} \ \sigma \in [-1,1], t \in \mathbb{R}. \quad (3.21)$$

**Proof.** It suffices to consider the case $\sigma = \mathcal{E}_2 = 0$ and $x \to +\infty$. Then (B.7), (3.8) and (3.11) imply

$$\phi_0(t,x) = -V_+^{\frac{1}{2}} \frac{1 + d_+(t) \int_x^\infty dx' [e^{2V_+^{\frac{1}{2}}(x-x')} + 1][V_2(t,x') - V_+] e^{V_+^{\frac{1}{2}} x'} \psi_{2,0}(t,x')}{1 - d_+(t) \int_x^\infty dx' [e^{2V_+^{\frac{1}{2}}(x-x')} - 1][V_2(t,x') - V_+] e^{V_+^{\frac{1}{2}} x'} \psi_{2,0}(t,x')},$$
$$(t,x) \in \mathbb{R}^2, \quad (3.22)$$

where

$$d_+(t) := e^{2V_+^{\frac{3}{2}} t}/2V_+^{\frac{1}{2}}, \ t \in \mathbb{R}. \quad (3.23)$$

Since

$$|e^{V_+^{\frac{1}{2}} x'} \psi_{2,0}(t,x')| \leq C(t), \ x' \geq 0, \ t \in \mathbb{R},$$
$$|e^{2V_+^{\frac{1}{2}}(x-x')} \pm 1| \leq 2, \quad x \leq x', \quad (3.24)$$

the integral terms in (3.22) behave like $\int_x^\infty \ldots = o(1)$ as $x \to +\infty$ and hence we may expand the denominator in (3.22) to obtain

$$\phi_0(t,x) \underset{x \to \infty}{=} -V_+^{\frac{1}{2}} \Big\{ 1 + 2d_+(t) \int_x^\infty dx' e^{2V_+^{\frac{1}{2}}(x-x')} [V_2(t,x') - V_+] e^{V_+^{\frac{1}{2}} x'} \psi_{2,0}(t,x')$$
$$+ O(w_+(x)^{-2}) \Big\}, \quad t \in \mathbb{R}. \quad (3.25)$$

Here we also used the monotonicity of $w_+$ in order to arrive at the estimate

$$\left| d_+(t) \int_x^\infty dx' \left[ e^{2V_+^{\frac{1}{2}}(x-x')} - 1 \right] [V_2(t,x') - V_+] e^{V_+^{\frac{1}{2}}x'} \psi_{2,0}(t,x') \right|^2$$

$$\leq \left[ c(t) \int_x^\infty dx' w_+(x')^{-1} w_+(x') |V_2(t,x') - V_+| \right]^2 \quad (3.26)$$

$$\leq c(t)^2 w_+(x)^{-2} \left[ \int_x^\infty dx' w_+(x') |V_2(t,x') - V_+| \right]^2, \, x \geq 0, t \in \mathbb{R}.$$

Finally we use the elementary fact that

$$f \in L^1((R,\infty)), \, R \in \mathbb{R} \text{ implies } g_\alpha \in L^1((R,\infty)), \quad (3.27)$$

where

$$g_\alpha(x) := \int_x^\infty dx' e^{\alpha(x-x')} f(x') \text{ for some } \alpha > 0. \quad (3.28)$$

Then (3.25) together with $w_+^{-1} \in L^1((0,\infty))$,

$$\int_x^\infty dx' e^{2V_+^{\frac{1}{2}}(x-x')} |V_2(t,x') - V_+| e^{V_+^{\frac{1}{2}}x'} |\psi_{2,0}(t,x')|$$

$$\leq D(t) w_+(x)^{-1} \int_x^\infty dx' e^{2V_+^{\frac{1}{2}}(x-x')} w_+(x') |V_2(t,x') - V_+|, \, t \in \mathbb{R}, x \geq 0, \quad (3.29)$$

and (3.27) proves (3.20).

Given Lemma 3.3, one can now prove Theorem 3.2 in analogy to Theorem 7.14 of [GSS].

**Remark 3.4.** A comparison of (3.17) and (3.18) shows that, whenever $\text{sgn}(\phi_{\sigma,-}) \neq \text{sgn}(\phi_{\sigma,+})$ then $\mathcal{E}_2 = 0$ and $\mathcal{E}_{1,\sigma} := \inf[\sigma(H_{1,\sigma})] > 0$ or $\mathcal{E}_2 > 0$ and $\mathcal{E}_{1,\sigma} = 0$. Hence either $H_2(t)$ is critical and $H_{1,\sigma}(t)$ is subcritical or vice versa. In particular, $H_{1,\sigma}(t)$ is critical for $\sigma \in (-1,1)$, $\mathcal{E}_2 > 0$. Only in the case where $\text{sgn}(\phi_{\sigma,-}) = \text{sgn}(\phi_{\sigma,+})$ one gets $\mathcal{E}_2 = \mathcal{E}_{1,\sigma} > 0$ and thus both $H_2$ and $H_{1,\sigma}$ are subcritical iff $\sigma = \pm 1$.

**Remark 3.5.** If $V_j, j = 1, 2$ in Lemma 2.12 actually satisfy (H.3.1) (i) then $\phi$ satisfies (H.3.1) (ii) and we obtain

$$\phi_+ - \phi_- = 2^{-1} \int_\mathbb{R} dx' [V_2(t,x') - V_1(t,x')], \, t \in \mathbb{R}, \quad (3.30)$$

$$\phi(t,x) = \phi_\pm - 2^{-1} \int_x^{\pm\infty} dx' [V_2(t,x') - V_1(t,x')], \ (t,x) \in \mathbb{R}^2, \qquad (3.31)$$

in addition to (2.39).

## 4. Soliton solutions of the mKdV-equation

In this section we review the derivation of soliton solutions of the mKdV-equation. Originally, these solutions have been constructed implicitly in [Ji-Mi] (as a by-product of their study of the modified Kadomtsev-Petviashvili equation) using vertex operator techniques and explictly in [Gro] on the basis of inverse scattering methods. Here we follow [GSS] exploiting the commutation approach outlined in Appendix A.

We start with an $N$-soliton solution $V_2(t,x)$ of the KdV-equation (2.1) given by the familar formula [Hir], [Zak]

$$V_2(t,x) = V_\infty - 2\partial_x^2 \ln\{\det[1 + C_{2,N,\pm}(t,x)]\}, \ N \in \mathbb{N}_0, \ (t,x) \in \mathbb{R}^2, \qquad (4.1)$$

$$V_{2,\pm} = \lim_{x \to \pm\infty} V_2(t,x) = V_\infty > 0,$$

where $V_\infty$ is $t$-independent,

$$C_{2,N,\pm}(t,x) = [c_{2,\pm,\ell,m}(t,x)]_{\ell,m=1}^N,$$

$$c_{2,\pm,\ell,m}(t,x) = (\kappa_\ell + \kappa_m)^{-1} c_{2,\pm,\ell}(0) c_{2,\pm,m}(0) e^{\mp(\kappa_\ell + \kappa_m)(x + 6V_\infty t)} e^{\pm 4(\kappa_\ell^3 + \kappa_m^3)t},$$
$$1 \leq \ell, m \leq N, \ N \in \mathbb{N}, \ (t,x) \in \mathbb{R}^2, \qquad (4.2)$$

$$c_{2,\pm,\ell}(0) > 0, \ 1 \leq \ell \leq N, \ 0 < \kappa_N < \kappa_{N-1} < \ldots < \kappa_1,$$

for $N \in \mathbb{N}$ and

$$C_{2,0,\pm}(t,x) = 0, \ (t,x) \in \mathbb{R}^2, \qquad (4.3)$$

for $N = 0$. Define

$$H_2(t) := -\partial_x^2 + V_2(t,.) \text{ on } H^2(\mathbb{R}), \ t \in \mathbb{R}, \qquad (4.4)$$

and let $f_{2,\pm}(t,z,x)$ be the Jost functions associated with $H_2(t)$ (cf. Appendix B). Then due to the fact that $V_{2,+} = V_{2,-} = V_\infty$, some of the formulas in Appendix B simplify. In particular, we have

$$k(z) := k_\pm(z) = (z - V_\infty)^{\frac{1}{2}}, \ \text{Im} k(z) \geq 0, \qquad (4.5)$$

and

$$\sigma_p(H_2(t)) = \begin{cases} \{\lambda_{2,\ell} = V_\infty - \kappa_\ell^2\}_{\ell=1}^N, & N \in \mathbb{N}, \\ \emptyset, & N = 0. \end{cases} \qquad (4.6)$$

As in Section 3 we need to distinguish between $\mathcal{E}_2 = 0$ and $\mathcal{E} > 0$.

$\mathcal{E}_2 = 0$:
In this case we necessarily have $N \geq 1$, $\lambda_{2,1} = 0$ and hence

$$V_\infty^{\frac{1}{2}} = \kappa_1. \tag{4.7}$$

In order to construct $H_{1,0}(t)$ we note that

$$H_2(t) = A_0(t) A_0(t)^*, \ t \in \mathbb{R}, \tag{4.8}$$

where

$$A_0(t) := \partial_x + \phi_0(t,.) \text{ on } H^1(\mathbb{R}), \ t \in \mathbb{R}, \tag{4.9}$$

and

$$\phi_0(t,x) = f_{2,\pm,x}(t,0,x)/f_{2,\pm}(t,0,x), \ (t,x) \in \mathbb{R}^2, \tag{4.10}$$

is uniquely defined since $H_2(t)$ is critical (cf. (3.7), (3.8)). Moreover,

$$\phi_{0,\pm} = \lim_{x \to \pm\infty} \phi_0(t,x) = \mp V_\infty^{\frac{1}{2}} = \mp \kappa_1 \tag{4.11}$$

by (3.17) and (4.7). Thus

$$0 \in \sigma_d(H_2(t)), \ 0 \notin \sigma_d(H_{1,0}(t)), \ t \in \mathbb{R}, \tag{4.12}$$

by (3.18). Hence

$$\sigma(H_{1,0}(t)) = \sigma(H_{2,0}(t)) \backslash \{0\},$$
$$\sigma_p(H_{1,0}(t)) = \begin{cases} \{\lambda_{2,\ell} = V_\infty - \kappa_\ell^2\}_{\ell=2}^N, & N \geq 2, \\ \emptyset, & N = 1; \end{cases} \ t \in \mathbb{R}. \tag{4.13}$$

Combining Remark B.6, (B.29) and (4.13), $V_{1,0}(t,x)$ must be a $(N-1)$-soliton solution of the KdV-equation explicitly given by

$$V_{1,0}(t,x) = V_\infty - 2\partial_x^2 \ln\{\det[1 + C_{1,N-1,\pm,0}(t,x)]\}, \ N \in \mathbb{N}, \ (t,x) \in \mathbb{R}^2,$$
$$V_{1,0,\pm} = \lim_{x \to \pm\infty} V_{1,0}(t,x) = V_\infty, \tag{4.14}$$

where

$$C_{1,N-1,\pm,0}(t,x) = [c_{1,\pm,0,\ell,m}(t,x)]_{\ell,m=2}^N,$$
$$c_{1,\pm,0,\ell,m}(t,x) = \left[\frac{(\kappa_1 \pm \kappa_\ell)(\kappa_1 \pm \kappa_m)}{(\kappa_1 \mp \kappa_\ell)(\kappa_1 \mp \kappa_m)}\right]^{\frac{1}{2}} c_{2,\pm\ell,m}(t,x), \tag{4.15}$$
$$2 \leq \ell, m \leq N, \ N \geq 2, \ (t,x) \in \mathbb{R}^2,$$

for $N \geq 2$ and

$$C_{1,0,0,\pm}(t,x) = 0, \ (t,x) \in \mathbb{R}^2, \tag{4.16}$$

for $N = 1$. It remains to compute $\phi_0(t,x)$. Before doing so we need

**Definition 4.1.** Assume (H.3.1) (ii). Then a solution $\phi$ of the mKdV-equation (2.2) is called an M-soliton solution, $M \in \mathbb{N}_0$ iff $\phi(t,x)$ is a reflectionless potential for the Dirac operator $Q(t)$ in (2.7) and $Q(t)$ has $M$(discrete) eigenvalues for some (and hence for all) $t \in \mathbb{R}$.

**Theorem 4.2.** Assume $\mathcal{E}_2 = 0$ and let

$$V_{j,(0)}(t,x) = V_\infty - 2\partial_x^2 \ln\left\{\det\left[1 + C_{j,\underset{(N-1)}{N},\pm,(0)}(t,x)\right]\right\}, \tag{4.17}$$

$$j = 1,2, \ N \in \mathbb{N}, \ (t,x) \in \mathbb{R}^2,$$

$$V_{j,(0),\pm} = V_\infty = \kappa_1^2$$

be the $N$ (resp. $N-1$) soliton solutions (4.1) (resp. (4.14)) of the KdV-equation. By construction, $V_2$ and $V_{1,0}$ are related to $\phi_0$ by Miura's transformation

$$V_{j,(0)}(t,x) = \phi_0(t,x)^2 + (-1)^j \phi_{0,x}(t,x), \ j = 1,2, \ (t,x) \in \mathbb{R}^2, \tag{4.18}$$

and one obtains

$$\phi_0(t,x) = \mp\kappa_1 - \partial_x \ln\left\{\det[1 + C_{2,N,\pm}(t,x)]/\det[1 + C_{1,N-1,\pm,0}(t,x)]\right\},$$
$$(t,x) \in \mathbb{R}^2, \tag{4.19}$$
$$\phi_{0,\pm} = \mp\kappa_1.$$

Moreover, $\phi_0(t,x)$ is a $(2N-1)$-soliton solution of the mKdV-equation. In particular, up to an overall sign ambiguity, the solutions (4.19) represent all reflectionless potentials of the associated Dirac operator $Q(t)$ in (2.7) under the assumption that $Q(0)$ has a zero eigenvalue, i.e., $0 \in \sigma_d(Q(0))$.

For the proof one only needs to combine Remark 2.13, Theorem 3.2, Remark 3.5, Theorem A.1 (vi), (vii) and Remark B.6.

Finally we consider the case

$\mathcal{E}_2 > 0$:
In this case (4.7) turns into

$$V_\infty = \lambda_{2,1} + \kappa_1^2 = \mathcal{E}_2 + \kappa_1^2. \tag{4.20}$$

Combining Remarks 2.13 and 3.4 we may confine ourselves to the cases $\sigma = \pm 1$ in Theorem 3.2. Writing

$$H_2(t) = A_\sigma(t)A_\sigma(t)^*, \ \sigma = \pm 1, \ t \in \mathbb{R},$$
$$A_\sigma(t) := \partial_x + \phi_\sigma(t,.) \text{ on } H^1(\mathbb{R}), \ \sigma = \pm 1, \ t \in \mathbb{R}, \tag{4.21}$$

we obtain

$$\phi_{\sigma,\pm} = \lim_{x \to \pm\infty} \phi_\sigma(t,x) = -\sigma V_\infty^{\frac{1}{2}}, \ \sigma = \pm 1, \tag{4.22}$$

and

$$\sigma(H_{1,\sigma}(t)) = \sigma(H_2(t)), \ \sigma = \pm 1, \quad (4.23)$$

$$\sigma_p(H_{1,\sigma}(t)) = \sigma_p(H_2(t)) = \begin{cases} \{\lambda_{2,\ell} = V_\infty - \kappa_\ell^2\}_{\ell=1}^N, & N \geq 1 \\ \emptyset, & N = 0 \end{cases} ; \sigma = \pm 1, \ t \in \mathbb{R}$$

by (3.17) and (3.18). Combining Remark B.6, (B.29) and (4.23), $V_{1,\sigma}(t,x)$, $\sigma = \pm 1$ now must be an $N$-soliton solution of the KdV-equation explicitly given by

$$V_{1,\sigma}(t,x) = V_\infty - 2\partial_x^2 \ln \det\{[1 + C_{1,N,\pm,\sigma}(t,x)]\}, \sigma = \pm 1, \ N \in \mathbb{N}_0,$$
$$(t,x) \in \mathbb{R}^2, \quad (4.24)$$
$$V_{1,\sigma,\pm} = \lim_{x \to \pm\infty} V_{1,\sigma}(t,x) = V_\infty, \ \sigma = \pm 1,$$

where

$$C_{1,N,\pm,\sigma}(t,x) = [c_{1,\pm,\sigma,\ell,m}(t,x)]_{\ell,m=1}^N,$$

$$c_{1,\pm\sigma,\ell,m}(t,x) = \left[\frac{\left(\sigma V_\infty^{\frac{1}{2}} \pm \kappa_\ell\right)\left(\sigma V_\infty^{\frac{1}{2}} \pm \kappa_m\right)}{\left(\sigma V_\infty^{\frac{1}{2}} \mp \kappa_\ell\right)\left(\sigma V_\infty^{\frac{1}{2}} \mp \kappa_m\right)}\right]^{\frac{1}{2}} c_{2,\pm,\ell,m}(t,x), \quad (4.25)$$

$$\sigma = \pm 1, \ 1 \leq \ell, m \leq N, \ N \in \mathbb{N}, (t,x) \in \mathbb{R}^2,$$

for $N \geq 1$ and
$$C_{1,0,\pm,\sigma}(t,x) = 0, \ \sigma = \pm 1, \ (t,x) \in \mathbb{R}^2, \quad (4.26)$$

for $N = 0$. It remains to compute $\phi_\sigma(t,x)$, $\sigma = \pm 1$.

**Theorem 4.3.** *Assume $\mathcal{E}_2 > 0$ and let*

$$V_{j,(\sigma)}(t,x) = V_\infty - 2\partial_x^2 \ln\{\det[1 + C_{j,N,\pm,(\sigma)}(t,x)]\},$$
$$j = 1, 2, \ \sigma = \pm 1, N \in \mathbb{N}_0, (t,x) \in \mathbb{R}^2, \quad (4.27)$$
$$V_{j,(\sigma),\pm} = V_\infty = \mathcal{E}_2 + \kappa_1^2$$

*be the $N$-soliton solutions (4.1) and (4.24) of the KdV-equation. By construction they are related to $\phi_\sigma$ by Miura's transformation*

$$V_{j,(\sigma)}(t,x) = \phi_\sigma(t,x)^2 + (-1)^j \phi_{\sigma,x}(t,x), \ j = 1, 2, \ (t,x) \in \mathbb{R}^2, \quad (4.28)$$

*and one obtains*

$$\phi_\sigma(t,x) = -\sigma V_\infty^{\frac{1}{2}} - \partial_x \ln\{\det[1 + C_{2,N,\pm}(t,x)]/\det[1 + C_{1,N,\pm,\sigma}(t,x)]\},$$
$$\sigma = \pm 1, \ N \in \mathbb{N}_0, \ (t,x) \in \mathbb{R}^2, \quad (4.29)$$
$$\phi_{\sigma,\pm} = -\sigma V_\infty^{\frac{1}{2}}, \ \sigma = \pm 1.$$

Moreover, $\phi_\sigma(t,x)$, $\sigma = \pm 1$ is a $2N$-soliton solution of the mKdV-equation. In particular, up to an overall sign ambiguity, the solutions (4.29) represent all reflectionless potentials of the associated Dirac operator $Q(t)$ in (2.7) under the assumption that $Q(0)$ has no eigenvalue zero, i.e., $0 \notin \sigma_d(Q(0))$.

The proof of Theorem 4.3 is analogous to that of Theorem 4.2. We conclude with

**Example 4.4.** $N = 1$, $\mathcal{E}_2 = 0$.

$$V_2(t,x) = \kappa_1^2 - 2\kappa_1^2[\cosh(\kappa_1 x + 2\kappa_1^3 t)]^{-2}, \ (t,x) \in \mathbb{R}^2,$$
$$V_\infty = \kappa_1^2, \quad c_{2,\pm,1}(0)^2 = 2\kappa_1, \quad \kappa_1 > 0. \tag{4.30}$$

Then
$$V_{1,0}(t,x) = \kappa_1^2,$$
$$\phi_0(t,x) = -\kappa_1 \tanh(\kappa_1 x + 2\kappa_1^3 t); \quad (t,x) \in \mathbb{R}^2. \tag{4.31}$$

## 5. Periodic solutions of the mKdV-equation

Here we apply Theorem 2.7 in the case where $V_2(t,x)$ is periodic with respect to $x$ and construct periodic solutions $\phi$ of the mKdV-equation.
We start with hypothesis

**(H.5.1).** Let $f \in C^\infty(\mathbb{R}^2)$ be real-valued and suppose there is an
$a > 0$ s.t. $f(t, x + a) = f(t,x)$, $(t,x) \in \mathbb{R}^2$.
Assume that $V$ satisfies (H.5.1) and define in $L^2(\mathbb{R})$

$$H(t) := -\partial_x^2 + V(t,.) \text{ on } H^2(\mathbb{R}), \ t \in \mathbb{R}. \tag{5.1}$$

Then the Floquet theory sketched in Appendix C applies to $H(t)$, $t \in \mathbb{R}$. In order to apply Theorem 2.7, we first need to clarify the time-dependence of the Floquet (Baker-Akhiezer) functions. Suppose $V$ satisfies (H.5.1) and the KdV-equation (2.1). We then define $c(t,z,x,x_0), s(t,z,x,x_0), \Delta(z), m_\pm(z), \phi_\pm(t,z,x_0), \mu_n(t,x)$ and $\psi_\pm(t,z,x,x_0)$ as in Appendix C replacing $V(x)$ by $V(t,x)$. Evidently $\Delta(z), m_\pm(z)$ are independent of $(t,x_0)$ by Theorem 2.2 (i), (C.6), C.22) and

$$\psi_\pm(t,z,x_0,x_0) = 1, \ (t,x_0) \in \mathbb{R}^2, \ z \in \hat{\Pi}_+(t,x_0) := \Pi_+ \backslash \{\mu_n(t,x_0)\}_{n \in \mathbb{N}}. \tag{5.2}$$

Since $\psi_\pm$ are normalized by (5.2), they will not satisfy (2.15) (with $z = k^2$) in general. In fact, we have

**Lemma 5.2.** Suppose $V$ satisfies (H.5.1) and $KdV(V) = 0$. Then

$$s_t(t,z,x,x_0) = 2[V(t,x) + 2z]s_x(t,z,x,x_0)$$
$$- [V_x(t,x) + V_x(t,x_0)]s(t,z,x,x_0) \tag{5.3}$$
$$- 2[V(t,x_0) + 2z]c(t,z,x,x_0), \ z \in \mathbb{C},$$

$$c_t(t,z,x,x_0) = 2[V(t,x) + 2z]c_x(t,z,x,x_0) - V_x(t,x)c(t,z,x,x_0)$$
$$+ V_x(t,x_0)c(t,z,x,x_0) \qquad (5.4)$$
$$+ \{V_{xx}(t,x_0) - 2[V(t,x_0) + 2z][V(t,x_0 - z]\}s(t,z,x,x_0), \ z \in \mathbb{C},$$

$$\phi_{\pm,t}(t,z,x_0) = 2[V(t,x_0) + 2z][V(t,x_0) - z] - V_{xx}(t,x_0)$$
$$+ \{2V_x(t,x_0) - 2[V(t,x_0) + 2z]\phi_\pm(t,z,x_0)\}\phi_\pm(t,z,x_0), z \in \hat{\Pi}_+(t,x_0), \qquad (5.5)$$

$$\psi_{\pm,t}(t,z,x,x_0) = 2[V(t,x) + 2z]\psi_{\pm,x}(t,z,x,x_0)$$
$$-\{V_x(t,x) - V_x(t,x_0) + 2[V(t,x_0) + 2z]\phi_\pm(t,z,x_0)\}\psi_\pm(t,z,x,x_0), \qquad (5.6)$$
$$z \in \hat{\Pi}_+(t,x_0); \ (t,x,x_0) \in \mathbb{R}^3.$$

**Proof.** Since $g := s_t - B_V s$ satisfies $Hg = zg$ by (2.10), we infer

$$g(t,z,x,x_0) = A(t,z,x_0)s(t,z,x,x_0) \qquad (5.7)$$
$$+ B(t,z,x_0)c(t,z,x,x_0), z \in \mathbb{C}, \ (t,x,x_0) \in \mathbb{R}^3.$$

This equation and its $x$-derivative taken at $x = x_0$ yield

$$A(t,z,x_0) = -V_x(t,x_0),$$
$$B(t,z,x_0) = -2[V(t,x_0) + 2z], \ z \in \mathbb{C}, \ (t,x) \in \mathbb{R}^2, \qquad (5.8)$$

since

$$s_t(t,z,x_0,x_0) = s_{tx}(t,z,x_0,x_0) = 0, \ z \in \mathbb{C}, (t,x_0) \in \mathbb{R}^2, \qquad (5.9)$$

by (C.2). Similarly one proves (5.4). Equations (5.5) and (5.6) then follow from the definitions in (C.24), (C.26).

Thus equation (5.6) shows that a certain $(t,z,x_0)$-dependent multiple of $\psi_\pm$ always satisfies (2.15).

Next we note that the Green's function $G(t,z,x,x')$ of $H(t)$ (i.e., the integral kernel of $(H(t) - z)^{-1}$) is given by

$$G(t,z,x,x') = W(\psi_+(t,z), \psi_-(t,z))^{-1} \begin{cases} \psi_-(t,z,x,x_0)\psi_+(t,z,x',x_0), \ x \leq x' \\ \psi_-(t,z,x',x_0)\psi_+(t,z,x,x_0), \ x \geq x', \end{cases}$$
$$z \in \Pi_+ \backslash \sigma(H(0))^\circ, \ Imz \geq 0, \quad (5.10)$$

($A^\circ$ the interior of $A \subseteq \mathbb{R}$) where we define $\psi_\pm(t,\lambda,x,x_0)$ on the cuts $\overline{\rho_0} = (-\infty, E_0]$, $\overline{\rho_n} = [E_{2n-1}, E_{2n}]$, $n \in I$ by

$$\psi_\pm(t,\lambda,x,x_0) := \lim_{\epsilon \downarrow 0} \psi_\pm(t, \lambda + i\epsilon, x, x_0), \ \lambda \in \overline{\rho_n}, n \in I_0, (t,x,x_0) \in \mathbb{R}^3, \quad (5.11)$$

in order to guarantee

$$\psi_\pm(t,z,.,x_0) \in L^2((0,\pm\infty)), \ z \in \Pi_+ \setminus \sigma(H(0)), \ \mathrm{Im}\, z \geq 0, \ (t,x_0) \in \mathbb{R}^2 \quad (5.12)$$

(see (C.32) and the paragraph following (C.15)). Introducing

$$\rho(t,z,x) := -2[\Delta(z)^2 - 1]^{\frac{1}{2}} G(t,z,x,x), \ z \in \mathbb{C}, \ (t,x) \in \mathbb{R}^2, \quad (5.13)$$

we obtain (see e.g., [Ge-Di])

$$\rho_{xx}(t,z,x)\rho(t,z,x) - (1/2)\rho_x(t,z,x)^2 - 2[V(t,x)-z]\rho(t,z,x)^2 \\ = -2[\Delta(z)^2 - 1], \ z \in \mathbb{C}. \quad (5.14)$$

Taking into account (C.30) and (E.5) we get

$$\rho(t,z,x) = s(t,z,x_0+a,x_0)\psi_-(t,z,x,x_0)\psi_+(t,z,x,x_0) \\ = s(t,z,x+a,x) = a \prod_{n \in \mathbb{N}} [\mu_n(t,x) - z](a^2/n^2\pi^2), \ z \in \mathbb{C}, \ (t,x) \in \mathbb{R}^2. \quad (5.15)$$

Equations (5.14) and (5.15) imply

$$\rho_x(t,\mu_n(t,x),x)^2 = 4[\Delta(\mu_n(t,x))^2 - 1], \ n \in \mathbb{N}, \ (t,x) \in \mathbb{R}^2, \quad (5.16)$$

and

$$\mu_{n,x}(t,x) = (2n^2\pi^2/a^3)[\Delta(\mu_n(t,x))^2 - 1]^{\frac{1}{2}} \\ \times \left\{ \prod_{\substack{m \in \mathbb{N} \\ m \neq n}} [\mu_m(t,x) - \mu_n(t,x)](a^2/m^2\pi^2) \right\}^{-1}, \ n \in I, \ (t,x) \in \mathbb{R}^2. \quad (5.17)$$

Similarly, using (5.6), (5.15) and (E.11)

$$\rho_t(t,\mu_n(t,x),x) = \\ 2[V(t,x) + 2\mu_n(t,x)]\rho_x(t,\mu_n(t,x),x), \ n \in I, (t,x) \in \mathbb{R}^2, \quad (5.18)$$

and

$$\mu_{n,t}(t,x) = 2[V(t,x) + 2\mu_n(t,x)]\mu_{n,x}(t,x) \quad (5.19) \\ = \left\{ 2E_\circ + 2\sum_{m \in I}[E_{2m-1} + E_{2m} - 2\mu_m(t,x)(1-\delta_{mn})] \right\} \mu_{n,x}(t,x), \\ n \in I, (t,x) \in \mathbb{R}^2.$$

Moreover,

$$\mu_n(t,x) = E_{2n-1} = E_{2n}, \ n \in \mathbb{N} \setminus I, \ (t,x) \in \mathbb{R}^2. \quad (5.20)$$

Now we are ready to apply Theorem 2.7. Suppose $V_2$ satisfies (H.5.1) and $\mathrm{KdV}(V_2) = 0$ and add the index 2 to all quantities defined in (5.1)-(5.20). Assuming that
$$\mathcal{E}_2 := \inf[\sigma(H_2(0))] \geq 0, \qquad (5.21)$$
we need to distinguish the cases $\mathcal{E}_2 = 0$ and $\mathcal{E}_2 > 0$ (cf. Remark 2.9). Similar to Section 3 we first start with the case

$\mathcal{E}_2 = 0$ (i.e., $H_2(t)$ is critical, $t \in \mathbb{R}$):
In this case $m_\pm(0) = 1$ and the Floquet solutions $\psi_{2,\pm}(t, 0, x, x_0)$ of $H_2(t)$ are linearly dependent and periodic with respect to $x$ with period $a > 0$ (see (C.27) and (C.30)). Since the second linearly independent solution of $H_2(t)\psi_2(t) = 0$ is of the type
$$xp_{2,1}(t,x) + p_{2,2}(t,x), \ p_{2,j}(t, x+a) = p_{2,j}(t,x), \ j = 1, 2, (t,x) \in \mathbb{R}^2, \qquad (5.22)$$
(see (C.32)) we infer that $H_2(t)$ is critical [Ge-Zh]. Thus there exists a unique, positive distributional solution $0 < \psi_{2,0} \in C^\infty(\mathbb{R}^2)$ of
$$\begin{aligned} H_2(t)\psi_{2,0}(t) = 0, \ 0 < \psi_{2,0}(t,.) \in L^\infty(\mathbb{R}), \\ \psi_{2,0}(t,x) = \psi_{2,\pm}(t, 0, x, x_0), \ (t,x) \in \mathbb{R}^2, \end{aligned} \qquad (5.23)$$
satisfying the time-evolution (5.6). According to (2.29) we then define
$$\phi_0(t,x) := \psi_{2,0,x}(t,x)/\psi_{2,0}(t,x), \ (t,x) \in \mathbb{R}^2. \qquad (5.24)$$
Next we turn to the case

$\mathcal{E}_2 > 0$ (i.e., $H_2(t)$ is subcritical, $t \in \mathbb{R}$):
In this case $H_2(t)$ is subcritical [Ge-Zh] and Floquet theory (cf. (C.31)) guarantees the existence of two linearly independent, positive distributional solutions $0 < \psi_{2,\pm} \in C^\infty(\mathbb{R}^2)$ of
$$\begin{aligned} H_2(t)\psi_{2,\pm}(t) = 0, \ 0 < \psi_{2,\pm}(t,.) \in L^\infty_{\mathrm{loc}}(\mathbb{R}), \\ \psi_{2,\pm}(t,x) = \lim_{\epsilon \downarrow 0} \psi_{2,\pm}(t, i\epsilon, x, x_0) \end{aligned} \qquad (5.25)$$
satisfying the time-evolution (5.6). In particular, one has (cf. (C.32))
$$\psi_{2,\pm}(t,x) \underset{x\to\pm\infty}{=} 0(\mathrm{e}^{\mp \kappa x}) \qquad (5.26)$$
for some $\kappa > 0$. In accordance with (2.29) we now define
$$\begin{aligned} \psi_{2,\sigma}(t,x) := (1/2)(1-\sigma)\psi_{2,-}(t,x) + (1/2)(1+\sigma)\psi_{2,+}(t,x), \\ \sigma \in [-1, 1], \ (t,x) \in \mathbb{R}^2, \end{aligned} \qquad (5.27)$$
(where $\sigma$ is $t$-independent) and
$$\phi_\sigma(t,x) := \psi_{2,\sigma,x}(t,x)/\psi_{2,\sigma}(t,x), \ \sigma \in [-1,1], \ (t,x) \in \mathbb{R}^2. \qquad (5.28)$$
By (C.32), only $\sigma = \pm 1$ yield spatially periodic functions $\phi_{\pm 1}$ in (5.28) and hence we restrict ourselves to these cases for the rest of this section. We have

**Theorem 5.3.** *[GSS] Assume $V_2$ satisfies (H.5.1) and $KdV(V_2) = 0$. Moreover, assume $\mathcal{E}_2 = \inf[\sigma(H_2(0))] \geq 0$. Then $\phi_\sigma$, $\sigma = 0, \pm 1$ (defined in (5.24) resp. (5.28)) and $V_{1,\sigma} = \phi_\sigma^2 - \phi_{\sigma,x}$, $\sigma = 0, \pm 1$ satisfy (H.5.1) and*

$$mKdV(\phi_\sigma) = 0, \quad KdV(V_{1,\sigma}) = 0, \quad \sigma = 0, \pm 1. \tag{5.29}$$

(Here the fact that $\psi_{2,0}, \psi_{2,\pm}$ satisfy the time-evolution (5.6) instead of (2.15) with $k = 0$ is irrelevant since by (5.6) a certain time-dependent multiple of $\psi_{2,0}, \psi_{2,\pm}$ (which drops out in the definitions (5.24), (5.28) for $\phi_0, \phi_{\pm 1}$) does satisfy (2.15) with $k = 0$.)

Moreover, formulas (C.9), (F.10), (G.4) and (G.7) yield

**Theorem 5.4.** *Assume $V_2$ satisfies (H.5.1) and $KdV(V_2) = 0$.*
(i) *If $\mathcal{E}_2 = 0$ then $\phi_0$, defined in (5.24), reads*

$$\phi_0(t, x) = (1/2)\partial_x \ln \left[ \prod_{n \in I} \mu_{2,n}(t, x)(a^2/n^2\pi^2) \right], \quad (t, x) \in \mathbb{R}^2. \tag{5.30}$$

*In the special case of finite genus $g < \infty$, (5.30) simplifies to*

$$\phi_0(t, x) = (1/2)\partial_x \ln \left[ \prod_{n \in I} \mu_{2,n}(t, x) \right], \quad (t, x) \in \mathbb{R}^2. \tag{5.31}$$

(ii) *If $\mathcal{E}_2 > 0$ then $\phi_{\pm 1}$, defined in (5.28), read*

$$\phi_{\pm 1}(t,x) = \left\{ \pm \tilde{R}(0)^{\frac{1}{2}} + (1/2)\partial_x \left[ \prod_{n \in I} \mu_{2,n}(t,x)(a^2/n^2\pi^2) \right] \right\}$$
$$\times \left\{ \prod_{n \in I} \mu_{2,n}(t,x)(a^2/n^2\pi^2) \right\}^{-1}, (t,x) \in \mathbb{R}^2. \tag{5.32}$$

*In the special case of finite genus $g < \infty$, (5.32) becomes*

$$\phi_{\pm 1}(t,x) = \left\{ \pm R_0(0)^{\frac{1}{2}} + (1/2)\partial_x \left[ \prod_{n \in I} \mu_{2,n}(t,x) \right] \right\}$$
$$\times \left\{ \prod_{n \in I} \mu_{2,n}(t,x) \right\}^{-1}, (t,x) \in \mathbb{R}^2. \tag{5.33}$$

*In (5.30)-(5.33), the $\mu_n$'s satisfy the first order systems (5.17) and (5.19).*

We conclude with the

**Example 5.5.** $g = 1$, $\mathcal{E}_2 = 0$.

$$V_2(t, x) = 2\mathcal{P}(x + 6\mathcal{P}(\omega)t + \omega') + \mathcal{P}(\omega), \quad (t, x) \in \mathbb{R}^2,$$
$$E_0 = 0, \quad E_1 = \mathcal{P}(\omega) - \mathcal{P}(\omega + \omega'), \quad E_2 = \mathcal{P}(\omega) - \mathcal{P}(\omega'). \tag{5.34}$$

Then
$$\phi_0(t,x) = \frac{1}{2}\frac{\mathcal{P}'(x+6\mathcal{P}(\omega)t+\omega')}{\mathcal{P}(x+6\mathcal{P}(\omega)t+\omega')-\mathcal{P}(\omega)}, \ (t,x) \in \mathbb{R}^2, \qquad (5.35)$$

where $\mathcal{P}(\cdot) := \mathcal{P}(\cdot,\omega,\omega')$ denotes the Weierstraß $\mathcal{P}$-function with halfperiods $\omega, \omega' (\omega > 0, -i\omega' > 0)$.

Formulas (5.31) and (5.33) may be expressed in terms of Riemann's theta function associated with the hyperelliptic Riemann surface $\mathcal{R}$ similar to the Its-Matveev formula in the KdV-case [It-Ma]. We shall report on that elsewhere. An approach to periodic solutions for the AKNS-system similar in spirit to Theorem 5.4 can be found in [De-Jo].

## Appendix A. Commutation formulas

We recall some of the abstract results based on commutation needed at various places in Sections 2-5.

Throughout this appendix let $\mathcal{H}$ be a separable, complex Hilbert space, $A$ a densely defined, closed linear operator in $\mathcal{H}$ and introduce

$$H_1 := A^*A, \ H_2 := AA^* \qquad (A.1)$$

in $\mathcal{H}$ and

$$Q := \begin{pmatrix} 0 & A^* \\ A & 0 \end{pmatrix} \text{ on } \mathcal{D}(A) \oplus \mathcal{D}(A^*) \qquad (A.2)$$

in $\mathcal{H} \oplus \mathcal{H}$. We state

**Theorem A.1.** *[Dei] (see also [GSS])*
 (i) $\sigma(H_1)\backslash\{0\} = \sigma(H_2)\backslash\{0\}$. (A.3)
 (ii) $H_1\psi_1 = E_1\psi_1, \ E_1 > 0, \ \psi_1 \in \mathcal{D}(H_1)$
 implies $(A\psi_1) \in \mathcal{D}(H_2), \ H_2(A\psi_1) = E_1(A\psi_1)$, (A.4)

 $H_2\psi_2 = E_2\psi_2, \ E_2 > 0, \psi_2 \in \mathcal{D}(H_2)$
 implies $(A^*\psi_2) \in \mathcal{D}(H_1), \ H_1(A^*\psi_2) = E_2(A^*\psi_2)$. (A.5)
 (iii) $A^*A|_{\text{Ker}(A)^\perp}$ is unitarily equivalent to $AA^*|_{\text{Ker}(A^*)^\perp}$.
 (iv) $Q^2 = \begin{pmatrix} H_1 & 0 \\ 0 & H_2 \end{pmatrix} = H_1 \oplus H_2$. (A.6)
 (v) $\sigma_3 Q \sigma_3 = -Q, \ \sigma_3 := \begin{pmatrix} 1 & 0 \\ 0 & -1 \end{pmatrix}$, (A.7)
 i.e., the spectrum $\sigma(Q)$ of $Q$ is symmetric with respect to zero on the real line.
 (vi) $H_1\psi_1 = E_1\psi_1, \ E_1 > 0, \ \psi_1 \in \mathcal{D}(H_1)$
 implies $\begin{pmatrix} \psi_1 \\ \pm E_1^{-1/2} A\psi_1 \end{pmatrix} \in \mathcal{D}(Q), \ Q\begin{pmatrix} \psi_1 \\ \pm E_1^{-1/2} A\psi_1 \end{pmatrix}$
 $= \pm E_1^{1/2} \begin{pmatrix} \psi_1 \\ \pm E_1^{-1/2} A\psi_1 \end{pmatrix}$ (A.8)
 and

$$H_2\psi_2 = E_2\psi_2,\ E_2 > 0,\ \psi_2 \in \mathcal{D}(H_2)$$
$$\text{implies } \begin{pmatrix} \pm E_2^{-1/2} A^*\psi_2 \\ \psi_2 \end{pmatrix} \in \mathcal{D}(Q),\ Q\begin{pmatrix} \pm E_2^{-1/2} A^*\psi_2 \\ \psi_2 \end{pmatrix}$$
$$= \pm E_2^{1/2} \begin{pmatrix} \pm E_2^{-1/2} A^*\psi_2 \\ \psi_2 \end{pmatrix} \tag{A.9}$$

(vii) $0 \in \sigma_p(Q)$ iff $0 \in \sigma_p(H_1)$ or $0 \in \sigma_p(H_2)$. (A.10)

(viii) $(Q - z)^{-1} = \begin{pmatrix} z(H_1 - z^2)^{-1} & A^*(H_2 - z^2)^{-1} \\ A(H_1 - z^2)^{-1} & z(H_2 - z^2)^{-1} \end{pmatrix}$,

$z^2 \in \mathbb{C}\setminus\{\sigma(H_1) \cup \sigma(H_2)\}$. (A.11)

## Appendix B. Spectral and scattering properties for one-dimensional Schrödinger and Dirac operators

We review basic spectral and scattering properties for one-dimensional Schrödinger operators needed in Sections 3 and 4.

We introduce hypothesis

**(H.B.1).**

(i) $V \in L^\infty(\mathbb{R})$ real-valued,
$\lim_{x \to \pm\infty} V(x) = V_\pm \in \mathbb{R},\ 0 < V_- \leq V_+$,
$\pm \int_0^{\pm\infty} dx(1 + |x|)|V(x) - V_\pm| < \infty$.

(ii) $\phi, \phi' \in L^\infty(\mathbb{R})$ real-valued,
$\lim_{x \to \pm\infty} \phi(x) = \phi_\pm \in \mathbb{R},\ 0 < \phi_-^2 \leq \phi_+^2$,
$\pm \int_0^{\pm\infty} dx(1 + |x|)[|\phi(x) - \phi_\pm| + |\phi'(x)|] < \infty$.

Assuming that $V, \phi$ satisfy (H.B.1) (i) resp. (H.B.1) (ii) we define in $L^2(\mathbb{R})$

$$H := -\frac{d^2}{dx^2} + V \text{ on } H^2(\mathbb{R}) \tag{B.1}$$

($H^m(\Omega),\ \Omega \subseteq \mathbb{R},\ m \in \mathbb{N}$ the standard Sobolev spaces) and in $L^2(\mathbb{R}) \otimes \mathbb{C}^2$

$$Q := \begin{pmatrix} 0 & -\frac{d}{dx} + \phi \\ \frac{d}{dx} + \phi & 0 \end{pmatrix} \text{ on } H^1(\mathbb{R}) \otimes \mathbb{C}^2. \tag{B.2}$$

Concerning spectral properties of $H$ and $Q$ we state

**Theorem B.2.** *[Co-Ka], [Da-Si], [Ges 2], [Ru-Bo]*

(i) *Assume (H.B.1) (i). Then*

$$\sigma_{ess}(H) = \sigma_{ac}(H) = [V_-, \infty),\ \sigma_{sc}(H) = \emptyset. \tag{B.3}$$

*Moreover, $H$ has simple spectrum in $(V_-, V_+)$ (iff $V_- < V_+$, if $V_- = V_+$ delete this assertion) and spectral multiplicity two in $(V_+, \infty)$. In addition, $H$ has finitely many simple eigenvalues in $(-\infty, V_-)$ and there are no eigenvalues embedded into the essential spectrum and no threshold eigenvalues, i.e.,*

$$\sigma_p(H) \cap [V_-, \infty) = \emptyset. \tag{B.4}$$

(ii) *Assume (H.B.1) (ii). Then*

$$\sigma_{\text{ess}}(Q) = \sigma_{ac}(Q) = (-\infty, -|\phi_-|] \cup [|\phi_-|, \infty), \ \sigma_{sc}(Q) = \emptyset. \tag{B.5}$$

*Moreover, $Q$ has simple spectrum in $(-|\phi_+|, -|\phi_-|) \cup (|\phi_-|, |\phi_+|)$ (iff $\phi_-^2 < \phi_+^2$, if $\phi_-^2 = \phi_+^2$ delete this assertion) and spectral multiplicity two in $(-\infty, -|\phi_+|) \cup (|\phi_+|, \infty)$. In addition, $Q$ has finitely many, simple eigenvalues in $(-|\phi_-|, |\phi_-|)$, symmetrically placed with respect to zero and there are no eigenvalues embedded into the essential spectrum and no threshold eigenvalues, i.e.,*

$$\sigma_p(Q) \cap \{(-\infty, -|\phi_-|] \cup [|\phi_-|, \infty)\} = \emptyset. \tag{B.6}$$

Finally, (A.6)-(A.11) hold for our concrete realization $Q$.
(Here $\sigma(.)$, $\sigma_{\text{ess}}(.), \sigma_{ac}(.), \sigma_{sc}(.), \sigma_p(.)$ and $\sigma_d(.)$ denote the spectrum, essential, absolutely continuous, singularly continuous, point and discrete spectrum respectively.)

Part (i) of this theorem follows e.g. from Jost function techniques (see e.g. [Ges 2]) with

$$f_\pm(z, x) = e^{\pm i k_\pm x} - \int_x^{\pm\infty} dx' k_\pm^{-1} \sin\left[k_\pm(x - x')\right] [V(x') - V_\pm] f_\pm(z, x'),$$

$$k_\pm(z) := (z - V_\pm)^{1/2}, \ \text{Im} k_\pm(z) \geq 0, \ z \in \mathbb{C}, x \in \mathbb{R}, \tag{B.7}$$

the Jost functions associated with $H$ and hence

$$H f_\pm(z) = z f_\pm, \ z \in \mathbb{C}, \tag{B.8}$$

in the sense of distributions. The eigenvalues $\lambda_\ell$ of $H$ in $(-\infty, V_-)$ are then determined by

$$W(f_-(\lambda_\ell), f_+(\lambda_\ell)) = 0. \tag{B.9}$$

Part (ii) is a consequence of part (i) above and of Theorem A.1 (iv)-(viii). Next we turn to scattering theory for $H$ and $Q$. We start with $H$.

**Theorem B.3.** *[Co-Ka], [Da-Si], [Ges 2], [Ru-Bo]* Assume (H.B.1) (i). Then the unitary scattering matrix $S_H(\lambda)$ in $\mathbb{C}^2$ for the triple $(H, H_0 + V_-, H_0 + V_+)$, $H_0 = -\dfrac{d^2}{dx^2}$ on $H^2(\mathbb{R})$ reads

(i) $\lambda > V_+$:

$$S_H(\lambda) = \begin{pmatrix} T_H(\lambda) & R_H^r(\lambda) \\ R_H^\ell(\lambda) & T_H(\lambda) \end{pmatrix}, \qquad (B.10)$$

where

$$\begin{aligned}
T_H(\lambda) &= 2i\,[k_+(\lambda)k_-(\lambda)]^{1/2}/W\,(f_-(\lambda),\ f_+(\lambda))\,, \\
R_H^\ell(\lambda) &= -W\,(g_-(\lambda), f_+(\lambda))/W\,(f_-(\lambda), f_+(\lambda))\,, \\
R_H^r(\lambda) &= -W\,(f_-(\lambda),\ g_+(\lambda))/W\,(f_-(\lambda), f_+(\lambda))\,, \\
k_\pm(\lambda) &= (\lambda - V_\pm)^{1/2} > 0,\ \lambda > V_+.
\end{aligned} \qquad (B.11)$$

(ii) $V_- < \lambda < V_+$ (if $V_- < V_+$):

$$S_H(\lambda) = -\overline{W(f_-(\lambda), f_+(\lambda))}/W(f_-(\lambda), f_+(\lambda)). \qquad (B.12)$$

Here $f_\pm(\lambda, x)$ are given by (B.7) and $g_\pm(\lambda, x)$ are defined by

$$g_\pm(z, x) = e^{\mp ik_\pm x} - \int_x^{\pm\infty} dx'\, k_\pm^{-1} \sin[k_\pm(x - x')]\,[V(x') - V_\pm]\, g_\pm(z, x'),$$

$$k_\pm(z) = (z - V_\pm)^{\frac{1}{2}},\ \mathrm{Im}\,k_\pm(z) \geq 0,\ z \in \mathbb{C},\ x \in \mathbb{R}. \qquad (B.13)$$

Next we turn to $Q$ and assume (H.B.1) (ii). Then the unitary scattering matrix $S_Q(E)$ in $\mathbb{C}^2$ for the triple $(Q, Q_{0,-}, Q_{0,+})$, $Q_{0,\pm} := \begin{pmatrix} 0 & -\dfrac{d}{dx} + \phi_\pm \\ \dfrac{d}{dx} + \phi_\pm & 0 \end{pmatrix}$ on $H^1(\mathbb{R}) \otimes \mathbb{C}^2$ reads

(i) $|E| > |\phi_+|$:

$$S_Q(E) = \begin{pmatrix} T_Q(E) & R_Q^r(E) \\ R_Q^\ell(E) & T_Q(E) \end{pmatrix}, \qquad (B.14)$$

where

$$\begin{aligned}
T_Q(E) &= \left\{2i[k_+(E^2)k_-(E^2)]^{\frac{1}{2}}/|E|\right\} W\,(\Psi_-(E), \Psi_+(E))^{-1}, \\
R_Q^\ell(E) &= -W\,(\Phi_-(E), \Psi_+(E))/W\,(\Psi_-(E), \Psi_+(E)), \\
R_Q^r(E) &= -W\,(\Psi_-(E), \Phi_+(E))/W\,(\Psi_-(E), \Psi_+(E)),\ |E| > |\phi_+|.
\end{aligned} \qquad (B.15)$$

(ii) $|\phi_-| < |E| < |\phi_+|$ (if $|\phi_-| < |\phi_+|$):

$$S_Q(E) = -\overline{W(\Psi_-(E), \Psi_+(E))}/W\Psi_-(E), \Psi_+(E)). \tag{B.16}$$

Here $W\left(\binom{a}{b}, \binom{c}{d}\right) := ad - bc$,

$$\Psi_\pm(E, x) := \begin{cases} \Psi_{1,\pm}(E, x), & E > |\phi_-|, \\ \Psi_{2,\pm}(-E, x), & E < -|\phi_-|, \end{cases}$$
$$\Phi_\pm(E, x) := \begin{cases} \Phi_{1,\pm}(E, x), & E > |\phi_-|, \\ \Phi_{2,\pm}(-E, x), & E < -|\phi_-|, \end{cases} \tag{B.17}$$

$$Q\Psi_\pm(E) = E\Psi_\pm(E), \ Q\Phi_\pm(E) = E\Phi_\pm(E), \ |E| > |\phi_-|, \tag{B.18}$$

$$\Psi_{1,\pm}(Z, x) := \begin{pmatrix} f_{1,\pm}(z, x) \\ Z^{-1}(Af_{1,\pm}(z))(x) \end{pmatrix},$$
$$\Psi_{2,\pm}(Z, x) := \begin{pmatrix} -Z^{-1}(A^* f_{2,\pm}(z))(x) \\ f_{2,\pm}(z, x) \end{pmatrix},$$
$$\Phi_{1,\pm}(Z, x) := \begin{pmatrix} g_{1,\pm}(z, x) \\ Z^{-1}(Ag_{1,\pm}(z))(x) \end{pmatrix}, \tag{B.19}$$
$$\Phi_{2,\pm}(Z, x) := \begin{pmatrix} -Z^{-1}(A^* g_{2,\pm}(z))(x) \\ g_{2,\pm}(z, x) \end{pmatrix},$$
$$Z^2 = z, \ z \in \mathbb{C}\setminus\{0\}, \ x \in \mathbb{R}.$$

$f_{j,\pm}(z, x), \ g_{j,\pm}(z, x)$ are the Jost functions associated with

$$H_j := -\frac{d^2}{dx^2} + V_j \text{ on } H^2(\mathbb{R}), \ j = 1, 2,$$
$$V_j(x) := \phi(x)^2 + (-1)^j \phi'(x), \ j = 1, 2, \ x \in \mathbb{R}, \tag{B.20}$$

and

$$A := \frac{d}{dx} + \phi \text{ on } H^1(\mathbb{R}) \tag{B.21}$$

such that

$$H_1 = A^*A, \ H_2 = AA^*, \ Q = \begin{pmatrix} 0 & A^* \\ A & 0 \end{pmatrix}. \tag{B.22}$$

Since $\phi$ is assumed to satisfy (H.B.1) (ii) we obtain

$$V_{j,\pm} := \lim_{x \to \pm\infty} V_j(x) = \phi_\pm^2. \tag{B.23}$$

This in turn yields (cf. also (A.4), (A.5))

$$f_{2,\pm}(z,x) = (\pm ik_\pm(z) + \phi_\pm)^{-1}(Af_{1,\pm}(z))(x),$$
$$f_{1,\pm}(z,x) = (\mp ik_\pm(z) + \phi_\pm)^{-1}(A^*f_{2,\pm}(z))(x), \quad \text{(B.24)}$$
$$z \in \mathbb{C}, \ x \in \mathbb{R}.$$

Moreover, applying the elementary identity

$$W(Af(z), Ag(z)) = zW(f(z), g(z)), \ z \in \mathbb{C}, \quad \text{(B.25)}$$

where $f(z,x)$, $g(z,x)$ are any distributional solutions of $H_1\psi_1(z) = z\psi_1(z)$, we infer

**Theorem B.4.** *[GSS] Assume $\phi$ satisfies (H.B.1) (ii) and denote by $S_j(\lambda) := S_{H_j}(\lambda)$ the scattering matrix associated with $(H_j, H_0 + \phi_-^2, H_0 + \phi_+^2)$, $j = 1, 2$. Then*

(i) $\lambda > \phi_+^2$:

$$T_1(\lambda) = [ik_-(\lambda) + \phi_-][ik_+(\lambda) + \phi_+]^{-1}T_2(\lambda),$$
$$R_1^\ell(\lambda) = [ik_-(\lambda) + \phi_-][-ik_-(\lambda) + \phi_-]^{-1}R_2^\ell(\lambda), \quad \text{(B.26)}$$
$$R_1^r(\lambda) = [-ik_+(\lambda) + \phi_+][ik_+(\lambda) + \phi_+]^{-1}R_2^r(\lambda).$$

(ii) $\phi_-^2 < \lambda < \phi_+^2$ (*if* $\phi_-^2 < \phi_+^2$):

$$S_1(\lambda) = [ik_-(\lambda) + \phi_-][-ik_-(\lambda) + \phi_-]^{-1}S_2(\lambda). \quad \text{(B.27)}$$

*In addition, let* $\lambda_\ell := \phi_\pm^2 - \kappa_{\pm,\ell}^2, \kappa_{\pm,\ell} > 0$, *denote the nonzero eigenvalues of $H_j$ and let*

$$c_{j,\pm,\ell} := \|f_{j,\pm}(\lambda_\ell,\cdot)\|_2^{-2}, \ j = 1,2, \quad \text{(B.28)}$$

*be the associated norming constants. Then*

$$c_{1,\pm,\ell} = \left[(\phi_\pm \pm \kappa_{\pm,\ell})(\phi_\pm \mp \kappa_{\pm,\ell})^{-1}\right]^{1/2} c_{2,\pm,\ell}. \quad \text{(B.29)}$$

Finally, taking into account the simple identities

$$W(\Psi_-(E), \Psi_+(E)) = W(f_{1,-}(E^2), f_{1,+}(E^2))\,|E|^{-1},$$
$$W(\Psi_-(E), \Phi_+(E)) = W(f_{1,-}(E^2), g_{1,+}(E^2))\,|E|^{-1},$$
$$W(\Phi_-(E), \Psi_+(E)) = W(g_{1,-}(E^2), f_{1,+}(E^2))\,|E|^{-1}, \quad \text{(B.30)}$$
$$W(\Phi_-(E), \Phi_+(E)) = W(g_{1,-}(E^2), g_{1,+}(E^2))\,|E|^{-1},$$
$$E > |\phi_-|,$$

and similarly for $E < -|\phi_-|$, we obtain

**Theorem B.5.** *[GSS] Assume (H.B.1) (ii) and let $S_Q(E)$ be the scattering matrix introduced in (B.14)-(B.16). Then*

$$S_Q(E) = \begin{cases} S_1(E^2), & E > |\phi_-|, \\ S_2(E^2), & E < -|\phi_-|. \end{cases} \quad (B.31)$$

**Remark B.6.** (B.26 and (B.31) prove, in particular, that if one of $V_1, V_2, \phi$ is a reflectionless potential (i.e., if one of $R_1^{\ell(r)}(\lambda)$, $R_2^{\ell(r)}(\lambda)$, $R_Q^{\ell(r)}(E)$ vanishes identically) then all three are reflectionless.

## Appendix C. Floquet theory

We review some of the basic results in the theory of periodic, one-dimensional Schrödinger and Dirac operators. General references for this material are e.g. [Eas], [Lev], [Le-Sa], [Ma-Wi], [Mar], [Mc-Tr 1,2], [NMPZ], [RS III].
Starting with Schrödinger operators we introduce

**(H.C.1).** Suppose $V \in C^\circ(\mathbb{R})$ is real-valued and periodic with period $a > 0$ and define in $L^2(\mathbb{R})$

$$H := -\frac{d^2}{dx^2} + V \text{ on } H^2(\mathbb{R}). \quad (C.1)$$

A fundamental system of distributional solutions for $H\psi = z\psi$, $z \in \mathbb{C}$ is given by solutions of the Volterra integral equations.

$$c(z, x, x_0) = \cosh\left[\sqrt{-z}(x - x_0)\right]$$
$$+ \int_{x_0}^{x} dx' (\sqrt{-z})^{-1} \sinh\left[\sqrt{-z}(x - x')\right] V(x') c(z, x', x_0),$$
$$s(z, x, x_0) = (\sqrt{-z})^{-1} \sinh\left[\sqrt{-z}(x - x_0)\right] \quad (C.2)$$
$$+ \int_{x_0}^{x} dx' (\sqrt{-z})^{-1} \sinh\left[\sqrt{-z}(x - x')\right] V(x') s(z, x' x_0),$$
$$z \in \mathbb{C}, \ x \in \mathbb{R}.$$

The corresponding fundamental matrix $\Phi(z, x, x_0)$ reads $(\prime \equiv \frac{d}{dx})$

$$\Phi(z, x, x_0) := \begin{pmatrix} c(z, x, x_0) & s(z, x, x_0) \\ c'(z, x, x_0) & s'(z, x, x_0) \end{pmatrix} \in SL(2, \mathbb{C}), \ z \in \mathbb{C}, \quad (C.3)$$

$$\Phi(z, x_0, x_0) = 1, \ z \in \mathbb{C},$$
$$\Phi(z, x, x_0) = \Phi(z, x, x_1)\Phi(z, x_1, x_0), \ z \in \mathbb{C}; \ x, x_0, x_1 \in \mathbb{R}, \quad (C.4)$$

and the discriminant (Floquet determinant) of $H$ is defined by

$$\Delta(z) := \frac{1}{2} Tr \left[ \Phi(z, x_0 + a, x_0) \right] \tag{C.5}$$
$$= [c(z, x_0 + a, x_0) + s'(z, x_0 + a, x_0)]/2, \ z \in \mathbb{C}.$$

We note
$$\frac{d}{dx_0} \Delta(z) = 0, \ z \in \mathbb{C}. \tag{C.6}$$

One has the product representation

$$\Delta(z)^2 - 1 = a^2 (E_0 - z) \prod_{n \in \mathbb{N}} \left[ (E_{2n-1} - z)(E_{2n} - z) a^4 / n^4 \pi^4 \right], z \in \mathbb{C}, \tag{C.7}$$
$$-\infty < E_0 < E_1 \leq E_2 < E_3 \leq E_4 < \ldots,$$

where $E_0, E_{4n-1}, E_{4n}, n \in \mathbb{N}$ are the zeroes of $\Delta(z) - 1$ and $E_{4n+1}, E_{4n+2}, n \in \mathbb{N}$, are the zeroes of $\Delta(z) + 1$ counting multiplicities.
The spectrum $\sigma(H)$ of $H$ is then characterized by

**Theorem C.2.** *Assume (H.C.1). Then the spectrum of $H$ is purely absolutely continuous of multiplicity two and has a band structure of the type*

$$\sigma(H) = \sigma_{ac}(H) = \{\lambda \in \mathbb{R} | \Delta(\lambda)| \leq 1\} = \bigcup_{n \in \mathbb{N}} \sigma_n,$$
$$\sigma_n := \left[ E_{2(n-1)}, E_{2n-1} \right], \ n \in \mathbb{N}, \tag{C.8}$$
$$\sigma_p(H) = \sigma_{sc}(H) = \emptyset.$$

Next we introduce the Riemann surface $\mathcal{R}$ associated with

$$R(z)^{1/2} := \left[ \Delta(z)^2 - 1 \right]^{1/2} \text{ resp.}$$
$$\tilde{R}(z)^{1/2} := \left[ (E_0 - z) \prod_{n \in I} (E_{2n-1} - z)(E_{2n} - z)(a^4/n^4\pi^4) \right]^{\frac{1}{2}}. \tag{C.9}$$

Denoting

$$\rho_0 = (-\infty, E_0), \ \rho_n := \begin{cases} (E_{2n-1}, E_{2n}), & E_{2n-1} < E_{2n} \\ \emptyset, & E_{2n-1} = E_{2n} \end{cases}, n \in \mathbb{N}, \tag{C.10}$$

we collect the spectral gaps of $H$ in

$$\rho_H := \mathbb{R} \backslash \sigma(H) = \bigcup_{n \in I_0} \rho_n, \tag{C.11}$$

where $I_0 := I \cup \{0\}$, $I \subseteq \mathbb{N}$ indexes the open spectral gaps of $H$, i.e., those $\rho_n$'s such that $\rho_n \neq \emptyset$ iff $n \in I_0$. Then $\mathcal{R}$ is realized as a branched Riemann surface in the canonical manner: Form two copies $\Pi_\pm$ of the cut plane

$$\Pi := \mathbb{C} \backslash \overline{\rho_H} \tag{C.12}$$

and paste them together such that the upper rims of the upper sheet $\Pi_+$ are connected with the lower rims of the lower sheet $\Pi_-$ and vice versa. The genus $g$ of $\mathcal{R}$ is then determined by

$$g = \text{card}(I) \qquad (0 \le g \le \infty). \tag{C.13}$$

If $g = \infty$, $\mathcal{R}$ is punctured at $\infty$ and represents a noncompact Riemann surface. If $g < \infty$ one adds the point $\{\infty\}$ to $\mathcal{R}$ and obtains a compact (hyperelliptic) Riemann surface. Let

$$\mathcal{B} := \{E_0, E_{2n-1}, E_{2n}, \ n \in I\} \tag{C.14}$$

then the set of branch points of $\mathcal{R}$ is given by $\mathcal{B}$ if $g = \infty$ and by $\mathcal{B} \cup \{\infty\}$ if $g < \infty$. Any point $P \ne \infty$ of $\mathcal{R}$ is represented by

$$P = \left(z, R(z)^{1/2}\right) \quad , z \in \mathbb{C}, \tag{C.15}$$

where $z$ denotes the projection of $P$ onto $\mathbb{C}$ and the value of $R(z)^{1/2}$ (for $P \in \mathcal{R}\setminus\mathcal{B}$) indicates on which sheet $P$ is lying on. The upper sheet $\Pi_+$ is characterized by the fact that we define $R(z)^{1/2}$ and $\tilde{R}(z)^{1/2}$ to be negative (resp. positive) on the upper rim (resp. lower rim) of the cut $\rho_0 = (-\infty, E_0)$ and analytically continue on $\Pi_+$ (the signs of $R(z)^{1/2}$, $\tilde{R}(z)^{1/2}$ on $\Pi_-$ are then just reversed).

The monodromy matrix $M(z, x_0)$ is defined by

$$M(z, x_0) := \Phi(z, x_0 + a, x_0) \in SL(2, \mathbb{C}), \ z \in \mathbb{C}, \tag{C.16}$$

and hence

$$\det[M(z, x_0)] = 1, \quad Tr[M(z, x_0)] = 2\Delta(z), \ z \in \mathbb{C}, \tag{C.17}$$

and

$$\Phi(z, x + a, x_0) = M(z, x)\Phi(z, x, x_0), z \in \mathbb{C}, \ x, x_0 \in \mathbb{R}, \tag{C.18}$$

by (C.3)-(C.5). The so called Floquet multipliers $m_\pm(z)$, the eigenvalues of $M(z, x_0)$, are then solutions of

$$m^2 - 2\Delta(z)m + 1 = 0, \ z \in \mathbb{C}. \tag{C.19}$$

This yields an analytic map

$$m : \mathcal{R} \to \mathbb{C}_\infty, \quad m(P) = \Delta(z) + R(z)^{1/2}, \ P = (z, R(z)^{1/2}) \tag{C.20}$$

($\mathbb{C}_\infty := \mathbb{C} \cup \{\infty\}$, the one point compactification of $\mathbb{C}$, homeomorphic to $S^2$) and hence $m_\pm(z)$ are given by

$$m_\pm(z) = \Delta(z) \pm [\Delta(z)^2 - 1]^{1/2}, \ z \in \Pi_+. \tag{C.21}$$

We have
$$\frac{d}{dx_0} m_\pm(z) = 0, \ z \in \Pi_+, \tag{C.22}$$

$$m_+(z)m_-(z) = 1, \ m_+(z) + m_-(z) = 2\Delta(z), \ z \in \Pi_+ \tag{C.23}$$

and

$$M(z, x_0) \begin{pmatrix} 1 \\ \phi_\pm(z, x_0) \end{pmatrix} = m_\pm(z) \begin{pmatrix} 1 \\ \phi_\pm(z, x_0) \end{pmatrix}, \ z \in \hat{\Pi}_+(x_0),$$
$$\phi_\pm(z, x_0) := [m_\pm(z) - c(z, x_0 + a, x_0)] s(z, x_0 + a, x_0)^{-1}, \tag{C.24}$$
$$z \in \hat{\Pi}_+(x_0) := \Pi_+ \backslash \{\mu_n(x_0)\}_{n \in \mathbb{N}},$$

where $\mu_n(x_0)$ are the zeroes of $s(z, x_0 + a, x_0)$, i.e.,

$$s(\mu_n(x_0), \ x_0 + a, x_0) = 0, \ n \in \mathbb{N}. \tag{C.25}$$

The Floquet solutions (resp. Baker-Akhiezer functions) of $H$ are then defined by

$$\psi_\pm(z, x, x_0) := c(z, x, x_0) + \phi_\pm(z, x_0) s(z, x, x_0), \ z \in \hat{\Pi}_+(x_0), \ x, x_0 \in \mathbb{R}. \tag{C.26}$$

They satisfy

$$\psi_\pm(z, x + a, x_0) = m_\pm(z)\psi_\pm(z, x, x_0), \ z \in \hat{\Pi}_+(x_0), \ x, x_0 \in \mathbb{R}, \tag{C.27}$$

by (C.23). Moreover, we have

$$\psi_\pm(z, x_0, x_0) = 1, \ z \in \mathbb{C}, \ x_0 \in \mathbb{R}, \tag{C.28}$$

$$\psi_+(\lambda, x, x_0) = \overline{\psi_-(\lambda, x, x_0)}, \ \lambda \in \sigma(H), \ x, x_0 \in \mathbb{R}, \tag{C.29}$$

$$W(\psi_-(z), \psi_+(z)) = [m_+(z) - m_-(z)] s(z, x_0 + a, x_0)^{-1}$$
$$= 2R(z)^{1/2} s(z, x_0 + a, x_0)^{-1}, \ z \in \hat{\Pi}_+(x_0), \tag{C.30}$$

and

$$H\psi_\pm(z, x, x_0) = z\psi_\pm(z, x, x_0), \ z \in \hat{\Pi}_+(x_0), \ x, x_0 \in \mathbb{R}, \tag{C.31}$$

in the sense of distributions. In particular, (C.30) shows that $\psi_-(z, x, x_0)$ and $\psi_+(z, x, x_0)$ are linearly dependent precisely at the band edges, i.e., at the branch points in $\mathcal{B}$. Moreover, one can show that $H\psi = \lambda\psi, \lambda \in \mathbb{R}$ has two linearly independent solutions of the type

$$\Delta(\lambda) > 1 \; (m_\pm(\lambda) = e^{\pm\kappa a}, \quad \kappa \in \mathbb{R}\backslash\{0\}): \; e^{\kappa x}p_1(x), \; e^{-\kappa x}p_2(x),$$
$$\Delta(\lambda) < -1 \; (m_\pm(\lambda) = -e^{\pm\kappa a}, \quad \kappa \in \mathbb{R}\backslash\{0\}): \; e^{\kappa x}p_1(x), \; e^{-\kappa x}p_2(x),$$
$$\Delta(\lambda) = 1 \; (m_\pm(\lambda) = 1):$$
$$\quad p_1(x), \; p_2(x) \text{ if } c'(\lambda, x_0 + a, x_0) = 0 = s(\lambda, x_0 + a, x_0),$$
$$\quad p_1(x), xp_1(x) + p_2(x) \text{ if } c'(\lambda, x_0 + a, x_0) \neq 0 \text{ or } s(\lambda, x_0 + a, x_0) \neq 0,$$
$$\Delta(\lambda) = -1 \; (m_\pm(\lambda) = -1):$$
$$\quad e^{i\pi x/a}p_1(x), \; e^{i\pi x/a}p_2(x) \text{ if } c'(\lambda, x_0 + a, x_0) = 0 = s(\lambda, x_0 + a, x_0),$$
$$\quad e^{i\pi x/a}p_1(x), \; e^{i\pi x/a}[xp_1(x) + p_2(x)] \text{ if } c'(\lambda, x_0 + a, x_0) \neq 0$$
$$\quad \text{or } s(\lambda, x_0 + a, x_0) \neq 0,$$
(C.32)

where $p_j(x+a) = p_j(x)$, $x \in \mathbb{R}$, $j = 1, 2$.

Finally we briefly discuss periodic Dirac operators. We introduce

**(H.C.3).** Suppose $\phi, \phi' \in C^0(\mathbb{R})$ are real-valued and periodic with period $a > 0$. Assuming (H.C.3) we define in $L^2(\mathbb{R})$

$$A := \frac{d}{dx} + \phi \text{ on } H^1(\mathbb{R}),$$
$$H_1 := A^*A = -\frac{d^2}{dx^2} + V_1, \; H_2 := AA^* = -\frac{d^2}{dx^2} + V_2, \quad \text{(C.33)}$$
$$V_j(x) = \phi(x)^2 + (-1)^j \phi'(x), \; j = 1, 2, \; x \in \mathbb{R},$$

and

$$Q := \begin{pmatrix} 0 & A^* \\ A & 0 \end{pmatrix} \text{ on } H^1(\mathbb{R}) \otimes \mathbb{C}^2 \quad \text{(C.34)}$$

in $L^2(\mathbb{R}) \otimes \mathbb{C}^2$. Denoting by $c_j(z, x, x_0)$, $s_j(z, x, x_0)$ the fundamental system (C.2) for $H_j$, $j = 1, 2$, a fundamental matrix $\Phi_Q(E, x, x_0)$ for $Q$ is defined by

$$\Phi_Q(E, x, x_0) = \begin{pmatrix} c_1(\lambda, x, x_0) - \phi(x_0)s_1(\lambda, x, x_0) & Es_1(\lambda, x, x_0) \\ E^{-1}[A(c_1 - \phi(x_0)s_1](\lambda, x, x_0) & (As_1)(\lambda, x, x_0) \end{pmatrix}, E > 0,$$

$$\Phi_Q(E, x, x_0) = \begin{pmatrix} -(A^*s_2)(\lambda, x, x_0) & -E^{-1}[A^*(c_2 + \phi(x_0)s_2)](\lambda, x, x_0) \\ Es_2(\lambda, x, x_0) & c_2(\lambda, x, x_0) + \phi(x_0)s_2(\lambda, x, x_0) \end{pmatrix},$$
$$E < 0,$$

$$\Phi_Q(E, x_0, x_0) = 1, \; E \in \mathbb{R}; \; \lambda = E^2, \; x, x_0 \in \mathbb{R}.$$
(C.35)

The discriminant $\Delta_Q$ of $Q$ is then defined by

$$\Delta_Q(E) = 2^{-1} Tr[\Phi_Q(E, x_0 + a, x_0)], \; E \in \mathbb{R}\backslash\{0\}, \quad \text{(C.36)}$$

and both, (C.35) and (C.36) extend to all $Z \in \mathbb{C}$ by analytic continuation. Denoting the discriminant and fundamental matrix of $H_j$ by $\triangle_j(z)$ and $\Phi_j(z,x,x_0)$, $j=1,2$ we infer that

$$c_2(z,x,x_0),$$
$$s_2(z,x,x_0),$$
$$c_1(z,x,x_0) = A^*\{z^{-1}\phi(x_0)c_2(z,x,x_0) - [1 - z^{-1}\phi(x_0)^2]s_2(z,x,x_0)\}, \quad \text{(C.37)}$$
$$s_1(z,x,x_0) = A^*\{z^{-1}c_2(z,x,x_0) + z^{-1}\phi(x_0)s_2(z,x,x_0)\}; \quad z \in \mathbb{C}\setminus\{0\},$$

are satisfying

$$H_j c_j(z) = z c_j(z), \; H_j s_j(z) = z s_j(z), \; j=1,2, \; z \in \mathbb{C}\setminus\{0\}, \qquad \text{(C.38)}$$

and

$$\Phi_j(z,x_0,x_0) = 1, \; z \in \mathbb{C}\setminus\{0\}. \qquad \text{(C.39)}$$

Thus (C.5) and (C.36) together with (A.3) and (C.8) (or Theorem A.1 (iii)) imply

**Theorem C.4.** *[GSS] Assume (H.C.3). Then $H_1$ and $H_2$ are isospectral*

$$\sigma(H_1) = \sigma(H_2) \qquad \text{(C.40)}$$

*and Theorem C.2 applies. Moreover,*

$$\triangle_Q(Z) = \triangle_1(z) = \triangle_2(z), \; Z^2 = z, \; z \in \mathbb{C}. \qquad \text{(C.41)}$$

Finally, Theorem A.1 and C.2 combined yield

**Theorem C.5.** *Assume (H.C.3). Then the spectrum $\sigma(Q)$ of $Q$ is purely absolutely continuous, symmetric with respect to zero and of multiplicity two*

$$\sigma(Q) = \sigma_{ac}(Q) = \{E \in \mathbb{R} | \; |\triangle_Q(E)| \le 1\} = \bigcup_{n \in \mathbb{Z}\setminus\{0\}} \sum_n,$$
$$\sum_n := \left[E_{2(n-1)}^{1/2}, E_{2n-1}^{1/2}\right], \; \sum_{-n} = -\sum_n, \; n \in \mathbb{N}, \qquad \text{(C.42)}$$
$$\sigma_p(Q) = \sigma_{sc}(Q) = \emptyset$$

with $E_n$, $n \in \mathbb{N}_0, E_0 \ge 0$ the zeroes of $\triangle(z) \pm 1$, where $\triangle(z) := \triangle_j(z)$, $z \in \mathbb{C}$ (cf. (C.41) and (C.7)).

## Appendix D. (Anti) periodic, Dirichlet, and Neumann spectra, Borg's theorem, FIT-formula

Here we first consider the operator $-\dfrac{d^2}{dx^2} + V$ on the periodicity interval $(x_0, x_0 + a)$ with various boundary conditions at $x_0, x_0 + a$ and then discuss

Borg's theorem and the FIT-formula. In addition to the general references quoted at the beginning of Appendix C we refer to [Po-Tr].

Throughout this appendix we shall assume

**(H.D.1).** $V \in C_a^\infty(\mathbb{R})$ real-valued.
(Here $C_a^\infty(\mathbb{R})$ denotes the $C^\infty(\mathbb{R})$-functions of period $a > 0$.)
Introducing $H^p$ in $L^2((x_0, x_0 + a))$ with periodic boundary conditions by

$$H^p := -\frac{d^2}{dx^2} + V,$$
$$\mathcal{D}(H^p) := \{g \in H^2((x_0, x_0 + a)) \mid g((x_0)_+) \qquad (D.1)$$
$$= g((x_0 + a)_-), \ g'((x_0)_+) = g'((x_0 + a)_-)\}$$

one obtains

$$\sigma(H^p) = \{E_0, \ E_{4n-1}, E_{4n}\}_{n \in \mathbb{N}}. \qquad (D.2)$$

Similarly, $H^{AP}$ in $L^2((x_0, x_0+a))$ with antiperiodic boundary conditions defined by

$$H^{AP} := -\frac{d^2}{dx^2} + V,$$
$$\mathcal{D}(H^{AP}) := \{g \in H^2((x_0, x_0 + a)) \mid g((x_0)_+) \qquad (D.3)$$
$$= -g((x_0 + a)_-), \ g'((x_0)_+) = -g'((x_0 + a)_-)\}$$

has spectrum

$$\sigma(H^{AP}) = \{E_{4n+1}, E_{4n+2}\}_{n \in \mathbb{N}}, \qquad (D.4)$$

where $E_n, n \in \mathbb{N}_o$ are given by (C.7). The corresponding Dirichlet and Neumann operators $H^D$ and $H^N$ in $L^2((x_0, x_0 + a))$ are then given by

$$H^D := -\frac{d^2}{dx^2} + V, \qquad (D.5)$$
$$\mathcal{D}(H^D) := \{g \in H^2((x_0, x_0 + a)) \mid g((x_0)_+) = 0 = g((x_0 + a)_-)\}$$

with simple spectrum

$$\sigma(H^D) = \{\mu_n(x_0)\}_{n \in \mathbb{N}}, \ \mu_n(x_0) \in [E_{2n-1}, E_{2n}], \ n \in \mathbb{N}, \qquad (D.6)$$

and

$$H^N := -\frac{d^2}{dx^2} + V, \qquad (D.7)$$
$$\mathcal{D}(H^N) := \{g \in H^2((x_0, x_0 + a)) \mid g'((x_0)_+) = 0 = g'((x_0 + a)_-)\}$$

with simple spectrum

$$\sigma(H^N) = \{\nu_n(x_0)\}_{n \in \mathbb{N}_o}, \ \nu_0(x_0) \leq E_0, \ \nu_n(x_0) \in [E_{2n-1}, E_{2n}], n \in \mathbb{N}. \qquad (D.8)$$

Denoting the zeroes of $\frac{d}{dz}\Delta(z)$ by

$$\{\lambda_n\}_{n\in\mathbb{N}} \tag{D.9}$$

one obtains

$$-\infty < \nu_0(x_0) \le E_0 < E_1 \le \mu_1(x_0), \nu_1(x_0), \lambda_1 \le E_2 < E_3 \le \mu_2(x_0), \nu_2(x_0),$$
$$\lambda_2 \le E_4 < E_5 \le \dots, \tag{D.10}$$

$$\sqrt{E_{\binom{2n}{2n-1}}} \underset{n\to\infty}{=} \frac{n\pi}{a} + \frac{1}{2n\pi}\int_{x_0}^{x_0+a} dx\, V(x) + 0(n^{-3}), \tag{D.11}$$

$$\sqrt{\mu_n(x_0), \nu_n(x_0), \lambda_n} \underset{n\to\infty}{=} \frac{n\pi}{a} + \frac{1}{2n\pi}\int_{x_0}^{x_0+a} dx\, V(x) + 0(n^{-3}) \tag{D.12}$$

and for the asymptotic widths of gaps [Hoc 1]

$$|E_{2n} - E_{2n-1}| \underset{n\to\infty}{=} 0(n^{-p}) \text{ for all } p > 0. \tag{D.13}$$

We also note the formulas

$$s(z, x_0 + a, x_0) = a \prod_{n\in\mathbb{N}} [\mu_n(x_0) - z](a^2/n^2\pi^2),\ z \in \mathbb{C}, \tag{D.14}$$

$$c'(z, x_0 + a, x_0) = a[\nu_0(x_0) - z] \prod_{n\in\mathbb{N}} [\nu_n(x_0) - z](a^2/n^2\pi^2),\ z \in \mathbb{C}, \tag{D.15}$$

$$\frac{d}{dz}\Delta(z) = -(a^2/2) \prod_{n\in\mathbb{N}} [\lambda_n - z](a^2/n^2\pi^2),\ z \in \mathbb{C}, \tag{D.16}$$

$$\frac{d}{dz}s(z, x_0 + a, x_0) =$$
$$- a\sum_{m=1}^{\infty}(a^2/m^2\pi^2) \prod_{\substack{n\in\mathbb{N}\\n\neq m}} [\mu_n(x_0) - z](a^2/n^2\pi^2), z \in \mathbb{C}, \tag{D.17}$$

$$\frac{d}{dz}s(\mu_n(x_0), x_0 + a, x_0) =$$
$$- (a^3/n^2\pi^2) \prod_{\substack{m\in\mathbb{N}\\m\neq n}} [\mu_m(x_0) - \mu_n(x_0)](a^2/m^2\pi^2),\ n \in \mathbb{N}. \tag{D.18}$$

The normalized Dirichlet $g_n(x,x_0)$ and Neumann $h_n(x,x_0)$ eigenfunctions of $H^D$ and $H^N$ are then given by

$$g_n(x,x_0) = c_n^D(x_0)s(\mu_n(x_0),x,x_0), \; n \in \mathbb{N}, \; x \in \mathbb{R}, \qquad (D.19)$$

$$h_n(x,x_0) = c_n^N(x_0)c(\nu_n(x_0),x,x_0), \; n \in \mathbb{N}_0, \; x \in \mathbb{R}, \qquad (D.20)$$

with the norming constants

$$[c_n^D(x_0)]^{-2} := \int_{x_0}^{x_0+a} dx\, s(\mu_n(x_0),x,x_0)^2$$
$$= s'(\mu_n(x_0),x_0+a,x_0)\frac{d}{dz}s(\mu_n(x_0),x_0+a,x_0), \; n \in \mathbb{N}. \qquad (D.21)$$

$$[c_n^N(x_0)]^{-2} := \int_{x_0}^{x_0+a} dx\, c(\nu_n(x_0),x,x_0)^2$$
$$= -c(\nu_n(x_0),x_0+a,x_0)\frac{d}{dz}c'(\nu_n(x_0),x_0+a,x_0), \; n \in \mathbb{N}_0. \qquad (D.22)$$

**Remark D.2.** In addition to (C.7) one also has the formulas

$$\Delta(z) - 1 = a^2(E_0 - z) \prod_{n \in \mathbb{N}} (E_{4n-1} - z)(E_{4n} - z)[a^4/(2n)^4\pi^4], \; z \in \mathbb{C}, \qquad (D.23)$$

$$\Delta(z) + 1 = \prod_{n \in \mathbb{N}_0} (E_{4n+1} - z)(E_{4n} - z)[a^4/(2n+1)^4\pi^4], \; z \in \mathbb{C}, \qquad (D.24)$$

which together with

$$\Delta(\lambda) \underset{\lambda \downarrow -\infty}{=} \cosh(|\lambda|^{1/2}a) \qquad (D.25)$$

prove that $\Delta(z)$ is determined by $\sigma(H^p)$ or by $\sigma(H^{AP})$ alone.

At a Dirichlet eigenvalue we have

$$s(\mu_n(x_0),x_0+a,x_0) = 0, \; n \in \mathbb{N}, \qquad (D.26)$$

and hence

$$c(\mu_n(x_0),x_0+a,x_0) = s'(\mu_n(x_0),x_0+a,x_0)^{-1}, \; n \in \mathbb{N}, \qquad (D.27)$$

since $W(c(z),s(z)) = 1$.

Given a fixed potential $V_0$ satisfying (H.D.1), we define the isospectral manifold $I(V_0)$ of $V_0$ by

$$I(V_0) := \{V \in C_a^\infty(\mathbb{R}) | E_n(V) = E_n(V_0), \; n \in \mathbb{N}_0\}. \qquad (D.28)$$

Borg's theorem then reads

**Theorem D.3.** *[Bor] The map*

$$I(V_0) \ni V \longrightarrow \{\mu_n(x_0),\ c_n^D(x_0)\}_{n\in\mathbb{N}} \tag{D.29}$$

*is injective, or equivalently, by (D.14), (D.21), the map*

$$I(V_0) \ni V \longrightarrow \{\mu_n(x_0),\ s'(\mu_n(x_0), x_0 + a, x_0)\}_{n\in\mathbb{N}} \tag{D.30}$$

*is injective.*

**Remark D.4.** Actually, $s'(\mu_n(x_0), x_0 + a, x_0)$ is determined from $\triangle(\mu_n(x_0))$ up to a sign ambiguity since (C.5) and (D.27) yield

$$2\triangle(\mu_n(x_0)) = s'(\mu_n(x_0), x_0 + a, x_0)^{-1} + s'(\mu_n(x_0), x_0 + a, x_0),\ n \in \mathbb{N}, \tag{D.31}$$

and thus

$$s'(\mu_n(x_0), x_0 + a, x_0) = \triangle(\mu_n(x_0)) \pm |\triangle(\mu_n(x_0))^2 - 1|^{1/2},\ n \in \mathbb{N}. \tag{D.32}$$

It is precisely this sign ambiguity that is fixed by the norming constants $c_n^D(x_0)$ since

$$s'(\mu_n(x_0), x_0 + a, x_0) =$$

$$\left[\frac{d}{dz} s(\mu_n(x_0), x_0 + a, x_0)\right]^{-1} \left[c_n^D(x_0)\right]^{-2},\ n \in \mathbb{N} \tag{D.33}$$

(cf. (D.21)). (However, this sign ambiguity is a genuine one only if $\mu_n(x_0) \in \rho_n$. In fact, if $\mu_n(x_0) \in \partial\overline{\rho_n} = \{E_{2n-1}, E_{2n}\}$, this ambiguity vanishes since then $\triangle(\mu_n(x_0))^2 = 1$.)

Thus one can rephrase Theorem D.3 as

**Theorem D.5.** *Assume (H.D.1). Then the map*

$$I(V_0) \ni V \longrightarrow \left\{\mu_n(x_0),\ \sigma_n(x_0)|\triangle(\mu_n(x_0))^2 - 1|^{1/2}\right\}_{n\in I} \tag{D.34}$$

*is injective, where* $\sigma_n(x_0) = \pm$ *denotes the sign of* $R(\mu_n(x_0))^{1/2} = [\triangle(\mu_n(x_0))^2 - 1]^{1/2}$, *i.e.,*

$$s'(\mu_n(x_0), x_0 + a, x_0) = \triangle(\mu_n(x_0)) + \sigma_n(x_0)|\triangle(\mu_n(x_0))^2 - 1|^{1/2},\ n \in I \tag{D.35}$$

*and* $I \subseteq \mathbb{N}$ *indexes all open gaps* $\rho_n$, $n \geq 1$ *of* $H$ *(i.e.,* $n \in I$ *iff* $\rho_n = (E_{2n-1}, E_{2n}) \neq \emptyset$*).*

**Remark D.6** Theorem D.5 implies in particular that the specification of $\sigma(H^D) = \{\mu_n(x_0)\}_{n\in\mathbb{N}}$ alone is insufficient to determine $V$ uniquely. In addition one must specify the sheet $\Pi_+$ or $\Pi_-$ on which the point $P_n := (\mu_n(x_0),$

$R(\mu_n(x_0))^{1/2}) \in \mathcal{R}$ lies (unless $\mu_n(x_0) \in \mathcal{B}$). By opening up the cuts $\overline{\rho_n}$, $n \in I$ on $\mathbb{C}$ to circles $S_n^1(V_0)$, the right hand side of (D.34) may be realized as a $g$-dimensional torus ($g = $ card $(I)$) $T^g(V_0)$

$$T^g(V_0) = \underset{n \in I}{\times} S_n^1(V_0), \quad S_n^1(V_0) := [E_{2n-1}, E_{2n}] \times \{+, -\}, \tag{D.36}$$

$$p_n(V) := (\mu_n(x_0), \sigma_n(x_0)) \in S_n^1(V_0), \sigma_n(x_0) = \operatorname{sgn} R(\mu_n(x_0))^{1/2}, \ n \in I,$$

and hence the map

$$I(V_0) \ni V \longrightarrow \{p_n(V)\}_{n \in I} \in T^g(V_0) \tag{D.37}$$

is injective. Actually, it has been proven by [Mc-Tr 1] that the map (D.37) is bijective.

The torus $T^g(V_0)$ can be explicitly realized by the Finkel, Isaacson, Trubowitz (FIT) formula.

**Theorem D.7.** *[FIT] (see also [Bu-Fi], [Iwa], [McK1]) Let $V \in I(V_0)$, $\hat{p}_j := (\hat{\mu}_j(x_0), \hat{\sigma}_j(x_0)) \in S_j^1(V_0)$, $j \in J_N \subseteq I$, card $(J_N) := N \leq g$. Define*

$$\hat{V}_N(x, \hat{p}_j, j \in J_N) := V(x) - 2\frac{d^2}{dx^2} \ln W\left(\psi_{\hat{\sigma}_{j_1}(x_0)}, g_{j_1}, \ldots, \psi_{\hat{\sigma}_{j_N}(x_0)}, g_{j_N}\right),$$
$$x \in \mathbb{R},$$
$$\tag{D.38}$$

*where $W(\ldots)$ denotes the Wronskian, $J_N = \{j_1, \ldots \leq j_N\}$, $j_1 < j_2 < \ldots < j_N$, and $\psi_{\hat{\sigma}_{j_n}(x_0)} = \psi_{\hat{\sigma}_{j_n}(x_0)}(\hat{\mu}_{j_n}(x_0), x, x_0)$ and $g_{j_n} = g_{j_n}(\hat{\mu}_{j_n}(x_0), x)$ are the Floquet functions (C.26) and Dirichlet eigenfunctions (D.19) of $H$ and $H^D$. Then $\hat{V}_N$ is the unique point in $I(V_0)$ with coordinates*

$$p_n(\hat{V}_N) = \begin{cases} \hat{p}_n, & n \in J_N, \\ p_n(V), & n \in I \backslash J_N. \end{cases} \tag{D.39}$$

*If $g = \infty$, $\hat{V}_N(x, \hat{p}_j, j \in J_N)$ and all its derivatives converge uniformly as $N \uparrow \infty$ to a potential $\hat{V}(x)$ with torus coordinates*

$$p_n(\hat{V}) = \begin{cases} \hat{p}_n, & n \in J, \\ p_n(V), & n \in I \backslash J \end{cases}, J = \bigcup_{N \in \mathbb{N}} J_N. \tag{D.40}$$

*Consequently, (D.37) is bijective.*

**Remark D.8.**
  (i) If $\hat{\mu}_{n_0}(x_0) \in \{E_{2n_0-1}, E_{2n_0}\}$ then the specification of $\hat{\sigma}_{n_0}(x_0)$ becomes superfluous (see Remark D.4) which is reflected in the fact that $\psi_+(\hat{\mu}_{n_0}(x_0), x, x_0) = \psi_-(\hat{\mu}_{n_0}(x_0), x, x_0)$ (cf. (C.30)) in this case.
  (ii) If $\hat{\mu}_{n_1}(x_0) = \mu_{n_1}(x_0)$, then $\psi_{\hat{\sigma}_{n_1}(x_0)}(z, x, x_0)$ has a pole at $z = \hat{\mu}_{n_1}(x_0)$ and the right hand side of (D.38) is defined by a limiting process. (This case occurs if one simply wants to flip the sign of $\sigma_j(x_0)$, or equivalently, if one changes $p_j(V)$ from one sheet to the other keeping the projection $\mu_j(x_0)$ of $p_j$ onto $\mathbb{C}$ fixed.)

## Appendix E. Trace relations

In this section we recall trace relations for $H^D$, i.e., the connections between $V(x_0)$ and $\sigma(H^D) = \{\mu_n(x_0)\}_{n \in \mathbb{N}}$ and present the first order differential system for $\mu_n(x_0 + t)$ corresponding to $V(x+t)$, $t \in \mathbb{R}$.

Assuming (H.D.1) throughout this section, we first consider translates of $V$

$$V^t(x) := V(x+t), \; (t,x) \in \mathbb{R}^2, \tag{E.1}$$

and add the suffix $t$ to all the quantities associated with

$$H^{D,t} := -\frac{d^2}{dx^2} + V(\cdot + t) \text{ on } \mathcal{D}(H^D), \; t \in \mathbb{R}, \tag{E.2}$$

such as $\mu_n^t(x_0)$, $c^t(z,x,x_0)$, $s^t(z,x,x_0)$ etc. The following quantities are invariant with respect to translations

$$\begin{aligned} E_n^t &= E_n, \; n \in \mathbb{N}_0, \\ \Delta^t(z) &= \Delta(z), \; z \in \mathbb{C}, \\ m_\pm^t(z) &= m_\pm(z), \; z \in \Pi_+; \; t \in \mathbb{R}, \end{aligned} \tag{E.3}$$

while

$$\Phi^t(z,x,x_0) = \Phi(z, x+t, x_0+t), z \in \mathbb{C}, \; (t,x,x_0) \in \mathbb{R}^3, \tag{E.4}$$

implies, in particular,

$$\begin{aligned} s^t(z, x_0+a, x_0) &= s(z, x_0+a, x_0)\psi_+(z, x_0+t, x_0)\psi_-(z, x_0+t, x_0) \\ &= s(z, x_0+t+a, \; x_0+t), \; z \in \mathbb{C}, (t,x_0) \in \mathbb{R}^2. \end{aligned} \tag{E.5}$$

Moreover,

$$\begin{aligned} \mu_n^t(x_0) &= \mu_n(x_0+t), \; n \in I, \\ E_{2n-1} &= \mu_n^t(x_0) = E_{2n}, \; n \in \mathbb{N}\setminus I. \end{aligned} \tag{E.6}$$

The $t$-dependence of $\mu_n^t(x_0)$ is governed by

**Theorem E.1.** *[Tru] Assume (H.D.1). Then the system*

$$\frac{d}{dt}\mu_n(x_0+t) = \frac{(2n^2\pi^2/a^3)[\Delta(\mu_n(x_0+t))^2 - 1]^{1/2}}{\prod\limits_{\substack{j \in \mathbb{N} \\ j \neq n}}[\mu_j(x_0+t) - \mu_n(x_0+t)](a^2/j^2\pi^2)}, t \in \mathbb{R}, n \in I, \tag{E.7}$$

*with the initial conditions*

$$(\mu_n(x_0), R(\mu_n(x_0))^{1/2}), \; n \in I, \text{ at } t=0 \tag{E.8}$$

has a unique solution $\mu_n(x_0 + \cdot) \in C_a^\infty(\mathbb{R})$ that does not pause at the simple eigenvalues $\{E_{2n-1}, E_{2n}\}$, $n \in I$. The sign $\sigma_n(x_0)$ of $[\Delta(\mu_n(x_0))^2 - 1]^{1/2}$ in (E.8) is chosen according to (D.35), i.e.,

$$[\Delta(\mu_n(x_0))^2 - 1]^{\frac{1}{2}} = \frac{d}{dz} s(\mu_n(x_0), x_0 + a, x_0)^{-1} 2c_n^D(x_0)^{-2} - \Delta(\mu_n(x_0))$$
$$= [s'(\mu_n(x_0), x_0 + a, x_0) - c(\mu_n(x_0), x_0 + a, x_0)]/2, \; n \in I. \quad \text{(E.9)}$$

**Remark E.2.** As $t$ runs through $[0, a)$, $p_n(t) := (\mu_n(x_0 + t)), \sigma_n(x_0 + t))$, $n \in I$ runs clockwise through the circle $S_n^1(V)$, changing sheets whenever it hits $E_{2n-1}, E_{2n}$, making $n$ complete revolutions.

The trace relations for $H^D$ finally read

**Theorem E.3.** [Fla], [Ge-Le], [Iwa], [Lev], [Le-Sa], [Mar], [Tru] Suppose $V$ satisfies (H.D.1). Then

$$V(x) = -\frac{1}{a} \int_{x_0}^{x_0+a} dy V(y) + 2 \sum_{n=1}^{\infty} \left[ (n^2\pi^2/a^2) - \frac{1}{a} \int_{x_0}^{x_0+a} dy V(y) - \mu_n(x) \right], \; x \in \mathbb{R},$$
(E.10)

$$V(x) = E_0 + \sum_{n \in I} [E_{2n-1} + E_{2n} - 2\mu_n(x)], \; x \in \mathbb{R}. \quad \text{(E.11)}$$

**Remark E.4.** Theorems E.1 and E.3 not only provide a solution of the inverse, periodic problem (i.e., to recover $V(x)$ given $(\mu_n(x_0), \sigma_n(x_0))$, $n \in I$, $\mu_n(x_0), n \in \mathbb{N}\setminus I$) but they also prove Theorem D.3 (resp. D.5) and the fact that the map (D.37) is bijective [Mc-Tr 1].

## Appendix F. Connections with a Riccati-type equation

Here we summarize some of the results in [Da-Ta] and [Bu-Fi].
We assume (H.D.1) throughout this section.
Given the definitions of $\phi_\pm(z, x_0)$ and $\psi_\pm(z, x, x_0)$ in (C.24) and (C.26), we also introduce

$$\phi_R(z, x_0) := 2^{-1} s(z, x_0 + a, x_0)^{-1} [s'(z, x_0 + a, x_0) - c(z, x_0 + a, x_0)],$$
$$z \in \hat{\mathbb{C}}(x_0) := \mathbb{C}\setminus\{\mu_n(x_0)\}_{n \in \mathbb{N}} \quad \text{(F.1)}$$

$$\phi_I(z, x_0) := -is(z, x_0 + a, x_0)^{-1}[\Delta(z)^2 - 1]^{1/2}, \; z \in \hat{\Pi}_+(x_0) \quad \text{(F.2)}$$
such that

$$\phi_\pm(z, x_0) = \phi_R(z, x_0) \pm i\phi_I(z, x_0), \; z \in \hat{\Pi}_+(x_0), \quad \text{(F.3)}$$

$$\phi_R(z, x_0) = -\frac{1}{2} \frac{d}{dx_0} \ln[\phi_I(z, x_0)], \; z \in \hat{\mathbb{C}}(x_0). \quad \text{(F.4)}$$

This yields the Riccati-type equation

$$\phi_\pm(z,x)^2 + \phi'_\pm(z,x) = V(x) - z, \quad z \in \hat{\mathbb{C}}(x). \quad (F.5)$$

Since also $\dfrac{d}{dx}\ln[\psi_\pm(z,x,x_0)]$ satisfies (F.5) we obtain

$$\phi_\pm(z,x) = \frac{d}{dx}\ln[\psi_\pm(z,x,x_0)], \quad z \in \hat{\Pi}_+(x), \quad (F.6)$$

(since it holds at $x = x_0$) and hence

$$\psi_\pm(z,x,x_0) = \exp\{\int_{x_0}^{x} dy\,\phi_\pm(z,y)\}, \quad z \in \hat{\Pi}_+(x_0). \quad (F.7)$$

Together with (F.3) and (F.4), (F.7) implies

$$\psi_-(z,x,x_0)\psi_+(z,x,x_0) = \frac{\phi_I(z,x_0)}{\phi_I(z,x)} = \frac{s(z,x+a,x)}{s(z,x_0+a,x_0)}, \quad z \in \hat{\mathbb{C}}(x_0). \quad (F.8)$$

We also note that

$$W(\psi_-(z),\psi_+(z)) = 2i\phi_I(z,x_0), \quad z \in \hat{\Pi}_+(x_0), \quad (F.9)$$

by (C.30) and (F.2). Combining (F.1)-(F.4) and (F.8) we finally obtain

$$\begin{aligned}\phi_\pm(z,x) &= \frac{\pm[\Delta(z)^2-1]^{1/2} + \frac{1}{2}\frac{d}{dx}\{s(z,x_0+a,x_0)\psi_-(z,x,x_0)\psi_+(z,x,x_0)\}}{s(z,x_0+a,x_0)\psi_-(z,x,x_0)\psi_+(z,x,x_0)}\\ &= \frac{\pm[\Delta(z)^2-1]^{1/2} + \frac{1}{2}\frac{d}{dx}\left\{a\prod_{n\in\mathbb{N}}[\mu_n(x)-z](a^2/n^2\pi^2)\right\}}{a\prod_{n\in\mathbb{N}}[\mu_n(x)-z](a^2/n^2\pi^2)},\\ &= \frac{\pm\tilde{R}(z)^{1/2} + \frac{1}{2}\frac{d}{dx}\prod_{n\in I}[\mu_n(x)-z](a^2/n^2\pi^2)}{\prod_{n\in I}[\mu_n(x)-z](a^2/n^2\pi^2)}, \quad z \in \hat{\Pi}_+(x),\end{aligned} \quad (F.10)$$

(cf. (D.14), (E.5) and (E.6)). It remains to study the behavior of $\phi_\pm(z,x_0)$ for $z$ near Dirichlet eigenvalues $\mu_n(x_0)$, $n \in \mathbb{N}$.

**Lemma F.1.** *[Bu-Fi] Assume (H.D.1) and let $\sigma \in \{+,-\}$. Then $\phi_\sigma(z,x)$, $x \in \mathbb{R}\setminus\{y \in \mathbb{R}|g_n(y,x_0)=0\}$ is continuous at points $z = \mu_n(x_0)$, $n \in \mathbb{N}$. In fact,*

$$\lim_{z\to\mu_n(x_0)} \phi_\sigma(z,x) = \frac{s'(\mu_n(x_0),x,x_0)}{s(\mu_n(x_0),x,x_0)} \quad (F.11)$$

$$= \frac{d}{dx}\ln[g_n(x,x_0)], \quad n \in \mathbb{N}, x \in \mathbb{R}\setminus\{y \in \mathbb{R}|g_n(y,x_0)=0\}.$$

Formula (F.10) together with (A.7), (C.42), (D.37) and Theorem E.1 also settles the isospectral manifold $I(\phi_0)$

$$I(\phi_0) := \{\phi \in C_a^\infty(\mathbb{R}) | E_n(\phi^2 \pm \phi') = E_n(\phi_0^2 \pm \phi_0'), \ n \in \mathbb{N}_0\} \quad (F.12)$$

for periodic potentials $\phi$ in the Dirac operator $Q$ given by (C.34). In fact, for $g \geq 1$, the maps

$$I(\phi_0) \ni \phi \longrightarrow (\{p_n(\phi^2 \pm \phi')\}_{n \in I}, \sigma) \in T^g(\phi_0^2 \pm \phi_0') \times \mathbb{Z}_2 \quad (F.13)$$
$$\text{if } 0 \in \sigma(Q), \ g \geq 1,$$

and

$$I(\phi_0) \ni \phi \longrightarrow (\{p_n(\phi^2 \pm \phi')\}_{n \in I}, \sigma, \sigma') \in T^g(\phi_0^2 \pm \phi_0') \times \mathbb{Z}_2 \times \mathbb{Z}_2 \quad (F.14)$$
$$\text{if } 0 \notin \sigma(Q), \ g \geq 1,$$

are bijective. Here $\sigma = \pm 1$ describes the overall sign ambiguity in $\phi$ (i.e., $\pm \phi(x)$ are isospectral due to (A.7)) and $\sigma' = \pm 1$ describes the $\pm$-sign in (F.10). (For $g = 0$, $I(\phi_0) = \{0\}$ for $E_0(\phi_0^2) = \phi_0^2 = 0$ and $I(\phi_0) = \{|\phi_0|, -|\phi_0|\}$ for $E_0(\phi_0^2) = \phi_0^2 > 0$.)

**Example F.2.** [Ges 1], [GSS] $g = 1$.
$Q$ has spectrum

$$\sigma(Q) = (-\infty, -E_2^{1/2}] \cup [-E_1^{1/2}, -E_0^{1/2}] \cup [E_0^{1/2}, E_1^{1/2}] \cup [E_2^{1/2}, \infty),$$
$$0 \leq E_0 < E_1 < E_2 \text{ iff } \phi(x) = (\sigma/2)\frac{\mathcal{P}'(x + \omega' + \alpha) - \sigma' \mathcal{P}'(b_0)}{\mathcal{P}(x + \omega' + \alpha) - \mathcal{P}(b_0)}, \quad (F.15)$$

where

$$\alpha \in \mathbb{R}, \ \sigma = \pm 1, \sigma' = \pm 1, \ E_0 = \mathcal{P}(b_0) - \mathcal{P}(\omega), E_1 = \mathcal{P}(b_0) - \mathcal{P}(\omega + \omega'),$$
$$E_2 = \mathcal{P}(b_0) - \mathcal{P}(\omega') \quad (F.16)$$

and $\mathcal{P}(.) := \mathcal{P}(.,\omega,\omega')$ denotes the Weierstraß $\mathcal{P}$-function with halfperiods $\omega, \omega'$ ($\omega > 0, -i\omega' > 0$).

Example F.2 extends Hochstadt's result (see Example F.3 below) in the context of periodic Schrödinger operators $H$ to the case of periodic Dirac operators $Q$.

**Example F.3.** [Hoc 3] (see also [Fla]) $g = 1$.
$H$ has spectrum

$$\sigma(H) = [E_0, E_1] \cup [E_2, \infty), \ E_0 < E_1 < E_2,$$
$$\text{iff } V(x) = 2\mathcal{P}(x + \omega' + \alpha) + C, \quad (F.17)$$

where

$$\alpha \in \mathbb{R}, \ E_0 = C - \mathcal{P}(\omega), E_1 = C - \mathcal{P}(\omega + \omega'), E_2 = C - \mathcal{P}(\omega'). \quad (F.18)$$

## Appendix G. The case $g < \infty$

Since many expressions in Appendices C-F considerably simplify in the case of finite genus $0 \leq g < \infty$, we list some of the relevant formulas in this appendix.

First of all (C.8) simplifies to

$$\sigma(H) = \begin{cases} \bigcup_{n \in I}[E_{2n-2}, E_{2n-1}] \cup [E_{2N_0}, \infty), & g \geq 1, \\ [E_0, \infty), & g = 0, \end{cases} \qquad (G.1)$$

$$g = \operatorname{card}(I) < \infty, \; N_0 := \max(I)$$

and

$$\rho_0 = (-\infty, 0), \; \rho_n = \emptyset, \; n \in \mathbb{N}\backslash I, \qquad (G.2)$$

$$E_{2n-1} = \mu_n(x) = \nu_n(x) = \lambda_n = E_{2n}, \; n \in \mathbb{N}\backslash I, \; x \in \mathbb{R}. \qquad (G.3)$$

The Riemann surface $\mathcal{R}$ becomes a hyperelliptic one of genus $g < \infty$, denoted by $\mathcal{R}_g$, associated with

$$R_0(z)^{1/2} := \left[(E_0 - z) \prod_{n \in I}(E_{2n-1} - z)(E_{2n} - z)\right]^{1/2} \qquad (G.4)$$

by splitting off the double zeroes (G.3) in $R(z)^{1/2} = [\Delta(z)^2 - 1]^{1/2}$ (cf. (C.7) and (C.9)). (F.2) and (F.8) turn into

$$\phi_I(z, x) = -i R_0(z)^{1/2} \prod_{n \in I}[\mu_n(x) - z]^{-1}, \; z \in \hat{\Pi}_+(x), \qquad (G.5)$$

$$\psi_-(z, x, x_0)\psi_+(z, x, x_0) = \prod_{n \in I}[\mu_n(x) - z][\mu_n(x_0) - z]^{-1}, \; z \in \hat{\mathbb{C}}(x_0). \qquad (G.6)$$

(G.5) and (G.6) together with (C.25) imply

**Lemma G.1.** *[Dub], [It-Ma], [NMPZ] Assume (H.D.1) and $g < \infty$. Then $\psi(., x, x_0) : \mathcal{R}_g \to \mathbb{C}_\infty$ has $g$ simple zeroes at $\{\mu_n(x)\}_{n \in I}$ and $g$ simple poles at $\{\mu_n(x_0)\}_{n \in I}$.*

Combining (F.3), (F.4) and (G.5), (F.10) becomes

$$\phi_\pm(z, x) = \frac{\pm R_0(z)^{1/2} + \frac{1}{2}\frac{d}{dx}\left\{\prod_{n \in I}[\mu_n(x) - z]\right\}}{\prod_{n \in I}[\mu_n(x) - z]}, \; z \in \hat{\Pi}_+(x). \qquad (G.7)$$

The system (E.7) turns into

$$\frac{d}{dt}\mu_n(x_0 + t) = \frac{2 R_0(\mu_n(x_0 + t))^{1/2}}{\prod_{\substack{j \in I \\ j \neq n}}[\mu_j(x_0 + t) - \mu_n(x_0 + t)]}, \; n \in I, t \in \mathbb{R}, \qquad (G.8)$$

(here $\prod_{\substack{j \in I \\ j \neq n}} \ldots \equiv 1$ for $I = \emptyset$, i.e., for $g = 0$).

Next we mention Hochstadt's theorem.

**Theorem G.2.** *[Hoc 2] (see also [McK 2], [Mc-Mo]) Assume $g = \text{card}(I) < \infty$ and let $\sum_s := \{E_0, E_{2n-1}, E_{2n}\}_{n \in I}$ be the simple periodic and antiperiodic eigenvalues of $H^P, H^{AP}$ (cf. (D.2), (D.4)). Then $\sum_s$ determines the double periodic and antiperiodic spectrum of $H$ (i.e., all degenerate eigenvalues of $H^P$ and $H^{AP}$) as well as the roots $\{\lambda_n\}_{n \in \mathbb{N}}$ of $\dfrac{d}{dz}\Delta(z)$. In particular, $\sum_s$ determines $\Delta(z)$, $z \in \mathbb{C}$.*

Finally, adopting the notation in (C.42), $\sigma(Q)$ becomes

$$\sigma(Q) = \bigcup_{n \in I} \sum\nolimits_n \cup \sum\nolimits_{-n} \cup (-\infty, -E_{2N_0}^{1/2}] \cup [E_{2N_0}^{1/2}, \infty). \tag{G.9}$$

**Acknowledgements.**

I would like to thank J.A. Goldstein, F. Kappel and W. Schappacher for their kind invitation to a most stimulating conference. I am indebted to D. Bollé for the great hospitality extended to me at the Institute for Theoretical Physics of the University of Leuven, Belgium during summer 1989, where most parts of this paper were written. Support by the Onderzoeksfonds of the University of Leuven during that period is gratefully acknowledged.

# References

[Ad-Mo] M. Adler, J. Moser, Commun. Math. Phys. **61**, 1 (1978).
[BDT] R. Beals, P. Deift, C. Tomei, *Direct and inverse scattering on the line*, Mathematical Surveys and Monographs **28**, AMS, Providence, RI, 1988.
[Bor] G. Borg, Acta Math. **78, 1** (1945).
[Bu-Fi] M. Buys, A. Finkel, J. Diff. Eqs. **55**, 257 (1984).
[Ch-Pe] S.S. Chern, C.-K. Peng, Manuscripta Math. **28, 207** (1979).
[Co-Ka] A. Cohen, T. Kappeler, Indiana Univ. Math. J. **34**, 127 (1985).
[Da-Ta] E. Date, S. Tanaka, Suppl. Progr. Theoret. Phys. **59**, 107 (1976).
[Da-Si] E.B. Davies, B. Simon, Commun. Math. Phys. **63**, 277 (1978).
[De-Jo] C. De Concini, R.A. Johnson, Ergod. Th. Dynam. Syst. **7**, 1 (1987).
[De-Sp] E.M. deJager, S. Spannenburg, J. Phys. **A18**, 2177 (1985).
[Dei] P. A. Deift, Duke Math. J. **45**, 267.
[Dub] B.A. Dubrovin, Funct. Anal. Appl. **9**, 215 (1975).
[Eas] M.S.P. Eastham, The Spectral Theory of Periodic Differential Equations, Scottish Academic Press, Edinburgh, 1973.
[FIT] A. Finkel, E. Isaacson, E. Trubowitz, SIAM J. Math. Anal **18**, 46 (1987).
[Fla] H. Flaschka, Arch. Rat. Mech. Anal. **59**, 293 (1975).
[Ge-Di] I.M. Gel'fand, L.A. Dikii, Russ. Math. Surv. **30:5**, 77 (1975).
[Ge-Le] I.M. Gel'fand, B.M. Levitan, AMS Transl. **1**, 253 (1955).

[Ges 1]  F. Gesztesy, *Some applications of commutation methods*, Schrödinger Operators, Lectures from the Nordic Summer School in Mathematics 1988, H. Holden, A. Jensen (eds.), Springer Lecture Notes in Physics **345**, Berlin, 1989, p. 93.

[Ges 2]  F. Gesztesy, *Scattering theory for one-dimensional systems with nontrivial spatial asymptotics*, Schrödinger Operators, Aarhus 1985, E. Balslev (ed.), Springer Lecture Notes in Mathematics **1218**, Berlin, 1986, p.93.

[Ge-Sc]  F. Gesztesy, W. Schweiger, *Rational KP and mKP-solutions in Wronskian form*, preprint, 1990.

[Ge-Si]  F. Gesztesy, B. Simon, *Constructing solutions of the mKdV-equation*, J. Funct. Anal. **89**, 53 (1990).

[Ge-Zh]  F. Gesztesy, Z. Zhao, *On critical and subcritical Sturm-Liouville operators*, J. Funct. Anal. (to appear).

[GSS]  F. Gesztesy, W. Schweiger, B. Simon, *Commutation methods applied to the mKdV-equation*, Trans. Amer. Math. Soc. (to appear).

[GHSS]  F. Gesztesy, H. Holden, E. Saab, B. Simon, *Explicit construction of solutions of the modified Kadomtsev-Petviashvili equation*, J. Funct. Anal. (to appear).

[GHSZ]  F. Gesztesy, H. Holden, B. Simon, Z. Zhao, *On the Toda and Kac-van Moerbeke systems*, preprint, 1990.

[Gro]  H. Gross, Lett. Math. Phys. **8**, 313 (1984).

[Har]  P. Hartman, *Ordinary Differential Equations*, $2^{nd}$ ed., Birkhäuser, Boston, 1982.

[Hir]  R. Hirota, Phys. Rev. Lett. **18**, 1192 (1971).

[Hoc 1]  H. Hochstadt, Proc. Amer. Math. Soc. **14**, 930 (1963).

[Hoc 2]  H. Hochstadt, Math. Z. **82**, 237 (1963).

[Hoc 3]  H. Hochstadt, Arch. Rat. Mech. Anal. **19**, 353 (1965).

[It-Ma]  A.R. Its, V.B. Matveev, Theoret. Math. Phys. **23**, 343 (1975).

[Iwa]  K. Iwasaki, Ann. Mat. Pura Appl. Ser. 4, **149**, 185 (1987).

[Ji-Mi]  M. Jimbo, T. Miwa, Publ. RIMS, Kyoto Univ. **19**, 943 (1983).

[Lax 1]  P.D. Lax, Commun. Pure Appl. Math. **21**, 467 (1968).

[Lax 2]  P.D. Lax, Lectures Appl. Math. **15**, 85 (1974).

[Lev]  B.M. Levitan, *Inverse Sturm-Liouville Problems*, VNU Science Press, Utrecht, 1987.

[Le-Sa]  B.M. Levitan, I.S. Sargsjan, *Introduction to Spectral Theory*, Translations of Mathematical Monographs **39**, AMS, Providence, RI, 1975.

[Ma-Wi]  W. Magnus, S. Winkler, *Hill's Equation*, Dover, New York, 1979.

[Mar]  V.A. Marchenko, *Sturm-Liouville Operators and Applications*, Birkhäuser, Basel, 1986.

[McK 1]  H.P. McKean, Commun. Pure Appl. Math. **38**, 669 (1985).

[McK 2]  H.P. McKean, *Integrable systems and algebraic curves*, Global Analysis, M. Grmela, J.E. Marsden (eds.), Springer Lecture Notes in Mathematics **755**, Berlin, 1979, p. 3.

[Mc-Mo]  H.P. McKean, P. vanMoerbeke, Invent. Math. **30**, 217 (1975).

[Mc-Tr1]  H.P. McKean, E. Trubowitz, Commun. Pure Appl. Math. **29**, 143 (1976).

[Mc-Tr2]  H.P. McKean, E. Trubowitz, Bull. Amer. Math. Soc. **84**, 1042 (1978).

[Miu]  R.M. Miura, J. Math. Phys. **9**, 1202 (1968).

[Mur]  M. Murata, Duke Math. J. **53**, 869 (1986).

[NMPZ]  S. Novikov, S.V. Manakov, L.P. Pitaevskii, V.E. Zakharov, *Theory of Solitons*, Consultants Bureau, New York, 1984.

[Po-Tr]  J. Pöschel, E. Trubowitz, *Inverse Spectral Theory*, Academic Press, Boston, 1987.

[RS III]  M. Reed, B. Simon, *Methods of Modern Mathematical Physics III, Scattering Theory*, Academic Press, New York, 1979.

[Ru-Bo]  S.N.M. Ruijsenaars, P.J.M. Bongaarts, Ann. Inst. H. Poincaré **A26**, 1 (1977).

[Sa-Zu]  D.H. Sattinger, V.D. Zurkovski, Physica **26D**, 225 (1987).

[Sim]  B. Simon, J. Funct. Anal. **40**, 66 (1981).

[Tan]  S. Tanaka, Publ. RIMS, Kyoto Univ. **8**, 429 (1972/73).

[Tru]  E. Trubowitz, Commun. Pure Appl. Math. **30**, 321 (1977).

[Wa-Es] H.D. Wahlquist, F.B. Estabrook, J. Math. Phys. **16**, 1 (1975).
[Zak] V.E. Zakharov, Sov. Phys. JETP **33**, 538 (1971).
[Zh-Ch] Yu-kun Zheng, W.L. Chan, J. Math. Phys. **29**, 308 (1988).


F. Gesztesy
Department of Mathematics
University of Missouri
Columbia, MO 65211, USA

# Integrodifferential Equations with Nondensely Defined Operators

Ronald Grimmer
Hetao Liu

Department of Mathematics, Southern Illinois University

## 1. Introduction

We are concerned with the equation

$$X'(t) = AX(t) + \int_0^t (F(t-s)A + K(t-s))X(s)\,ds + f(t), \qquad (1.1)$$
$$X(0) = X$$

in a Banach space **X** with $A$ a not necessarily densely defined linear closed operator defined on **X**, while $F$ and $K$ are bounded linear operators on **X**.

There are many studies of (1.1) when $A$ generates a $C_0$-semigroup, eg. [4], [6-9], [11-12], but [2] is the only previous study of an integrodifferential equation involving non-densely defined operators with which we are familiar. The initial value problem

$$X'(t) = AX(t) + f(t), \qquad (1.2)$$
$$X(0) = X$$

with $A$ not necessarily densely defined has been studied by Da Prato and Sinestrari [5], Arendt [1] and Thieme [14]. Specifically, it was shown in [1] and in [5]

---

This work was partially supported by NSF.

that if $A$ satisfies resolvent estimates of the Hille - Yosida type, then $X \in D(A)$ and $AX + f(0) \in \overline{D(A)}$ implies that (1.2) has a unique classical solution.

Also, it is shown in [5] that if $A$ satisfies the estimates of Hille - Yosida type then the equation

$$X(t) - X = A \int_0^t X(s)\,ds + \int_0^t f(s)\,ds \qquad (1.3)$$

has a unique $\overline{D(A)}$-valued solution for any $X \in \overline{D(A)}$ and f locally integrable. (Solutions of (1.3) are called integral solutions of (1.2)).

For more general integrodifferential equations of the form

$$\begin{aligned} U'(t) &= \int_0^t AU(t-s)\,d\eta(s), \\ U(0) &= X, \end{aligned} \qquad (1.4)$$

where $\eta$ is of bounded variation and is exponentially bounded, Arendt and Kellerman, [2], developed the theory of an n-times integrated solution family so that under some Hille-Yosida type conditions involving the generalized resolvent $(\lambda - \hat{\eta}(\lambda)A)^{-1}$ (with $\hat{\eta}$ being the Laplace transform of $\eta$) it is proved that there is an n-times integrated solution family for (1.4). Hence for appropriate $(X, f)$, (1.4) has a unique classical solution. Integral solutions of (1.4) were not studied in [2].

In this paper, we shall only consider n-times integrated semigroups with $n = 1$, that is integrated semigroups, and their application to (1.1). In Section 2, we recognize that the integral solution of (1.2) agrees in some sense with the weak solution of Ball [3]. This enables us to prove that the existence and uniqueness of solutions to (1.3) and that $\lambda \|(\lambda - A)^{-1}\| \leq C < \infty$, $\lambda > \omega$, is also sufficient for $A$ to satisfy the resolvent estimates of Hille - Yosida type. Thus one can get a generation theorem for integrated semigroups analogous to the generation theorem for semigroups. In Section 3, we look at (1.1) in the strong form. We assume that $A$ satisfies the Hille - Yosida type estimates and that $f$ is continuously differentiable. Then we can prove under certain smoothness conditions on $F(\cdot)$ and $K(\cdot)$ that (1.1) has a unique classical solution iff

$$X \in D(A), \quad AX + f(0) \in \overline{D(A)}. \qquad (1.5)$$

That is, we get the same condition as required for (1.2). In doing this we use methods as in [12] and [8] to write (1.1) as

$$\begin{aligned} \begin{pmatrix} U(t) \\ \phi_t \end{pmatrix}' &= \begin{pmatrix} A & \delta \\ F(\cdot)A + K(\cdot) & D \end{pmatrix} \begin{pmatrix} U(t) \\ \phi_t \end{pmatrix} + \begin{pmatrix} f(t) \\ 0 \end{pmatrix} \\ &= \mathcal{A} \begin{pmatrix} U(t) \\ \phi_t \end{pmatrix} + \begin{pmatrix} f(t) \\ 0 \end{pmatrix}, \\ \begin{pmatrix} U(0) \\ \phi_0 \end{pmatrix} &= \begin{pmatrix} X \\ 0 \end{pmatrix} \end{aligned} \qquad (1.6)$$

and prove that $\mathcal{A}$ generates an integrated semigroup in an appropriate product space.

Note that in case $F(t)$ and $K(t)$ are scalar functions, (1.4) is more general than (1.1). The difference is that for this particular problem (1.1), if we set it into the form of (1.4) and try to get the classical solutions, then we must check a Hille - Yosida type condition involving the generalized resolvent to see whether it generates an n-times integrated solution family. This is, in general, not easy. Also, the sufficient conditions on initial data and $f$ for (1.4) to have a classical solution (when there is an n-times integrated solution family) in [2] are

$$X \in D(A^{n+1}), \quad f \in C^{n+1}([0,T], \mathbf{X}), \quad f^{(k)}(0) \in D(A^{n-k}), \\ k = 0, 1, \ldots, n-1, \quad n \geq 0. \tag{1.7}$$

We only impose resolvent conditions on the operator $A$, and require the condition (1.5) which is weaker than (1.7) with n = 1.

In Section 4, we look at (1.1) in the integrated form

$$X(t) - X = A \int_0^t X(s)\,ds + \int_0^t (F(t-s)A + K(t-s)) \int_0^s X(u)\,du\,ds \\ + \int_0^t f(s)\,ds. \tag{1.8}$$

Note that Thieme [14, section 6] studied the integrated solutions of (1.2) by integrating (1.2) twice to get

$$V(t) - tX = A \int_0^t V(s)\,ds + \int_0^t (t-s)f(s)\,ds \tag{1.9}$$

and proved that if $A$ generates a non-degenerate integrated semigroup $S(t)$, then the unique solution of (1.9) for any $X \in \mathbf{X}$ is given by

$$V(t) = S(t)X + \int_0^t S(t-u)f(u)\,du.$$

In addition, Thieme has shown (taking $f = 0$) that $A$ generates a non-degenerate integrated semigroup iff for any $X \in \mathbf{X}$,

$$V(t) = A \int_0^t V(s)\,ds + tX, \quad t \geq 0,$$

has a unique solution $V(t)$ such that $\|V(t)\| \leq C(t)\|X\|$ with some $C(t) > 0$ not depending on $X$. Since we consider the integral solution of (1.1), (that is only integrate (1.1) once), and we want our integrated semigroup to satisfy a Lipschitz condition in order to use a perturbation result we cannot use the results of [14] here. However, in Section 4 we will see that if the strong derivative

$F'(t)$ exists the results of Section 2 will show that $A$ satisfies the Hille - Yosida estimates iff $\lambda \|(\lambda - A)^{-1}\| \leq C < \infty$, $\lambda > \omega$, and (1.8) has a unique continuous $\overline{D(A)}$-valued solution for any $X \in \overline{D(A)}$ and $f$ locally integrable. (Solutions of (1.8) are called integral solutions of (1.1).) In doing this we write (1.1) as the system

$$\begin{pmatrix} X(t) \\ V(t) \end{pmatrix}' = \begin{pmatrix} A & F(0) \\ A & 0 \end{pmatrix} \begin{pmatrix} X(t) \\ V(t) \end{pmatrix}$$
$$+ \int_0^t \begin{pmatrix} K(t-s) & F'(t-s) \\ 0 & 0 \end{pmatrix} \begin{pmatrix} X(s) \\ V(s) \end{pmatrix} ds + \begin{pmatrix} f(t) \\ 0 \end{pmatrix},$$
$$\begin{pmatrix} X(0) \\ V(0) \end{pmatrix} = \begin{pmatrix} X \\ 0 \end{pmatrix} \tag{1.10}$$

and apply the results in Section 2 to (1.10).

Finally, in Section 5 we show that the one-dimensional wave equation with Dirichlet boundary conditions can be written as a $2 \times 2$ system considered on $C[0,1] \times C[0,1]$ and that when considered abstractly, the associated operator generates an integrated semigroup. Thus one can apply this theory to certain equations arising from problems in viscoelasticity. For further results concerning symmetric hyperbolic systems we refer to [10].

## 2. Generation of Integrated Semigroups

Consider now the equation

$$W'(t) = GW(t) + q(t) , W(0) = W, \tag{2.1}$$

in a Banach space $\mathbf{K}$ with $q$ locally integrable and recall the following well known results (cf. Pazy [13] and Ball [3] ).

**Theorem 2.1.** *Let $G$ be a densely defined closed linear operator. Then the following statements are equivalent:*

  a) *There exist $M \geq 1$, and $\alpha \in \mathbb{R}$ such that $\lambda > \alpha$ implies $\lambda \in \rho(A)$ and*

$$\|(\lambda - G)^{-n}\| \leq M/(\lambda - \alpha)^n, \ \forall \lambda > \alpha, \ n = 1, 2, 3, \ldots$$

  b) *$G$ generates a $C_0$ semigroup.*
  c) *(2.1) has a unique continuous weak solution $W(t)$ for each $W \in \mathbf{K}$ and $q : [0, \infty) \to \mathbf{K}$ locally integrable. That is, for each $V \in D(G^*)$, $\langle W(t), V \rangle$ is absolutely continuous and*

$$\frac{d}{dt}\langle W(t), V \rangle = \langle W(t), G^*V \rangle + \langle q(t), V \rangle, \text{ a.e.}, W(0) = W.$$

  *where $G^*$ is the adjoint of $G$ and $\langle \cdot, \cdot \rangle$ is the pairing between $\mathbf{K}$ and its dual $\mathbf{K}^*$.*

We now must formally define integrated semigroups.

**Definition 2.2.** *A family $S(t)$, $t \geq 0$, of bounded linear operators on a Banach space $\mathbf{K}$ is called an integrated semigroup iff the following properies are satisfied:*
  a) $S(t)S(\tau) = \int_0^t (S(\tau+r) - S(r))\, dr$.
  b) *For any $X \in \mathbf{K}$, $S(t)X$ is a continuous function of $t \geq 0$ with values in $\mathbf{K}$.*
  c) $S(0) = 0$.

**Definition 2.3.** *An integrated semigroup $S$ is called non-degenerate if and only if $S(t)X = 0$ for all $t \geq 0$ implies that $X = 0$.*

**Definition 2.4.** *An integrated semigroup $S$ is said to be of type $(M, \omega)$ iff for $t, r \geq 0$*

$$\|S(t+r) - S(t)\| \leq M \int_t^{t+r} e^{ws}\, ds.$$

**Definition 2.5.** *The generator $G$ of $S$ is defined by: Let $X, Y \in \mathbf{K}$, then $X \in D(G)$ and $GX = Y$ iff $S(\cdot)X \in C^1([0,\infty), \mathbf{K})$ and $S'(t)X - X = S(t)Y$.*

**Definition 2.6.** *The part $G_0$ of $G$ in $\overline{D(G)}$ is*

$$G_0 = G \text{ on } D(G_0) = \{W \in D(G) | GW \in \overline{D(G)}\}.$$

**Remark.** Note that the integrated semigroup as we have defined it is the once integrated semigroup of Arendt [1].

We now can state the result parallel to Theorem 2.1.

**Theorem 2.7.** *The following three statements are equivalent:*
  a) *$G$ is a (not necessarily densely defined) linear closed operator and there exist constants $M \geq 1$, $\alpha \in \mathbb{R}$, such that $\lambda > \alpha \Rightarrow \lambda \in \rho(G)$, and*

$$\|(\lambda - G)^{-n}\| \leq M/(\lambda - \alpha)^n, \forall \lambda > \alpha,\ n = 1, 2, 3, \ldots$$

  b) *$G$ generates a non-degenerate integrated semigroup $S(t)$ of type $(M, \alpha)$.*
  c) *For $\lambda$ sufficiently large, $\lambda \in \rho(G)$ and $\limsup_{\lambda \to \infty} \lambda \|(\lambda - G)^{-1}\| < \infty$. In addition, for any $W \in \overline{D(G)}$ and $q : [0, \infty) \to \mathbf{K}$ locally integrable, (2.1) has a unique continuous $\overline{D(G)}$-valued 'integral solution' $W(t)$, i.e., for $t \geq 0$, $W(t) \in \overline{D(G)}$ and*

$$W(t) - W = G \int_0^t W(s)\, ds + \int_0^t q(s)\, ds.$$

**Proof.** (a) $\iff$ (b) is proved in Arendt [1, Theorem 4.1 with n = 0]. (a) $\Rightarrow$ (c) is proved in Da Prato and Sinestrari [5, Corollary 7.3]. Thus, we need only show (c) $\Rightarrow$ (b).

Suppose that (c) is valid. For $X \in D(G)$,

$$\|\lambda(\lambda - G)^{-1}X - X\| = \|(\lambda - G)^{-1}GX\|$$

so

$$\|\lambda(\lambda - G)^{-1}X - X\| \to 0, \text{ as } \lambda \to \infty.$$

As $\limsup_{\lambda \to \infty} \lambda \|(\lambda - G)^{-1}\| < \infty$, we conclude $\|\lambda(\lambda - G)^{-1}X - X\| \to 0$, as $\lambda \to \infty$, for $X \in \overline{D(G)}$. Hence we see that $G_0$ is densely defined in $\overline{D(G)}$. Note that $G_0$ is a closed operator in $\overline{D(G)}$ so the adjoint $G_0^*$ of $G_0$ is well defined on $(\overline{D(G)})^*$.

Next, for $W \in \overline{D(G)}$, $q : [0, \infty) \to \overline{D(G)} \subset \mathbf{K}$ locally integrable, we may denote the unique $\overline{D(G)}$ - valued integral solution of (2.1) as $W(t) = W(t, W, q)$, then $\int_0^t W(s)ds \in D(G_0)$ and, hence, for each $V \in D(G_0^*)$,

$$\langle W(t), V \rangle = \langle W, V \rangle + \langle G \int_0^t W(s)\, ds, V \rangle + \langle \int_0^t q(s)\, ds, V \rangle$$
$$= \langle W, V \rangle + \int_0^t \langle W(s), G_0^* V \rangle\, ds + \int_0^t \langle q(s), V \rangle\, ds.$$

Therefore $\langle W(t), V \rangle$ is absolutely continuous and

$$\frac{d}{dt} \langle W(t), V \rangle = \langle W(t), G_0^* V \rangle + \langle q(t), V \rangle \text{ a.e.,}$$
$$W(0) = W$$

so $W(t)$ is the weak solution of (2.1) in $\overline{D(G)}$. To prove the uniqueness of $W(t)$, let $V(t) = V(t, W, q)$ be another weak solution corresponding to $(W, q)$, then

$$\langle W(t) - V(t) \rangle = \langle \int_0^t (W(s) - V(s))\, ds, G_0^* V \rangle, \ \forall\, V \in D(G_0^*),$$

and, hence, [3],

$$\int_0^t (W(s) - V(s))\, ds \in D(G_0)$$

and

$$G_0 \int_0^t (W(s) - V(s))\, ds = W(t) - V(t).$$

Thus, $W(t) - V(t) \equiv 0$ by the assumption that (2.1) has a unique integral solution. Now it follows from [3] that $G_0$ generates a $C_0$-semigroup on $\overline{D(G)}$. Therefore (b) is valid as we know that if $G_0$ generates a semigroup on $\overline{D(G)}$ and $\limsup_{\lambda \to \infty} \lambda \|(\lambda - G)^{-1}\| < \infty$ then (b) is valid, [15]. In particular, if $T_0(t)$ is the semigroup generated on $\overline{D(A)}$ by $G_0$ then $S_0(t)$ defined on $\overline{D(A)}$ by

$$S_0(t) = \int_0^t T_0(s)\,ds$$

can be extended to **K** by

$$S(t) = (\lambda - G_0)S_0(t)(\lambda - G)^{-1}.$$

**Remark.** In $\overline{D(G)}$ 'integral solution' in the sense of Da Prato and Sinestrari [5] agrees with the 'weak solution' of Ball [3]. Thus, Theorem 2.7 is an extension of Theorem 2.1 to the integrated semigroup setting.

Note that when $G$ is densely defined, $G = G_0$, so we have:

**Corollary 2.8.** *Let $G$ be a densely defined closed linear operator. Then the following are equivalent:*

  a) *$G$ generates a $C_0$ semigroup.*
  b) *For any $W \in \mathbf{K}$ and $q : [0, \infty) \to \mathbf{K}$ locally integrable,*

$$W(t) - W = G\int_0^t W(s)\,ds + \int_0^t q(s)\,ds$$

*has a unique continuous solution in* **K**.

## 3. Classical solutions of (1.1)

In this section we look at (1.1) in the strong form and write (1.1) as (1.6). We first prove that $\mathcal{A}$ generates an integrated semigroup so that with appropriate conditions on the initial data and function $f$, (1.6) has a classical solution. Next we build some equivalent relations between (1.1) and (1.6) which enable us to obtain classical solutions of (1.1).

Define $\mathcal{F}$ to be the space of bounded uniformly continuous functions on $[0, \infty)$ into **X** with the usual sup norm and define $\mathcal{Z} = \mathbf{X} \times \mathcal{F}$ with the usual norm. Also, define $\delta : \mathcal{F} \to \mathbf{X}$ by $\delta f = f(0)$, and the operators $F$ and $K$ on **X** into $\mathcal{F}$ by $(F(\cdot)X)(\theta) = F(\theta)X$, $(K(\cdot)X)(\theta) = K(\theta)X$, $\theta \geq 0$, $X \in \mathbf{X}$. Finally, $\mathcal{D} = \frac{d}{d\theta}$ is the generator of the $C_0$ translation semigroup $f(\cdot) \to f(t + \cdot)$ on $\mathcal{F}$.

**Assumption 3.1.** *$A : D(A) \subset \mathbf{X} \to \mathbf{X}$ is linear and closed, $(D(A)$ may not be dense in **X**), and there are $\omega \in \mathbb{R}$ and $M \geq 1$ such that $\lambda > \omega \Rightarrow \lambda \in \rho(A)$ and*

$$\|(\lambda - A)^{-n}\| \leq M/(\lambda - \omega)^{-n}, \ \forall \lambda > \omega, \ n = 1, 2, 3, \ldots$$

**Assumption 3.2.** For $t \geq 0$, $F(t)$ and $K(t)$ are bounded linear operators on **X**. For $X \in \mathbf{X}$, $F(\cdot)X \in D(\mathcal{D})$, $K(\cdot)X \in \mathcal{F}$.

**Definition 3.3.** We say that $(U(t), \phi_t)$ is a classical solution of (1.6) if $(U(t), \phi_t)$ is continuously differentiable in $t$, $(U(t), \phi_t) \in D(\mathcal{A})$ and (1.6) is satisfied for $t \geq 0$.

**Theorem 3.4.** Let Assumptions 3.1 and 3.2 be satisfied and $f : [0, \infty) \to \mathbf{X}$ be continuously differentiable. Then (1.6) has a classical solution iff

$$\begin{pmatrix} X \\ 0 \end{pmatrix} \in D(A), \quad A\begin{pmatrix} X \\ 0 \end{pmatrix} + \begin{pmatrix} f(0) \\ 0 \end{pmatrix} \in \overline{D(A)}.$$

**Proof.** Let Assumptions 3.1 and 3.2 be satisfied and consider

$$\mathcal{A} = \begin{pmatrix} A & \delta \\ F(\cdot)A + K(\cdot) & \mathcal{D} \end{pmatrix}$$
$$= \begin{pmatrix} I & 0 \\ F(\cdot) & I \end{pmatrix} \begin{pmatrix} A & 0 \\ 0 & \mathcal{D} \end{pmatrix} \begin{pmatrix} I & 0 \\ -F(\cdot) & I \end{pmatrix} + \begin{pmatrix} 0 & \delta \\ \mathcal{D}F(\cdot) + K(\cdot) & 0 \end{pmatrix}$$
$$= E_1 + E_2.$$

As $\mathcal{D}$ generates a $C_0$ semigroup, it's easy to show that $\text{diag}(A, \mathcal{D})$ satisfies Hille - Yosida type estimates. Thus, because $F(\cdot)$ is a bounded operator on $\mathbf{X} \to \mathcal{F}$ by the Uniform Boundedness Principle, we see that $E_1$ satisfies Hille - Yosida type estimates also. It follows from the Closed Graph Theorem that $E_2$ is a bounded operator and so a standard argument from perturbation theory shows that $\mathcal{A}$ satisfies the Hille - Yosida estimates. It now follows from Theorem 2.7 that $\mathcal{A}$ generates a non-degenerate integrated semigroup on $\mathcal{Z}$. The result now follows from [1] and [5].

Next we define **Y** to be the Banach space formed from $D(A)$ with the graph norm and use the following definition of a classical solution of (1.1).

**Definition 3.5.** $X(t)$ is a classical solution of (1.1) if $X(\cdot) \in C([0, \infty), \mathbf{Y}) \cap C^1([0, \infty), \mathbf{X})$ and (1.1) is satisfied for $t \geq 0$.

**Theorem 3.6.** Let Assumptions 3.1 and 3.2 be satisfied and let $f : [0, \infty) \to \mathbf{X}$ be continuously differentiable. Then (1.1) has a classical solution iff $X \in D(A)$ and $AX + f(0) \in \overline{D(A)}$. Moreover, (1.1) is wellposed. That is, $X \in D(A)$ and $AX + f(0) \in \overline{D(A)}$ implies that (1.1) has a unique classical solution $X(t, X, f)$. For each $T > 0$, there is a constant $M(T)$ so that for $t \in [0, T]$,

$$\|X(t, X, f)\| \leq M(\|X\| + \int_0^t \|f(s)\| \, ds).$$

**Proof.** One argues as in [9] that (1.6) has a classical solution iff (1.1) has a classical solution and invokes Theorem 3.4.

## 4. Integral solutions of (1.1)

In this section we look at (1.1) in the integrated form (1.8). Using the results obtained in Section 2 and assuming that the strong derivative $F'(t)$ exists, we will prove that $A$ satisfies resolvent estimates of the Hille - Yosida type iff (1.8) has a unique continuous $\overline{D(A)}$ - valued solution for any $X \in \overline{D(A)}$ and $f$ locally integrable. First we consider the case when $F(\cdot) \equiv 0$ and use (1.6) in this case. Then (1.8) and (1.6) become

$$X(t) - X = A \int_0^t X(s)\, ds + \int_0^t K(t-s) \int_0^s X(u)\, du\, ds + \int_0^t f(s)\, ds, \quad (4.1)$$

and

$$\begin{pmatrix} U(t) \\ \phi_t \end{pmatrix} - \begin{pmatrix} X \\ \phi_0 \end{pmatrix} = \begin{pmatrix} A & \delta \\ K(\cdot) & \mathcal{D} \end{pmatrix} \int_0^t \begin{pmatrix} U(s) \\ \phi_s \end{pmatrix} ds + \int_0^t \begin{pmatrix} q(s) \\ 0 \end{pmatrix} ds. \quad (4.2)$$

**Definition 4.1.** *Let $f : [0, \infty) \to \mathbf{X}$ be locally integrable and $X \in \overline{D(A)}$. Then $X(t)$ is said to be an integral solution of (1.1) if $X(\cdot) : [0, \infty) \to \mathbf{X}$ is continuous and satisfies (1.8) for $t \geq 0$.*

**Lemma 4.2.** *The following two statements are equivalent:*
  a) *For $X \in \overline{D(A)}$, $f : [0, \infty) \to \mathbf{X}$ locally integrable, (4.1) has a unique continuous $\overline{D(A)}$ - valued solution.*
  b) *For $(X, \phi_0) \in \overline{D(A)} \times \mathcal{F}$ and $(q, 0) : [0, \infty) \to \mathcal{Z}$ locally integrable, (4.2) has a unique continuous $\overline{D(A)} \times \mathcal{F}$ - valued solution.*

**Proof.** First assume b). For $X \in \overline{D(A)}$, $f : [0, \infty) \to \mathbf{X}$ locally integrable, we have $(X, 0) \in \overline{D(A)} \times \mathcal{F}$, and $(f, 0) : [0, \infty) \to \mathcal{Z}$ locally integrable. By b) we know that (4.2) has a unique continuous $\overline{D(A)} \times \mathcal{F}$ - valued solution such that

$$\begin{aligned} U(t) - X &= A \int_0^t U(s)\, ds + \delta \int_0^t \phi_s\, ds + \int_0^t f(s)\, ds, \\ \phi_t &= \mathcal{D} \int_0^t \phi_s\, ds + \int_0^t K(\cdot) U(s)\, ds. \end{aligned} \quad (4.3)$$

Note that $\mathcal{D}$ generates the $C_0$ translation semigroup $T(t)$. Then from (4.3) and [14],

$$\begin{aligned} \int_0^t \phi_s\, ds &= \int_0^t \left( \int_0^{t-s} T(r)\, dr \right) K(\cdot) U(s)\, ds \\ &= \int_0^t \int_0^{t-s} K(r + \cdot) U(s)\, dr\, ds \\ &= \int_0^t K(r + \cdot) \int_0^{t-r} U(s)\, ds\, dr \\ &= \int_0^t K(t - \rho + \cdot) \int_0^{\rho} U(s)\, ds\, d\rho. \end{aligned}$$

Also,

$$\delta \int_0^t \phi_s \, ds = \delta \int_0^t K(t-s+\cdot) \int_0^s U(r) \, dr \, ds$$
$$= \int_0^t K(t-s) \int_0^s U(r) \, dr \, ds$$

thus,

$$U(t) - X = A \int_0^t U(s) \, ds + \int_0^t K(t-s) \int_0^s U(r) \, dr \, ds + \int_0^t f(s) \, ds$$

and therefore $U(t)$ is a solution of (4.1).

If $V(t)$ is another solution of (4.1) corresponding to $(X, f)$, then $W = U - V$ satisfies

$$W(t) = A \int_0^t W(s) \, ds + \int_0^t K(t-s) \int_0^s W(r) \, dr \, ds.$$

Recall that $\mathcal{D}$ is the generator of a $C_0$ semigroup. Thus, there exists $\psi_t$ such that

$$\psi_t = \mathcal{D} \int_0^t \psi_s \, ds + \int_0^t K(\cdot) W(s) \, ds$$
$$= \mathcal{D} \int_0^t \psi_s \, ds + K(\cdot) \int_0^t W(s) \, ds.$$

Thus,

$$\begin{pmatrix} W(t) \\ \psi_t \end{pmatrix} = \begin{pmatrix} A & \delta \\ K(\cdot) & \mathcal{D} \end{pmatrix} \int_0^t \begin{pmatrix} W(s) \\ \psi_s \end{pmatrix} ds$$

and therefore by the uniqueness assumption for solutions of (4.2), $W(t) = U(t) - V(t) = 0$, $t \geq 0$, so the solution of (3.1) is unique.

Now assume (a). Let $(X, \phi_0) \in \overline{D(A)} \times \mathcal{F}$ and $q : [0, \infty) \to \mathcal{Z}$ be locally integrable. Then (a) implies that

$$U(t) - X = A \int_0^t U(s) \, ds + \int_0^t K(t-s) \int_0^s U(r) \, dr \, ds$$
$$+ \int_0^t (\phi_0(s) + q(s)) \, ds,$$
$$\phi_t - \phi_0 = \mathcal{D} \int_0^t \phi_s \, ds + \int_0^t K(\cdot) U(s) \, ds$$

has a solution $(U(t), \phi_t)$. Again, as $\mathcal{D}$ generates a $C_0$ translation semigroup $T(t)$,

$$\int_0^t \phi_s \, ds = \int_0^t T(s) \phi_0 \, ds + \int_0^t \left( \int_0^{t-s} T(r) \, dr \right) K(\cdot) U(s) \, ds$$
$$= \int_0^t \phi_0(s + \cdot) \, ds + \int_0^t K(t-s+\cdot) \int_0^s U(r) \, dr \, ds$$

and
$$\delta \int_0^t \phi_s\, ds = \int_0^t \phi_0(s)\, ds + \int_0^t K(t-s) \int_0^s U(r)\, dr\, ds$$
and, hence, $(U(t), \phi_t)$ is a solution of (4.2). The uniqueness of the solution follows easily.

Applying Theorem 2.7 to (4.2) we get

**Theorem 4.3.** *The following three statements are equivalent:*

a) *$A$ is a (not necessarily densely defined) linear closed operator and there exist constants $\alpha \in \mathbb{R}$, $M \geq 1$, such that $\lambda > \alpha \Rightarrow \lambda \in \rho(A)$ and*

$$\|(\lambda - A)^{-n}\| \leq M/(\lambda - \alpha)^n, \ \forall \lambda > \alpha, \ n = 1, 2, 3, \ldots$$

b) *$A$ generates a non-degenerate integrated semigroup $S(t)$ of type $(M, \alpha)$.*

c) *For $\lambda$ sufficiently large, $\lambda \in \rho(A)$ and $\limsup_{\lambda \to \infty} \lambda \|(\lambda - A)^{-1}\| < \infty$. In addition, for any $X \in \overline{D(A)}$ and $f : [0, \infty) \to \mathbf{X}$ locally integrable,*

$$X(t) - X = A \int_0^t X(s)\, ds + \int_0^t K(t-s) \int_0^s X(r)\, dr\, ds + \int_0^t f(s)\, ds$$

*has a unique continuous $\overline{D(A)}$ - valued (integral) solution whenever $K(\cdot)$ satifies Assumption 3.2.*

**Proof.** We know (a) $\iff$ (b) already. Note that (c) $\Rightarrow$ (a) by taking $K(\cdot) = 0$ and using Theorem 2.7. Thus we only need to prove that (a) $\Rightarrow$ (c). Now assume (a) is valid. Then from the discussion in Section 3,

$$G = \begin{pmatrix} A & \delta \\ K(\cdot) & D \end{pmatrix}$$

satisfies condition (a) of Theorem 2.7. Hence, Theorem 2.7((a) $\Rightarrow$ (c)) implies that (4.2) has a unique continuous $\overline{D(A)} \times \mathcal{F}$ - valued solution for each choice of initial data. Therefore (c) is true by applying Lemma 4.2.

Now we are in position to get the integral solution of (1.1), i.e., the solution of (1.8). We formally define $V(t)$ by

$$V'(t) = AX(t), \qquad V(0) = 0,$$

and rewrite (1.1)

$$X(t)' = AX(t) + \int_0^t F(t-s) V'(s)\, ds + \int_0^t K(t-s) X(s)\, ds + f(t),$$
$$V'(t) = AX(t).$$

As we assume that the strong derivative $F'(t)$ exists for $t \geq 0$, integration by parts with $V(0) = 0$ yields an equation whose integrated form is

$$\begin{pmatrix} X(t) \\ V(t) \end{pmatrix} - \begin{pmatrix} X \\ 0 \end{pmatrix} = \begin{pmatrix} A & F(0) \\ A & 0 \end{pmatrix} \int_0^t \begin{pmatrix} X(s) \\ V(s) \end{pmatrix} ds + \int_0^t \begin{pmatrix} f(s) \\ 0 \end{pmatrix} ds$$
$$+ \int_0^t \begin{pmatrix} K(t-s) & F'(t-s) \\ 0 & 0 \end{pmatrix} \int_0^s \begin{pmatrix} X(r) \\ V(r) \end{pmatrix} dr\, ds. \quad (4.4)$$

We now build some relationships between (1.8) and (4.4).

**Lemma 4.4.** *(1.8) has a unique $\overline{D(A)}$ - valued solution iff (4.4) has a unique $\overline{D(A)} \times X$ - valued solution.*

**Proof.** Integrate by parts.

Now assume that Assumptions 3.1 and 3.2 are valid and consider

$$A_1 = \begin{pmatrix} A & 0 \\ A & 0 \end{pmatrix}, \quad A_2 = \begin{pmatrix} 0 & F(0) \\ 0 & 0 \end{pmatrix}.$$

Then for $\lambda > \omega$, $\lambda > 0$,

$$\lambda(\lambda - A_1)^{-1} = \begin{pmatrix} \lambda(\lambda - A)^{-1} & 0 \\ \lambda(\lambda - A)^{-1} - I & I \end{pmatrix}.$$

An easy estimate then shows that $\limsup_{\lambda \to \infty} \lambda \|(\lambda - A_1)^{-1}\| < \infty$.

Now consider

$$\begin{pmatrix} Y(t) \\ W(t) \end{pmatrix}' = \begin{pmatrix} A & 0 \\ A & 0 \end{pmatrix} \begin{pmatrix} Y(t) \\ W(t) \end{pmatrix} + \begin{pmatrix} f(t) \\ q(t) \end{pmatrix}$$
$$\begin{pmatrix} Y(0) \\ W(0) \end{pmatrix} = \begin{pmatrix} Y \\ W \end{pmatrix}.$$

Its integrated form is

$$\begin{pmatrix} Y(t) \\ W(t) \end{pmatrix} - \begin{pmatrix} Y \\ W \end{pmatrix} = \begin{pmatrix} A & 0 \\ A & 0 \end{pmatrix} \int_0^t \begin{pmatrix} Y(s) \\ W(s) \end{pmatrix} ds + \int_0^t \begin{pmatrix} f(s) \\ q(s) \end{pmatrix} ds. \quad (4.5)$$

Note that as $A$ satisfies Assumption 3.1, the first equation of (4.5) has a unique $\overline{D(A)}$ - valued solution $Y(t)$ so by setting

$$W(t) = W + A \int_0^t Y(s)\, ds + \int_0^t q(s)\, ds$$

we get a unique $\overline{D(A)} \times X$ - valued solution of (4.5). Now applying Theorem 2.7 $((c) \Rightarrow (a))$ to (4.5) we see that $A_1$ satisfies Hille - Yosida resolvent estimates.

Then from the discussion of Section 3, $A_1 + A_2$ also satisfies Hille - Yosida estimates as $A_2$ is bounded. We next note that

$$\begin{pmatrix} K(\cdot) & F'(\cdot) \\ 0 & 0 \end{pmatrix}$$

is bounded. Applying Theorem 4.3 $((a) \Rightarrow (c))$ we conclude that (4.4) has a unique continuous $\overline{D(A)} \times X$ - valued solution and, hence, by Lemma 4.4 we have:

**Theorem 4.5.** *The following are equivalent:*

a) *A is a (not necessarily densely defined) linear closed operator and there exist constants $\alpha \in \mathbb{R}$ and $M \geq 1$, such that $\lambda > \alpha \Rightarrow \lambda \in \rho(A)$ and*

$$\|(\lambda - A)^{-n}\| \leq M/(\lambda - \alpha)^n, \quad \forall \lambda > \alpha, \ n = 1, 2, 3, \ldots$$

b) *A generates a non-degenerate integrated semigroup $S(t)$ of type $(M, \omega)$.*

c) *For $\lambda$ sufficiently large, $\lambda \in \rho(A)$, and $\limsup_{\lambda \to \infty} \lambda \|(\lambda - A)^{-1}\| < \infty$. In addition, for any $X \in \overline{D(A)}$ and $f : [0, \infty) \to \mathbf{X}$ locally integrable,*

$$X(t) - X = A \int_0^t X(s)\,ds + \int_0^t (F(t-s)A + K(t-s)) \int_0^s X(r)\,dr\,ds + \int_0^t f(s)\,ds$$

*has a unique continuous $\overline{D(A)}$ - valued solution whenever $F$ and $K$ satisfy Assumption 3.2.*

## 5. An example

In this section we shall examine a special case from [10]. Consider the wave equation

$$U_{tt}(x,t) = U_{xx}(x,t), \quad 0 \leq x \leq 1, \ t \geq 0,$$

with initial data given for $U$, $U_t$ and Dirichlet boundary conditions. In order to apply the theory of integrated semigroups we define the new variables $V = U_t$ and $W = U_x$ then, we can write the system as

$$\begin{pmatrix} V \\ W \end{pmatrix}_t = \begin{pmatrix} 0 & \partial_x \\ \partial_x & 0 \end{pmatrix} \begin{pmatrix} V \\ W \end{pmatrix}$$

with initial data and boundary conditions

$$V(x,0) = H(x), \quad W(x,0) = F(x),$$

$$V(x,t) = 0, \ t \geq 0, \ x = 0, 1.$$

We now diagonalize to get the system

$$\begin{pmatrix} V \\ W \end{pmatrix}_t = \begin{pmatrix} \partial_x & 0 \\ 0 & -\partial_x \end{pmatrix} \begin{pmatrix} V \\ W \end{pmatrix}$$

with initial data and boundary conditions

$$V(x,0) = (H(x) + F(x))/2, \quad W(x,0) = (H(x) - F(x))/2,$$

$$V(x,t) + W(x,t) = 0, \ t \geq 0, \ x = 0, 1.$$

We now define **X** to be $C[0,1] \times C[0,1]$ and

$$D(A) = \{\begin{pmatrix} V \\ W \end{pmatrix} \in C^1 \times C^1 : V(x) + W(x) = 0, x = 0, 1\},$$

$$A = \begin{pmatrix} \partial_x & 0 \\ 0 & -\partial_x \end{pmatrix} \tag{5.1}$$

It is clear that $D(A)$ is not dense in **X**. To obtain a Hille - Yosida estimate for the resolvent of $A$ consider

$$\begin{aligned} \lambda V - V' &= f, \\ \lambda W + W' &= g. \end{aligned} \tag{5.2}$$

In order to fully take advantage of the diagonal form of $A$ we define the norm on **X** by

$$\left\| \begin{pmatrix} V \\ W \end{pmatrix} \right\| = \max\{\|V\|_\infty, \|W\|_\infty\}.$$

Now suppose that $\left\| \begin{pmatrix} V \\ W \end{pmatrix} \right\| = \|V\|_\infty$. If the max of $|V(x)|$ occurs at an interior point $z \in (0,1)$, then $V'(z) = 0$ implies

$$\lambda V(z) = f(z) \quad \text{and} \quad \|V\|_\infty \le \|f\|_\infty/\lambda, \ \lambda > 0.$$

If the max of $|V|$ occurs at 0 we may assume $V(0) > 0$ (by considering $-\begin{pmatrix} V \\ W \end{pmatrix}$ if necessary). Then $V'(0) \le 0$ and (5.2) yield $\lambda V(0) \le f(0)$. Finally, if max $|V(x)|$ occurs at $x = 1$ we make use of the boundary condition $V(1) + W(1) = 0$ to see that we may consider $\|W\|_\infty$ instead of $\|V\|_\infty$. In this case, we assume $W(1) > 0$ so $W'(1) \ge 0$. This and (5.2) give $\lambda W(1) \le g(1)$. In any case, we get

$$\left\| \begin{pmatrix} V \\ W \end{pmatrix} \right\| \le \left\| \begin{pmatrix} f \\ g \end{pmatrix} \right\|/\lambda.$$

The case where we must consider $\|W\|_\infty$ is clearly the same and we have

**Theorem 5.1.** *The operator $A$ associated with the wave equation given by (5.1) generates a non-degenerate integrated semigroup of type (1,0) on* $\mathbf{X} = C[0,1] \times C[0,1]$.

One can actually show that much more general symmetric hyperbolic systems will lead to integrated semigroups and we refer to [10]. It is an easy thing now to construct an integrodifferential equation in **X** which fits our discussion and which can be easily related to problems in viscoelasticity, for example.

For additional examples the reader is referred to Da Prato and Sinestrari [5] and Thieme [14].

## References

[1] W. Arendt, *Vector - valued Laplace transforms and Cauchy problems*, Israel J. Math. **59** (1987), 327–352.

[2] W. Arendt and H. Kellerman, *Integrated solutions of Volterra integrodifferential Equations and applications*, Semesterbericht Funktionalanalysis **12** (1987), 141 –166.

[3] J. M. Ball, *Strongly continuous semigroups, weak solutions and the variation of constants formula*, Proc. Amer. Math. Soc. **63** (1977), 370 – 373.

[4] G. Chen and R. Grimmer, *Semigroups and integral equations*, J. Integral Eq. **2** (1980), 133–154.

[5] G. Da Prato and E. Sinestrari, *Differential operators with non-dense domain*, Ann. Scuola Norm. Sup. Pisa **14(2)** (1987), 285–344.

[6] W. Desch, R. Grimmer and W. Schappacher, *Some considerations for linear integrodifferential equations*, J. Math. Anal. Appl. **104** (1984), 219–234.

[7] W. Desch and W. Schappacher, *A semigroup approach to integrodifferential equations in Banach space*, J. Integal Eq. **10** (1985), 99–110.

[8] R. Grimmer, *Resolvents for integral equations in abstract spaces*, Evolution equations and their applications, F. Kappel and W. Schappacher, (eds.), Pittman, Boston, 1982, pp. 101–120.

[9] R. Grimmer, *Resolvent operators for integral equations in a Banach space*, Trans. Amer. Math. Soc. **273** (1982), 333–349.

[10] R. Grimmer and E. Sinestrari, *Integrated semigroups and symmetric hyperbolic systems*, preprint.

[11] R.K. Miller, *An integrodifferential equation for rigid heat conductors with memory*, J. Math. Anal. Appl. **66** (1978), 313–332.

[12] R.K. Miller, *Volterra integral equations in a Banach space*, Funkcial. Ekvac. **18** (1975), 163–193.

[13] A. Pazy, *Semigroups of linear operators and applications to partial differential equations*, Springer Verlag, New York, 1983.

[14] H. Thieme, *Integrated semigroups and integrated solutions to abstract Cauchy problems*, preprint.

[15] H. Thieme, *personal communication*.

[16] H. Thieme, *Semiflows generated by Lipschitz perturbations of non-densely defined operators*, preprint.

Ronald Grimmer
Hetao Liu
Department of Mathematics
Southern Illinois University
Carbondale, Illinois 62901, USA

# On Nodes of Local Solutions to Schrödinger Equations

M. HOFFMANN-OSTENHOF[1]
T. HOFFMANN-OSTENHOF

Institut für Mathematik, Universität Wien
Institut für Theoretische Chemie, Universität Wien

We shall consider the behaviour of a real valued solution $\psi$ of a Schrödinger equation in the neighbourhood of a zero. Let

$(-\Delta + V)\psi = 0$ in $\Omega$,

where $\Omega$ is a domain in $\mathbb{R}^n$, $n \geq 2$ and $V \in C^\infty(\Omega)$ with $V$ real valued. (1)

Without loss we assume $\mathcal{O} \in \Omega$ and $\psi(\mathcal{O}) = 0$.

Note that $\psi \in C^\infty(\Omega)$ by elliptic regularity. Further let $\varepsilon > 0$, $B_\varepsilon = \{x \in \mathbb{R}^n | |x| < \varepsilon\}$ such that $B_\varepsilon \subset \Omega$. In the following we present some recent results on the behaviour of the nodal set and the nodal domains of $\psi$ in $B_\varepsilon$. Most of these results have just appeared in [9]. To prove some of these results we suitably adapt techniques which have been developed in [7] to investigate the nodal behaviour of local solutions of Schrödinger equations in a neighbourhood of infinity. Those aspects just indicated in [9] are given in detail here.

Let $\mathcal{N}_\varepsilon = \{x \in B_\varepsilon | \psi(x) = 0\}$, a component $D$ of $B_\varepsilon \setminus \mathcal{N}_\varepsilon$ will be called a *local nodal domain (l.n.d.)* of $\psi$ in $B_\varepsilon$ and we define

$$\mathcal{D}_\varepsilon = \{\text{l.n.d. } D \text{ of } \psi \text{ in } B_\varepsilon \text{ with } \mathcal{O} \in \partial D\}.$$

Let $\mathcal{C}_\varepsilon = \{x \in B_\varepsilon | \psi(x) = 0, \nabla \psi(x) = 0\}$. It is known (see [2,3,5]) that under the above assumptions the manifold $\mathcal{N}_\varepsilon \setminus \mathcal{C}_\varepsilon$ is as regular as the solution $\psi$,

---

[1] Partially supported by "Fonds zur Förderung der wissenschaftlichen Forschung in Österreich", Projekt Nr. P7011 at the Institut für Theoretische Physik.

and that the Hausdorff dimension of $\mathcal{C}_\varepsilon \leq n - 2$. So clearly the case $\mathcal{O} \in \mathcal{C}_\varepsilon$ is the interesting one, and given $D \in \mathcal{D}_\varepsilon$ one may ask whether it satisfies an interior cone condition. Furthermore we shall also deal with the question of the cardinality of $\mathcal{D}_\varepsilon$. To investigate such problems we rely heavily on a result of Bers [1]:

**Proposition 1.** *Let $\psi$ satisfy (1). Then there exists a harmonic homogeneous polynomial $P_M(x) \not\equiv 0$ of degree $M \geq 1$ such that for $0 < \nu < 1$*

$$\frac{\partial^\ell (\psi - P_M)(x)}{\partial x_1^{i_1} \ldots \partial x_n^{i_n}} = O(|x|^{M-\ell+\nu}) \quad \text{for } |x| \to 0 \tag{2}$$

*for $\ell = 0, 1, \ldots M$, where $\sum_{j=1}^n i_j = \ell$.*

Using polar coordinates $x = ry$ with $r = |x|$ and $y = x/|x| \in S^{n-1}$, $S^{n-1}$ the $n-1$-dimensional unit sphere, we can write $P_M(ry) = r^M Y_M(y)$ with $Y_M$ some surface harmonic, and (2) implies

$$r^{-M} \psi(ry) \to Y_M(y) \quad \text{for } r \to 0, \quad \text{for } y \in S^{n-1}. \tag{3}$$

We denote the nodal set of $Y_M$ with $\mathcal{N}(Y_M)$ and the set of nodal domains of $Y_M$ with $\mathcal{U}(Y_M)$. (The components of $S^{n-1} \setminus \mathcal{N}(Y_M)$ are called nodal domains of $Y_M$.)

For dimension $n = 2$ Cheng [3] showed that the nodal lines of $\psi$ look locally as the nodal lines of $P_M$, which are straight lines intersecting in $\mathcal{O}$ and forming an equiangular system. So for $\varepsilon$ small, $\mathcal{D}_\varepsilon$ and $\mathcal{U}(Y_M)$ have the same number of elements.

For dimensions $n \geq 3$ the situation is more delicate as can be seen from the following harmonic function in $\mathbb{R}^3$

$$\psi(x_1, x_2, x_3) = x_1 x_2 + x_1^2 x_3 - x_3^2/3$$

which has a zero in $\mathcal{O}$ of order 2. The corresponding harmonic homogeneous polynomial is $P_2(x_1, x_2, x_3) = x_1 x_2$, so obviously $\#\mathcal{U}(Y_2) = 4$. On the other hand studying the intersection of the nodal set of $\psi$ with planes $x_3 = c$ with $c < 0, = 0$ and $> 0$, it is easily seen that $\psi$ has only 2 nodal domains.

To state our results we need the following definition: For $D \in \mathcal{D}_\varepsilon$, $D$ arbitrary but fixed, let

$$S(r) = \{y \in S^{n-1} | ry \in D\} \quad \text{for } 0 < r < \varepsilon.$$

Further we denote $|\{\cdot\}| = \int_{\{\cdot\}} d\sigma$, with $d\sigma$ the surface measure on $S^{n-1}$.

**Theorem 1.** Let $n \geq 3$, suppose $\psi$ satisfies (1), and let $Y_M$ denote the surface harmonic for which (3) holds. Let $D \in \mathcal{D}_\varepsilon$, $D$ arbitrary but fixed, with $S(r)$, $0 < r < \varepsilon$ given as above.

Let $\mathcal{M} \subset S^{n-1}$ denote the union of all nodal domains $U$ of $Y_M$ with the property that there exists a sequence $\{r_m y_m\}$ with $r_m y_m \in D$ for all $m$, $r_m \to 0$ and $y_m \to \bar{y}$ for $m \to \infty$ for some $\bar{y} \in U$.

Then $\mathcal{M} \neq \emptyset$ and $|\mathcal{M} \setminus S(r) \cup S(r) \setminus \mathcal{M}| \to 0$ for $r \to 0$.

This result, which implies $|S(r)| \to |\mathcal{M}| \neq 0$ for $r \to 0$ in particular rules out that $S(r)$ "shrinks" for $r \to 0$ into a subset of $\mathcal{N}(Y_M)$. There are some rather immediate consequences:

**Corollary.**
(i) There exists a cone $K$ with vertex $\mathcal{O}$ and $K \subset D$.
(ii) $\#\mathcal{D}_\varepsilon \leq \#\mathcal{U}(Y_M)$, furthermore $\#\mathcal{D}_\varepsilon$ is constant for $\varepsilon$ small enough. ($\#\{\cdot\}$ denotes the cardinality of $\{\cdot\}$.)
(iii) Let $\psi_0^2(r) = \int_{S(r)} \psi^2 d\sigma$ and $\psi_{av}^2(r) = \int_{S^{n-1}} \psi^2 d\sigma$, then

$$r^{-M}\psi_0 \to (\int_{\mathcal{M}} Y_M^2 d\sigma)^{1/2} > 0$$

and

$$\psi_0/\psi_{av} \to (\int_{\mathcal{M}} Y_M^2 d\sigma / \int_{S^{n-1}} Y_M^2 d\sigma)^{1/2} \quad \text{for } r \to 0.$$

These findings show that the local properties of $\psi$ in the neighborhood of a zero are determined to a certain extent by global properties of the nodal set of the corresponding surface harmonic.

Of course it would be desirable to study the local behaviour of the nodal domains of $\psi$ with weaker regularity assumptions on $V$. It would be also of interest to extend the foregoing results appropriately to the case where the Laplacian is replaced by more general elliptic operators.

Let us sketch the main idea of the

**Proof of Theorem 1.** (Compare also [7] and [9].) The difficult part of the proof is to verify that $\mathcal{M} \neq \emptyset$. For this purpose we investigate the asymptotic behaviour of $\psi_0(r) = (\int_{S(r)} \psi^2 d\sigma)^{1/2}$ for $r \to 0$ and proceed similarly as in [7] where we studied the asymptotics of a solution $\psi$ of a Schrödinger equation for $r \to \infty$. We suppose indirectly that $\mathcal{M} = \emptyset$, which implies

$$|S(r)| \to 0 \quad \text{for } r \to 0. \tag{6}$$

Let for $0 < r < \varepsilon$, $\varepsilon$ small

$$\lambda^2(r) = \inf_{\varphi \in C_0^\infty(S(r))} \int |L\varphi|^2 d\sigma / \int |\varphi|^2 d\sigma$$

where $-L^2$ denotes the Laplace-Beltrami operator on $S^{n-1}$, then we obtain from (6)
$$\lambda^2(r) \to \infty \quad \text{for } r \to 0. \tag{7}$$

The proof is essentially based on the following two lemmas:

**Lemma 1.** *Let $\tilde{\psi}_0 = r^{(n-1)/2}\psi_0$, then $\tilde{\psi}_0$ satisfies*
$$\left(-\frac{d^2}{dr^2} + \inf_{y \in S^{n-1}} V + \frac{(n-1)(n-3)}{4r^2} + \frac{\lambda^2(r)}{r^2}\right)\tilde{\psi}_0 \leq 0 \quad \text{for } 0 < r < \varepsilon \tag{8}$$

*in the distributional sense.*

The proof of Lemma 1 is the same as of Lemma 3.1 in [7].

**Lemma 2.** *For some $C > 0$*
$$\tilde{\psi}_0 \geq C\left(\frac{r}{\lambda(r)}\right)^{2\gamma} \quad \text{for } r < \varepsilon$$
$$\text{with } 2\gamma = M + \frac{n-1}{2}. \tag{9}$$

The proof of Lemma 2 is analogous to that of Lemma 3.2 in [7] (where $v(r)$ and $A(y)$ must be replaced by $r^M$ and $Y_M$ resp.). Since it is rather involved we shall sketch the main idea here later on.

We take into account (7) and obtain from inequality (8) by standard comparison techniques that
$$\tilde{\psi}_0 = O(r^m) \quad \text{for } r \to 0 \text{ for all } m \in \mathbb{N}. \tag{10}$$

On the other hand combination of (8) and (9) yields
$$-\tilde{\psi}_0'' + \tilde{\psi}_0^{1-2\alpha} \leq 0 \quad \text{for } 0 < r < R_\alpha \tag{11}$$

for some $\alpha < (2\gamma)^{-1}$ and some $R_\alpha < \varepsilon$.

Define for $R$ small
$$\phi(r) = e^{-\alpha\mu(r)} \quad \text{with } \mu(r) = \int_r^R \tilde{\psi}_0(t)^{-\alpha} dt,$$

then it is easily seen that for small $r$
$$-\phi'' + \tilde{\psi}_0^{-2\alpha}\phi = \alpha\tilde{\psi}_0^{-1-\alpha}\tilde{\psi}_0'\phi \geq 0. \tag{12}$$

(11) and (12) imply again via comparison techniques that $\tilde{\psi}_0 \phi^{-1} \leq const$ for $r \to 0$. Therefore
$$\alpha^{-1}(e^{-\alpha\mu})' = \psi_0^{-\alpha}\phi^\alpha \geq const > 0 \text{ for small } r.$$

Integration finally implies

$$e^{-\alpha\mu(r)} \geq const \cdot r \text{ for } r \text{ small}$$

which contradicts (10), hence $\mathcal{M} \neq \emptyset$.

We finally give the main ideas of the

**Proof of Lemma 2.** For convenience we assume that $S(r)$ "shrinks" for $r \to 0$ into a single point $\bar{y} \in S^{n-1}$. The general case which is a little more involved can be treated in a similar manner. (The techniques in [7] can be easily adapted.)

Let $B_\varepsilon(\bar{y})$ denote the geodesic disk centered in $\bar{y}$ with radius $\varepsilon > 0$ small. Then due to our assumption $S(r) \subset B_\varepsilon(\bar{y})$ for $r < r_\varepsilon$ for $r_\varepsilon$ small enough. Using local coordinates $\xi = \phi(y)$, $y \in B_\varepsilon(\bar{y})$, $\xi \in \mathbb{R}^{n-1}$ as introduced in section 2 of [7] and taking into account (6) it is easily seen that

$$\lambda^2(r) \geq C_\varepsilon \inf_{f \in W_0^{1,2}(G(r))} \frac{\|\nabla f\|^2}{\|f\|^2} \text{ for } r \leq r_\varepsilon \qquad (13)$$

for some $0 < C_\varepsilon < 1$, where $G(r) \equiv \phi(S(r))$, and $\|\cdot\|^2 = \int_{G(r)} |\cdot|^2 d\xi$.

Next we apply a result of Davies (Theorem 1.5.3 in [4]) which tells us that

$$\inf_{f \in W_0^{1,2}(G(r))} (\|\nabla f\|^2/\|f\|^2) \geq \frac{n-1}{4} q_\infty(r)^{-2}. \qquad (14)$$

Thereby $q_\infty(r)$ denotes the $L^\infty$-norm on $G(r)$ of $q(\xi)$ which is defined by

$$q(\xi)^{-2} = \int_{S^{n-1}} d(\xi, e)^{-2} d\mu(e)$$

($\mu(e)$ denotes the normalized surface measure on $S^{n-2}$) and where

$$d(\xi, e) = \inf\{|t| | \xi + te \notin G(r)\} \quad \forall \xi \in G(r), \forall e \in S^{n-2}.$$

Combining (13) and (14) we obtain

**Proposition 2.** *For some* $c(n, \varepsilon) > 0$

$$\lambda(r) \geq c(n, \varepsilon) q_\infty(r)^{-1} \quad \text{for small } r. \qquad (15)$$

Obviously this implies that it suffices to show

$$\tilde{\psi}_0 \geq const(r q_\infty(r))^{2\gamma} \quad \text{for small } r \qquad (16)$$

instead of (9). To prove (16) we use the following one-dimensional inequality:

**Proposition 3.** *Let $f \in C^M(J)$, $M \geq 1$, $J \subset \mathbb{R}$ a bounded interval with length $|J|$, $f$ bounded in $\bar{J}$ and $\inf_{t \in J} |f^{(M)}(t)| \equiv m_M > 0$, then*

$$\int_J |f| dt \geq 3^{-(M+M^2)/2} |J|^{M+1} m_M. \tag{17}$$

This inequality is an immediate consequence of Prop. 4.3 in [7], which itself is rather elementary.

Suppose without loss that $\psi > 0$ in $D$. Making use of Proposition 1 it can be shown that for $\nu > 0$ small and $r \leq \bar{r}$ small there exists $\bar{\xi}(r) \in G(r)$ and $J_{\nu,r} \subset S^{n-2}$ with $|J_{\nu,r}| \geq c_0(\nu) > 0$ such that

$$\left.\begin{array}{l} g(r,t) \equiv r^{-M} \int_{J_{\nu,r}} \psi(r\phi^{-1}(\bar{\xi}(r) + te)) d\mu(e) \\ \text{satisfies for } t \in [0, \nu q_\infty(r)] \text{ and } r \leq \bar{r} \\ g \geq 0, \left|\dfrac{\partial^k}{\partial t^k} g\right| \geq const > 0 \text{ for some } k \leq M. \end{array}\right\} \tag{18}$$

Now we identify in Proposition 3 for each $r$, $f$ with $t^{n-2}g(r,t)$ and $J$ with $[0, \nu q_\infty(r)]$, and obtain

$$\int_0^{\nu q_\infty(r)} t^{n-2} g(r,t) dt \geq const \, q_\infty(r)^{k+n-1}. \tag{19}$$

Let $\mathcal{C}_{\nu,r} = \{\xi \in \mathbb{R}^{n-1} | \xi = \bar{\xi}(r) + te, t \in [0, \nu q_\infty(r)], e \in J_{\nu,r}\}$, then $|\mathcal{C}_{\nu,r}| \leq const \, q_\infty(r)^{n-1}$, and we conclude by Cauchy-Schwarz that

$$\int_0^{\nu q_\infty(r)} t^{n-2} g(r,t) dt \leq const (\int_{\mathcal{C}_{\nu,r}} \psi^2(r\phi^{-1}(\xi)) d\xi)^{1/2} q_\infty(r)^{(n-1)/2}. \tag{20}$$

(19) and (20) together finally imply (16).

In connection with the results given here more detailed questions about the local behaviour of nodal sets arise in a natural way. For instance one might ask whether $S(r)$ is connected for $r > 0$ sufficiently small. This we could not answer, but for dimension $n = 3$ we obtained more detailed results about the set $\mathcal{M}$.

**Theorem 2.** *Let $n = 3$ and suppose that the assumptions of Theorem 1 hold. Let $\{r_m y_m\}$ be a sequence with $r_m y_m \in D$, $\forall m$ and $r_m \to 0$, $y_m \to \bar{y}$ for $m \to \infty$ for some $\bar{y} \in S^{n-1}$, then $\bar{y} \in \overline{\mathcal{M}}$. Furthermore $\overline{\mathcal{M}}$ is connected ($\overline{\mathcal{M}}$ denotes the closure of $\mathcal{M}$).*

**Proof of Theorem 2.** The case $Y_M(\bar{y}) \neq 0$ is trivial, so let $Y_M(\bar{y}) = 0$, and without loss we assume $\psi > 0$ in $D$.

(a) $\nabla P_M(\bar{y}) \neq 0$

Using the fact that there are only finitely many zeros of order greater one of $Y_M$ we choose a geodesic disk $U_\varepsilon(\bar{y}) \subset S^2$ centered in $\bar{y}$ with radius $\varepsilon$, $\varepsilon$ small enough, where the nodal line of $Y_M$ can be represented by a $C^1$-function $f : J \to S^2$, $J$ a compact interval, with $f'(s) \neq 0$, and

$$U_\varepsilon(\bar{y}) \setminus \mathcal{N}(Y_M) = U^+ \cup U^- \text{ with } U^+, U^- \text{ disjoint domains,} \qquad (21)$$
$$Y_M > 0 \text{ in } U^+ \text{ and } Y_M < 0 \text{ in } U^-.$$

Clearly there is a unique $D^+ \in \mathcal{D}_\varepsilon$ (and due to Theorem 1 a corresponding $\mathcal{M}_{D^+}$), where $U^+$ is a subset of $\mathcal{M}_{D^+}$. So if $D^+ = D$, then $\bar{y} \in \overline{\mathcal{M}}$. Now let $D^+ \cap D = \emptyset$: Define for $\delta > 0$ small

$$\mathcal{Z}_\delta = \{y \in U_\varepsilon(\bar{y})|\ \inf_{s \in J} |y - f(s)| < \delta\}$$

and denote by $\gamma_m$, for $m \geq m_\delta$ ($m_\delta$ large enough) the geodesic on $S^2$ which hits the nodal line $f$ orthogonally and passes through $y_m$, with $|\gamma_m(t)| = |\gamma_m'(t)| = 1$. Since $y_m \in S(r_m) \cap \mathcal{Z}_\delta$ for $m \geq m_\delta$ and $U^+ \setminus \mathcal{Z}_\delta \subset S^+(r)$ for $r \leq r_{m_\delta}$, Rolle's theorem implies for each $m$ that for some $\bar{y}_m \equiv \gamma_m(t_m) \in S(r_m) \cap \gamma_m \cap \mathcal{Z}_\delta$

$$\left.\frac{d}{dt}\psi(r_m \gamma_m(t))\right|_{t=t_m} = 0.$$

This together with Proposition 1 leads to

$$\gamma_m'(t_m) \cdot \nabla P_M(\bar{y}_m) = O(r_m^\nu) \text{ for large } m.$$

Assuming that $\gamma_m'(t_m) \to \tilde{y}$ for $m \to \infty$ for some $\tilde{y} \in S^2$ (otherwise choose a convergent subsequence) the above leads for $m \to \infty$ to

$$\tilde{y} \cdot \nabla P_M(\bar{y}) = 0. \qquad (22)$$

On the other hand it is not difficult to see from the above construction that $\tilde{y} \cdot \bar{y} = \tilde{y} \cdot f'(\bar{s}) = 0$, where $f(\bar{s}) \equiv \bar{y}$. Hence due to (21) $\tilde{y} \cdot \nabla P_M(\bar{y}) \neq 0$ contradicting (22). Therefore $D^+ = D$ and $\bar{y} \in \overline{\mathcal{M}}$.

(b) $\nabla P_M(\bar{y}) = 0$

Suppose indirectly $\bar{y} \notin \overline{\mathcal{M}}$. Choose geodesic disks $U_{\varepsilon'}'(\bar{y})$, $U_\varepsilon(\bar{y})$ with $0 < \varepsilon' < \varepsilon$, with $\varepsilon$ so small that $\bar{y}$ is the only zero of $Y_M$ of order greater one in $U_\varepsilon(\bar{y})$ and $\overline{U}_\varepsilon(\bar{y}) \cap \overline{\mathcal{M}} = \emptyset$. Further let $y' \in \mathcal{M}$ be arbitrary but fixed. Let $B_m = \{x \in \mathbb{R}^3 | |x| < r_m\}$ and suppose without loss $r_m \downarrow 0$ for $m \to \infty$. Since for $m \geq m_\varepsilon$, $m_\varepsilon$ large enough, $y_m \in S(r_m) \cap U_{\varepsilon'}$, $r_m y' \in D$ and $B_m \cap D$ is connected, we can find a path in $B_{m-1} \cap D$ connecting $r_m y_m$ with $r_m y'$ with the property that for some $\rho_m < r_{m-1}$ there is a $\tilde{y}_m \in S(\rho_m) \cap U_\varepsilon \setminus U_{\varepsilon'}$. Choosing a subsequence $\{\tilde{y}_{m(n)}\}$ which converges to some $\tilde{y} \in S^2$ for $n \to \infty$ we obtain $\rho_{m(n)} \to 0$ and $\tilde{y} \in \overline{U_\varepsilon \setminus U_{\varepsilon'}}$. But if $Y_M(\tilde{y}) \neq 0$ clearly $\tilde{y} \in \overline{\mathcal{M}}$, and if $Y_M(\tilde{y}) = 0$, then $\nabla P_M(\tilde{y}) \neq 0$ and (a) implies $\tilde{y} \in \overline{\mathcal{M}}$. This is a contradiction.

That $\overline{\mathcal{M}}$ is connected is proven indirectly following essentially the idea of the proof of (b).

Though we could not prove it we believe that Theorem 2 holds also for dimensions $> 3$.

As already noted the methods for the main part of the proof of Theorem 1 have been developed in [7] to investigate the asymptotic behaviour of nodes of solutions of Schrödinger equations in exterior domains for dimension $n \geq 3$. We remark that there the situation is much more complex than here, even for the 2-dimensional case it is rather delicate (see [6]). For a survey on these results see also [8].

## References

[1] L. Bers, *Local behaviour of solutions of general linear elliptic equations*, Comm. Pure Appl. Math. **8** (1955), 473–496.

[2] L.A. Caffarelli and A. Friedman, *Partial regularity of the zero-set of solutions of linear and superlinear elliptic equations*, J. Differential Equations **60** (1985), 420-433.

[3] S.Y. Cheng, *Eigenfunctions and nodal sets*, Comment. Math. Helvetici **51** (1976), 43-55.

[4] E.B. Davies, *Heat Kernels and Spectral Theory*, Cambridge Tracts in Mathematics 92, Cambridge University Press 1989.

[5] R. Hardt and L. Simon, *Nodal sets for solutions of elliptic equations*, Preprint (1988).

[6] M. Hoffmann-Ostenhof, *Asymptotics of the nodal lines of solutions of 2-dimensional Schrödinger equations*, Math. Z. **198** (1988), 161-179.

[7] M. Hoffmann-Ostenhof and T. Hoffmann-Ostenhof, *On the asymptotics of nodes of $L^2$-solutions of Schrödinger equations in dimensions $\geq 3$*, Commun. Math. Phys. **117** (1988), 49-77.

[8] M. Hoffmann-Ostenhof and T. Hoffmann-Ostenhof, *Asymptotic properties of nodes of solutions to Schrödinger equations*, Symposium "Partial Differential Equations", Holzhau 1988, Teubner-Texte zur Mathematik, Bd. 112, 157-162.

[9] M. Hoffmann-Ostenhof and T. Hoffmann-Ostenhof, *On the local behaviour of nodes of solutions of Schrödinger equations in dimensions $\geq 3$*, Commun. P.D.E. **15**(4) (1990), 435-451.

M. Hoffmann-Ostenhof
Institut für Mathematik
Universität Wien
A-1090 Wien, Strudlhofgasse 4, Austria

T. Hoffmann-Ostenhof
Institut für Theoretische Chemie
Universität Wien
A-1090 Wien, Währinger Straße 17, Austria

# On Integro-Differential Equations with Weakly Singular Kernels

KAZUFUMI ITO
FRANZ KAPPEL

Center for Applied Mathematical Sciences, University of Southern California
Institut für Mathematik, Universität Graz

## 1. Introduction

Consider the integro-differential equation of the form

$$\frac{d}{dt}\left(\int_{-r}^{0} g(\theta) x(t+\theta)\, d\theta\right) = f(t), \quad t \geq 0,$$
$$x(\theta) = \phi(\theta), \quad \theta < 0,$$

where $0 < r \leq \infty$ and $f \in L^1_{\mathrm{loc}}(0, \infty)$. Equations of this type appear as models of certain aeroelastic systems [BCH]. Also Abel's equation can be written in this form [IT]. In this paper we develop a semigroup theoretic treatment for these equations (in case of Abel's equation see [BHS] and [KZ]). In [BI] the case was considered when the kernel $g$ satisfies

$$g > 0 \quad \text{on} \quad (-r, 0), \tag{A.1}$$

$$g \in H^1_{\mathrm{loc}}(-r, 0) \quad \text{and} \quad \dot{g} \geq 0 \quad \text{on} \quad (-r, 0), \tag{A.2}$$

$$g \in L^1(-r, 0). \tag{A.3}$$

If there is no danger of confusion we shall denote with $D$ the operator of differentiation. It has been shown in [BI] that the linear operator $A$ in the Hilbert

---

Research supported in part by NSF under grant UINT-8521208 (K. I.) and by FWF (Austria) under grant P6005 (F. K.).

space $L_g^2(-r,0)$ (the weighted $L_2$-space on $(-r,0)$ with weight $g$) defined by

$$A\phi = D\phi, \quad \phi \in \operatorname{dom} A,$$
$$\operatorname{dom} A = \{\phi \in L_g^2(-r,0) \mid \phi \text{ is locally absolutely continuous in } (-r,0)$$
$$\text{with } D\phi \in L_g^2 \text{ and } \int_{-r}^0 g(\theta)D\phi(\theta)\,d\theta = 0\}$$

generates a $C_0$-semigroup $S(t)$, $t \geq 0$, on $L_g^2(-r,0)$.

The objective of this paper is twofold. First, we investigate the case when $g(\theta) = |\theta|^{-p}$, $0 < p < 1$ and $r < \infty$. We show that there exists a $t_0 > 0$ (depending on $r$ and $p$) such that the semigroup is differentiable for $t \geq t_0$. Next, the case when $r = \infty$ and $g \in L^1(-1,0)$ but $g \notin L^1(-\infty,0)$ is considered. We establish that $A$ generates a $C_0$-semigroup on $L_g^1(-\infty,0)$.

## 2. Differentiability of the solution semigroup

Assume that $g$ satisfies (A1)–(A3). In order to compute the resolvent we consider the equation

$$(\lambda I - A)\phi = f \in L_g^2, \quad \phi \in \operatorname{dom} A,$$

which is equivalent to

$$\lambda \phi - D\phi = f, \tag{2.1}$$

$$\int_{-r}^0 g(\theta)(D\phi)(\theta)\,d\theta = 0. \tag{2.2}$$

From (2.1) we get

$$\phi(\theta) = e^{\lambda \theta}\phi(0) + \int_\theta^0 e^{\lambda(\theta-\xi)} f(\xi)\,d\xi. \tag{2.3}$$

For $\operatorname{Re}\lambda > 0$ we have $\int_{-r}^0 g(\theta)|e^{\lambda\theta}|^2\,d\theta \leq \int_{-r}^0 g(\theta)\,d\theta < \infty$ and

$$\int_{-r}^0 g(\theta)\left|\int_\theta^0 e^{\lambda(\theta-\xi)} f(\xi)\,d\xi\right|^2 d\theta \leq \int_{-r}^0 \left|\int_\theta^0 e^{(\theta-\xi)\operatorname{Re}\lambda} \sqrt{g(\xi)}|f(\xi)|d\xi\right|^2 d\theta$$

$$\leq \int_{-r}^0 \left(\int_\theta^0 e^{(\theta-\xi)\operatorname{Re}\lambda}d\xi\right) \int_\theta^0 e^{(\theta-\xi)\operatorname{Re}\lambda} g(\xi)|f(\xi)|^2 d\xi\, d\theta$$

$$\leq \frac{1}{\operatorname{Re}\lambda} \int_{-r}^0 \left(\int_{-r}^\xi e^{(\theta-\xi)\operatorname{Re}\lambda}d\theta\right) g(\xi)|f(\xi)|^2 d\xi$$

$$\leq \frac{1}{(\operatorname{Re}\lambda)^2} \int_{-r}^0 g(\xi)|f(\xi)|^2 d\xi.$$

Thus, $\phi \in L_g^2$. Substituting (2.3) into (2.2) we obtain

$$\Delta(\lambda)\phi(0) = -\int_{-r}^{0} g(\theta)(D\psi)(\theta)\,d\theta = \int_{-r}^{0} g(\theta)\bigl(f(\theta) - \lambda\psi(\theta)\bigr)\,d\theta, \qquad (2.4)$$

where

$$\Delta(\lambda) = \lambda \int_{-r}^{0} e^{\lambda\theta} g(\theta)\,d\theta \qquad (2.5)$$

and

$$\psi(\theta) = \int_{\theta}^{0} e^{\lambda(\theta-\xi)} f(\xi)\,d\xi. \qquad (2.6)$$

Hence, we can state

**Lemma 2.1.** *$\lambda \in \rho(A) \cap \mathbb{C}^+$ if and only if $\Delta(\lambda) \neq 0$. Moreover, if $r$ is finite, then $\sigma(A)$ is only point spectrum and $\lambda \in \sigma(A)$ if and only if $\Delta(\lambda) = 0$. If $\lambda \in \rho(A)$,*

$$(\lambda I - A)^{-1} f = e^{\lambda\theta}\phi(0) + \psi(\theta), \qquad (2.7)$$

*where $\phi(0)$ satisfies (2.4) and $\psi$ is given by (2.6).*

Throughout the rest of the discussions in this section, we assume that $r$ is finite and $g(\theta) = |\theta|^{-p}$, $0 < p < 1$.

**Lemma 2.2.** *If $\psi$ is given by (2.6), then for $\operatorname{Re}\lambda \leq 0$*

$$\left| \int_{-r}^{0} |\theta|^{-p} (D\psi)(\theta)\,d\theta \right| \leq \frac{1}{\sqrt{1+p}} \left(\frac{1+p}{1-p}\right) r^{(1-p)/2} e^{-(r\operatorname{Re}\lambda)} \|f\|_{L_g^2}.$$

**Proof.** First note that for $\varepsilon > 0$,

$$\int_{-r}^{-\varepsilon} g(\theta)(D\psi)(\theta)\,d\theta = -g(-r)\psi(-r) + g(-\varepsilon)\psi(-\varepsilon) - \int_{-r}^{-\varepsilon} \dot{g}(\theta)\psi(\theta)\,d\theta. \qquad (2.8)$$

We have the estimate

$$|g(-r)\psi(-r)| \leq g(-r) \Bigl(\int_{-r}^{0} \frac{d\xi}{g(\xi)}\Bigr)^{1/2} \Bigl(\int_{-r}^{0} g(\xi)|f(\xi)|^2\,d\xi\Bigr)^{1/2} e^{-r\operatorname{Re}\lambda}$$

$$\leq \frac{1}{\sqrt{1+p}} r^{(1-p)/2} e^{-r\operatorname{Re}\lambda} \|f\|_{L_g^2}$$

and similarly,

$$|g(-\varepsilon)\psi(-\varepsilon)| \leq \frac{\varepsilon^{(1-p)/2}}{\sqrt{1+p}} e^{-\varepsilon\operatorname{Re}\lambda} \|f\|_{L_g^2} \to 0 \quad \text{as} \quad \varepsilon \to 0^+.$$

Moreover,

$$\left|\int_{-r}^{-\varepsilon} \dot{g}(\theta)\psi(\theta)\,d\theta\right| \le \int_{-r}^{-\varepsilon} \dot{g}(\theta)\left(\int_{\theta}^{0} \frac{d\xi}{g(\xi)}\right)^{1/2}\left(\int_{\theta}^{0} g(\xi)|f(\xi)|^2 d\xi\right)^{1/2} d\theta\, e^{-r\mathrm{Re}\,\lambda}$$

$$\le \frac{p}{\sqrt{1+p}} \int_{-r}^{-\varepsilon} |\theta|^{-(1+p)/2} d\theta\, e^{-r\mathrm{Re}\,\lambda} \|f\|_{L_g^2}$$

$$\le \frac{p}{\sqrt{1+p}} \frac{2}{1-p} r^{(1-p)/2} e^{-r\mathrm{Re}\,\lambda} \|f\|_{L_g^2}.$$

Hence, the estimate follows from (2.8). □

Similarly, using Lemma 2.1, we have

**Lemma 2.3.** For $\mathrm{Re}\,\lambda \le 0$,

$$\|(\lambda I - A)^{-1} f\|_{L_g^2} \le \left(\frac{1}{\sqrt{1-p}} r^{(1-p)/2} |\phi(0)| + \frac{r}{\sqrt{2(1+p)}} \|f\|_{L_g^2}\right) e^{-r\mathrm{Re}\,\lambda}.$$

**Proof.** Note that

$$\int_{-r}^{0} g(\theta) |e^{\lambda\theta}|^2 d\theta \le \frac{r^{1-p}}{1-p} e^{-2r\mathrm{Re}\,\lambda}$$

and

$$\int_{-r}^{0} g(\theta)|\psi(\theta)|^2 d\theta \le \int_{-r}^{0} g(\theta)\left(\int_{\theta}^{0} \frac{d\xi}{g(\xi)}\right)\left(\int_{\theta}^{0} g(\xi)|f(\xi)|^2 d\xi\right) d\theta\, e^{-2r\mathrm{Re}\,\lambda}$$

$$\le \int_{-r}^{0} \frac{|\theta|}{1+p} d\theta\, \|f\|_{L_g^2}^2 e^{-2r\mathrm{Re}\,\lambda} = \frac{r^2}{2(1+p)} \|f\|_{L_g^2}^2 e^{-2r\mathrm{Re}\,\lambda}.$$

Thus, the estimate follows from (2.7). □

Next we calculate lower bounds for $|\Delta(\lambda)|$.

**Lemma 2.4.** For each $p \in (0,1)$ there exist positive constants $c_p$ and $C_p$ such that

$$|\Delta(\lambda)| \ge C_p |\mathrm{Im}\,\lambda|^p - c_p e^{-r\mathrm{Re}\,\lambda} \quad \text{for } \mathrm{Re}\,\lambda < 0,\ \left|\frac{\mathrm{Re}\,\lambda}{\mathrm{Im}\,\lambda}\right| \le \frac{p}{2\pi} \text{ and } |\mathrm{Im}\,\lambda| \ge \frac{2\pi}{r}, \tag{2.9}$$

and

$$|\Delta(\lambda)| \ge C_p |\lambda|\,|\mathrm{Im}\,\lambda|^p \quad \text{for } \mathrm{Re}\,\lambda \ge 0,\ \left|\frac{\mathrm{Re}\,\lambda}{\mathrm{Im}\,\lambda}\right| \le \frac{1-p}{2\pi} \text{ and } |\mathrm{Im}\,\lambda| \ge \frac{2\pi}{r}. \tag{2.10}$$

**Proof.** For $\lambda = \alpha + i\omega$ with $\omega \ge 0$ we have

$$\int_{-r}^{0} |\theta|^{-p} e^{\lambda\theta} d\theta = \int_{-r}^{0} (-\theta)^{-p} e^{\alpha\theta} e^{i\omega\theta} d\theta$$

$$= \int_{-r}^{0} F(\theta) \cos\omega\theta\, d\theta + i \int_{-r}^{0} F(\theta) \sin\omega\theta\, d\theta,$$

where $F(\theta) = (-\theta)^{-p} e^{\alpha\theta}$. Since $F'(\theta) = (-\theta)^{-1-p} e^{\alpha\theta}(p - \alpha\theta)$, $F$ is monotonically increasing on $(-\infty, 0)$ in case $\alpha \geq 0$. For $\alpha < 0$ the function $F$ is decreasing on $(-\infty, \theta_0)$ and increasing on $(\theta_0, 0)$, where $\theta_0 = \frac{p}{\alpha}$. Assume that $\theta_0 \in (-r, 0)$ and let

$$J = \int_{-r}^{0} F(\theta) \sin \omega\theta \, d\theta = \int_{-r}^{\theta_0} + \int_{\theta_0}^{0} =: J_1 + J_2.$$

Let $j_0$ be the largest integer such that $\omega r > j_0 \pi$ and $j_1$ be the smallest integer such that $j_1 \pi > \omega |\theta_0|$. Then, we have

$$J_1 = \sum_{k=0}^{j_0 - j_1 + 1} a_k \quad \text{with} \quad a_k = \int_{I_k} F(\theta) \sin \omega\theta \, d\theta,$$

where $I_0 = [-r, -\frac{j_0 \pi}{\omega}]$, $I_k = [-(\frac{j_0 - k + 1}{\omega})\pi, -(\frac{j_0 - k}{\omega})\pi]$, $1 \leq k \leq j_0 - j_1$, and $I_{j_0 - j_1 + 1} = [-(\frac{j_1}{\omega})\pi, \theta_0]$. Observe that $(-1)^k a_k$, $k \geq 0$, have the same sign and that $|a_k|$, $k \geq 1$, is monotonically decreasing. Thus it is not difficult to show that

$$|J_1| \leq \max(|a_0|, |a_1|) \leq \frac{2}{\omega} r^{-p} e^{-r \operatorname{Re} \lambda}.$$

Similarly, one can show that $\tilde{J}_1 = \int_{-r}^{\theta_0} F(\theta) \cos \omega\theta \, d\theta$ satisfies

$$|\tilde{J}_1| \leq \frac{2}{\omega} r^{-p} e^{-r \operatorname{Re} \lambda}.$$

Next, we consider the integrals on $(\theta_0, 0)$,

$$J_2 = \int_{\theta_0}^{0} F(\theta) \sin \omega\theta \, d\theta = \sum_{k=1}^{j_1} b_k \quad \text{with} \quad b_k = \int_{\tilde{I}_k} F(\theta) \sin \omega\theta \, d\theta,$$

where $\tilde{I}_k = [-(\frac{k}{\omega})\pi, -(\frac{k-1}{\omega})\pi]$, $k = 1, \ldots, j_1 - 1$, and $\tilde{I}_{j_1} = [\theta_0, -(\frac{j_1 - 1}{\omega})\pi]$. Note that $(-1)^k b_k$, $k \geq 0$, have the same sign and that $|b_k|$, $k \geq 1$, is monotonically decreasing. Thus in case $\theta_0 \leq -\frac{2\pi}{\omega}$

$$|J_2| \geq |b_0 + b_1| = \left| \int_0^{\frac{2\pi}{\omega}} F(-s) \sin \omega s \, ds \right| = |\omega|^{p-1} \rho(-\alpha/\omega),$$

where $\rho(x) = \int_0^{2\pi} \sigma^{-p} e^{-x\sigma} \sin \sigma \, d\sigma$. Since

$$\rho'(x) = \int_0^{2\pi} \sigma^{-p+1} e^{x\sigma} \sin \sigma \, d\sigma = \int_0^{\pi} (\sigma^{1-p} - (\sigma + \pi)^{1-p} e^{x\pi}) e^{x\sigma} \sin \sigma \, d\sigma,$$

we see that $\rho'(x) < 0$ certainly for $x > \frac{p-1}{2\pi}$ (a sharper bound would be $\frac{p-1}{\pi} \ln 2$). Thus $\rho(x)$ is monotonically decreasing on $(\frac{p-1}{2\pi}, \infty)$.

If $\theta_0 \leq -r$ then we put $J_1 = \tilde{J}_1 = 0$ and $J_2 = \int_{-r}^0 F(\theta)\sin\omega\theta\, d\theta$. As long as $\frac{2\pi}{\omega} \leq r$ we obtain the same estimate for $|J_2|$.

We first consider $\lambda = \alpha + i\omega$ with $\alpha < 0$ and $-\frac{\alpha}{\omega} \leq \frac{p}{2\pi}$ and $\omega \geq \frac{2\pi}{r}$. Then $\theta_0 = \frac{p}{\alpha} \leq -\frac{2\pi}{\omega}$ and $-\frac{2\pi}{\omega} \geq -r$. Hence,

$$|J_2| \geq C_p|\operatorname{Im}\lambda|^{p-1} \quad \text{for } \operatorname{Re}\lambda < 0, \quad \left|\frac{\operatorname{Re}\lambda}{\operatorname{Im}\lambda}\right| \leq \frac{p}{2\pi},$$

where $C_p = \rho(\frac{p}{2\pi})$. From $|\Delta(\lambda)| \geq (|J_2| - (|J_1|^2 + |\tilde{J}_1|^2)^{1/2}) \geq |\omega||J_2| - |\lambda|(|J_1|^2 + |\tilde{J}_1|^2)^{1/2}$ we see that (2.9) holds with $c_p = 2\sqrt{2}\,r^{-p}\sqrt{1 + (\frac{p}{2\pi})^2}$.

Next suppose that $\alpha \geq 0$, $\frac{\alpha}{\omega} \leq \frac{1-p}{2\pi}$ and $\omega \geq \frac{2\pi}{r}$. Then

$$|\Delta(\lambda)| \geq |\lambda|\left|\int_{-r}^0 F(\theta)\sin\omega\theta\, d\theta\right|$$
$$\geq |\lambda||J_2| \geq \rho(0)|\lambda||\operatorname{Im}\lambda|^{p-1},$$

i.e., (2.10) holds because $\rho(0) > C_p$. $\square$

Now, one can state

**Theorem 2.5.** *There exist positive constants $a$, $\gamma$ and $M$ such that*

$$\|(\lambda I - A)^{-1}f\|_{L_g^2} \leq M|\operatorname{Im}\lambda|^p \|f\|_{L_g^2}$$

*provided that $r\operatorname{Re}\lambda \geq a - p\ln|\operatorname{Im}\lambda|$ and $\operatorname{Re}\lambda \leq \gamma$. The semigroup $S(t)$ generated by $A$ is differentiable for $t > t_0$ with $t_0 = \frac{2+p}{p}r$.*

**Proof.** Let $\phi = (\lambda I - A)^{-1}f$, $f \in L_g^2$. Then, by Lemma 2.1, $\phi(0) = -\frac{1}{\Delta(\lambda)}\int_{-r}^0 g\, D\psi\, d\theta$. The estimates of Lemma 2.4 show that there exist constants $\gamma > 0$ and $a > 0$ such that

$$|\Delta(\lambda)| \geq \frac{1}{2}C_p|\operatorname{Im}\lambda|^p \quad \text{for } \lambda \text{ with } \operatorname{Re}\lambda \leq \gamma \text{ and } r\operatorname{Re}\lambda \geq a - p\ln|\operatorname{Im}\lambda|.$$

Thus we get from Lemma 2.2

$$|\phi(0)| \leq \frac{1}{\sqrt{1+p}}\frac{1+p}{1-p}r^{(1-p)/2}e^{-r\operatorname{Re}\lambda}\frac{2}{C_p}|\operatorname{Im}\lambda|^p\|f\|_{L_g^2}$$
$$\leq \frac{1}{\sqrt{1+p}}\frac{1+p}{1-p}r^{(1-p)/2}\frac{2}{C_p}e^{-a}\|f\|_{L_g^2}.$$

Therefore it follows from Lemma 2.3 that

$$\|(\lambda I - A)^{-1}f\|_{L_g^2} \leq \left(\frac{1}{\sqrt{1-p^2}}\frac{1+p}{1-p}r^{1-p}\frac{2}{C_p}e^{-a} + \frac{r}{\sqrt{2(1+p)}}\right)e^{-r\operatorname{Re}\lambda}\|f\|_{L_g^2}$$
$$\leq M|\operatorname{Im}\lambda|^p\|f\|_{L_g^2}$$

for some appropriately chosen constant $M > 0$.

Note that for $\lambda = \alpha + i\tau$ satisfying $\alpha r = a - p\ln|\tau|$

$$\|\lambda e^{\lambda t}(\lambda I - A)^{-1}\| \leq M|\tau|^{p+1}e^{at/r}|\tau|^{-pt/r} = Me^{at/r}|\tau|^{1+p-pt/r}.$$

Hence, using the arguments in the proof of Theorem 2.47 in [P, p. 54–57] one can show that $t \to S(t)\phi$, $\phi \in L_g^2$, is differentiable for $t > t_0 = \frac{2+p}{p}r$. $\square$

## 3. A well-posedness result for non-integrable kernels

Throughout this section, we assume $r = \infty$ and that $g$ satisfies the following assumptions:

(i)   $g$ is locally absolutely continuous on $(-\infty, 0)$,
(ii)  $g > 0$ on $(-\infty, 0)$, $\dot{g} \geq 0$ a. e. on $(-\infty, 0)$,
(iii) $\lim\limits_{\theta \to 0^-} g(\theta) = \infty$ and $\int_{-1}^{0} g(\theta)\, d\theta < \infty$.

We do not assume that $\int_{-\infty}^{0} g(\theta)\, d\theta < \infty$. Consider the linear operator $A$ on $L^1_g(-\infty, 0)$ defined by

$$\text{dom}\, A = \{\phi \in L^1_g \mid D\phi \in L^1_g \text{ and } \int_{-\infty}^{0} g(\theta)(D\phi)(\theta)\, d\theta = 0\},$$
$$A\phi = D\phi \quad \text{for } \phi \in \text{dom}\, A.$$

**Lemma 3.1.** *The operator $A$ is dissipative.*

**Proof.** We need to show that

$$\tau_-(\phi, A\phi) \leq 0 \quad \text{for all } \phi \in \text{dom}\, A,$$

where $\tau_-(\cdot, \cdot) : L^1_g \times L^1_g \to \mathbb{R}$ is given by (see [M])

$$\tau_-(\phi, \psi) = \lim_{t \to 0^+} t^{-1}\big(\|\phi\|_{L^1_g} - \|\phi - t\psi\|_{L^1_g}\big)$$
$$= \int_{Y_1(\phi)} (\text{sgn}\,\phi(\theta))\psi(\theta) g(\theta)\, d\theta - \int_{Y_0(\phi)} |\psi(\theta)| g(\theta)\, d\theta,$$

where

$$Y_1(\phi) = \{\theta \in (-\infty, 0] \mid \phi(\theta) \neq 0\},$$
$$Y_0(\phi) = \{\theta \in (-\infty, 0] \mid \phi(\theta) = 0\}.$$

Since $\phi \in \text{dom}\, A$ is continuous on $(-\infty, 0]$, $Y_1(\phi)$ is a countable union of open intervals (open with respect to $(-\infty, 0]$). These intervals are of the form $(\alpha, \beta)$ with $-\infty < \alpha < \beta < 0$, $(\alpha, 0]$ or $(\alpha, 0)$ with $-\infty < \alpha < 0$, or $(-\infty, 0)$ or $(-\infty, 0]$. For intervals of the form $(\alpha, \beta)$ with $-\infty < \alpha < \beta < 0$ we have $\phi(\alpha) = \phi(\beta) = 0$ and

$$\int_\alpha^\beta (\text{sgn}\,\phi(\theta))\dot\phi g(\theta)\, d\theta = -\int_\alpha^\beta \dot{g}(\theta)|\phi(\theta)|\, d\theta. \tag{3.1}$$

For intervals $(-\infty, \beta)$ with $\beta < 0$ we have $\phi(\beta) = 0$ and

$$\int_{-\infty}^{\beta} \bigl(\operatorname{sgn} \phi(\theta)\bigr) \dot{\phi}(\theta) g(\theta)\, d\theta = \lim_{R \to \infty} \left( -g(-R)|\phi(-R)| - \int_{-R}^{\beta} \dot{g}(\theta)|\phi(\theta)|\, d\theta \right). \quad (3.2)$$

We will consider the following three cases for $Y_1(\phi)$.
**Case 1.** $Y_1(\phi) = (-\infty, 0)$ or $(-\infty, 0]$. In this case

$$\tau_-(\phi, A\phi) = \operatorname{sgn} \phi \int_{-\infty}^{0} g(\theta) \dot{\phi}(\theta)\, d\theta = 0.$$

**Case 2.** $Y_1(\phi)$ contains only intervals of the form $(\alpha, \beta)$ with $-\infty < \alpha < \beta < 0$ or $(-\infty, \beta)$ with $\beta < 0$. Then, obviously (because $\dot{g} \geq 0$ a. e.) formulae (3.1) and (3.2) imply $\tau_-(\phi, A) \leq 0$.

**Case 3.** $Y_1(\phi)$ contains an interval of the form $(\alpha, 0)$ or $(\alpha, 0]$ with $-\infty < \alpha < 0$. Since $\phi \in \operatorname{dom} A$, we obtain

$$\int_{\alpha}^{0} \bigl(\operatorname{sgn} \phi(\theta)\bigr) \dot{\phi}(\theta) g(\theta)\, d\theta = \operatorname{sgn}_{(\alpha, 0)} \phi \int_{\alpha}^{0} \dot{\phi}(\theta) g(\theta)\, d\theta$$

$$= -\operatorname{sgn}_{(\alpha, 0)} \phi \int_{-\infty}^{\alpha} g(\theta) \dot{\phi}(\theta)\, d\theta$$

$$= -\operatorname{sgn}_{(\alpha, 0)} \phi \int_{Y_0(\phi)} \dot{\phi}(\theta) g(\theta)\, d\theta - \operatorname{sgn}_{(\alpha, 0)} \phi \int_{Y_1(\phi) \cap (-\infty, \alpha)} \dot{\phi}(\theta) g(\theta)\, d\theta.$$

Therefore,

$$\tau_-(\phi, A\phi) = -\int_{Y_0(\phi)} \bigl(|\dot{\phi}(\theta)| \pm \dot{\phi}(\theta)\bigr) g(\theta)\, d\theta$$

$$+ \int_{Y_1(\phi) \cap (-\infty, \alpha)} \bigl(\operatorname{sgn} \phi(\theta) \pm 1\bigr) \dot{\phi}(\theta) g(\theta)\, d\theta.$$

Here $Y_1(\phi) \cap (-\infty, \alpha)$ only contains intervals as considered under Case 2. Thus, we have

$$\int_{\alpha}^{\beta} \bigl(\operatorname{sgn} \phi(\theta) \pm 1\bigr) \dot{\phi}(\theta) g(\theta)\, d\theta = -\int_{\alpha}^{\beta} \dot{g}(\theta) \bigl(|\phi(\theta)| \pm \phi(\theta)\bigr)\, d\theta \leq 0$$

and

$$\int_{-\infty}^{\beta} \bigl(\operatorname{sgn} \phi(\theta) \pm 1\bigr) \dot{\phi}(\theta) g(\theta)\, d\theta$$

$$= \lim_{R \to \infty} \left( -g(-R)\bigl(|\phi(-R)| \pm \phi(-R)\bigr) - \int_{-R}^{\beta} \dot{g}(\theta)\bigl(|\phi(\theta)| \pm \phi(\theta)\bigr)\, d\theta \right) \leq 0.$$

Hence, we have shown that also in this case $\tau_-(\phi, A\phi) \leq 0$. $\square$

**Lemma 3.2.** dom $A$ is dense in $L^1_g(-\infty, 0)$.

**Proof.** Obviously $C^\infty$-functions with compact support in $(-\infty, 0]$ are dense in $L^1_g$. Let $\phi$ be a $C^\infty$-function with supp $\phi \subset [-\tau, -\varepsilon]$. Define, for $n$ sufficiently large,
$$\phi_n(\theta) = \begin{cases} \phi(\theta) & \text{for } \theta < -\frac{1}{n}, \\ n(\theta + \frac{1}{n})a_n & \text{for } -\frac{1}{n} \le \theta \le 0. \end{cases}$$

Obviously, $\phi_n$ is locally absolutely continuous with $\phi_n, \dot\phi_n \in L^1_g$. If we put
$$a_n = -\int_{-\tau}^{-\varepsilon} g(\theta)\dot\phi(\theta)\,d\theta \left(n\int_{-\frac{1}{n}}^0 g(\theta)\,d\theta\right)^{-1},$$

then also $\int_{-\infty}^0 g(\theta)\dot\phi_n(\theta)\,d\theta = 0$, and therefore $\phi_n \in \text{dom}\,A$. We put $\alpha = \left|\int_{-\tau}^{-\varepsilon} g(\theta)\dot\phi(\theta)\,d\theta\right|$. Then

$$\|\phi - \phi_n\|_{L^1_g} = \alpha \int_{-\frac{1}{n}}^0 (\theta + \frac{1}{n})\,d\theta \left(\int_{-\frac{1}{n}}^0 g(\theta)\,d\theta\right)^{-1} \le \frac{\alpha}{n} \to 0$$

as $n \to \infty$. $\square$

**Lemma 3.3.** For all $\lambda > 0$
$$(\lambda I - A)\text{dom}\,A = L^1_g.$$

**Proof.** For $\lambda > 0$ and $f \in L^1_g$ we get from $(\lambda I - D)\phi = f$ the representation

$$\phi(\theta) = e^{\lambda\theta}\phi(0) + \int_\theta^0 e^{\lambda(\theta-\tau)}f(\tau)\,d\tau, \quad \theta \le 0.$$

Obviously, $\phi$ is locally absolutely continuous. We shall prove that $\phi \in L^1_g(-\infty, 0)$. Since $\lambda > 0$, we have

$$\int_{-\infty}^0 g(\theta)e^{\lambda\theta}\,d\theta \le g(-1)\int_{-\infty}^{-1} e^{\lambda\theta}\,d\theta + \int_{-1}^0 g(\theta)\,d\theta < \infty$$

and, by Fubini's theorem,

$$\int_{-\infty}^0 g(\theta)\left|\int_\theta^0 e^{\lambda(\theta-\tau)}f(\tau)\,d\tau\right|d\theta$$
$$\le \int_{-\infty}^0 |f(\tau)|\int_{-\infty}^\tau g(\theta)e^{\lambda(\theta-\tau)}\,d\theta\,d\tau \le \frac{1}{\lambda}\int_{-\infty}^0 g(\tau)|f(\tau)|\,d\tau < \infty,$$

where we also have used the monotonicity of $g$. This proves $\phi \in L_g^1$. Since $D\phi = \lambda\phi - f$, we also get $D\phi \in L_g^1$. Finally, $\phi(0)$ is determined so that $\int_{-\infty}^0 g(\theta)(D\phi)(\theta)\,d\theta = 0$, which leads to

$$\phi(0) = -\frac{1}{\lambda \int_{-\infty}^0 g(\theta)e^{\lambda\theta}d\theta} \int_{-\infty}^0 g(\theta)\Big(\lambda \int_\theta^0 e^{\lambda(\theta-\tau)}f(\tau)\,d\tau - f(\theta)\Big)d\theta.$$

Here, $\int_{-\infty}^0 g(\theta)e^{\lambda\theta}d\theta > 0$ for $\lambda > 0$. Thus, $\phi \in \operatorname{dom} A$ and $(\lambda I - A)\phi = f$ for given $f \in L_g^1$. □

By Lemmas 3.1 – 3.3 the assumptions of the Lumer-Phillips theorem [P] are satisfied. Thus we have

**Theorem 3.4.** *The operator $A$ defined by (3.1) and (3.2) generates a strongly continuous semigroup $S(t)$, $t \geq 0$, on $L_g^1(-\infty, 0)$.*

## References

[BCH]  J. A. Burns, E. M. Cliff and T. L. Herdman, *A state space model for an aeroelastic system*, Proc. 22nd IEEE Conference on Decision and Control.

[BHS]  J. A. Burns, T. L. Herdman and H. W. Stech, *Linear functional differential equations as semigroups on product speces*, SIAM J. Math. Analysis **14** (1983), 98–116.

[BI]  J. A. Burns and K. Ito, *On well-posedness of solutions to integro-differential equations of neutral type in a weighted $L_2$-space*, preprint 1989.

[IT]  K. Ito and J. Turi, *Numerical method for a class of singular integro-differential equation based on semigroup approximation*, submitted.

[KZ]  F. Kappel and K. P. Zhang, *On neutral functional differential equations with nonatomic difference operator*, J. Math. Anal. Appl. **113** (1986), 311–343.

[M]  R. H. Martin, Jr., *Nonlinear Operators and Differential Equations in Banach Spaces*, J. Wiley, New York, 1976.

[P]  A. Pazy, *Semigroup of Linear Operators and Applications to Partial Differnetial Equations*, Springer-Verlag, New York, 1983.


Kazufumi Ito
Center for Applied Mathematical Sciences
University of Southern California
Los Angeles, CA 90089-1113, USA


Franz Kappel
Institut für Mathematik
Universität Graz
Heinrichstraße 36, A8010 Graz, Austria

# Ground States of Semi–Linear Diffusion Equations

HANS G. KAPER
MAN KAM KWONG

Mathematics and Computer Science Division, Argonne National Laboratory

## 1. Statement of results

Let $f$ be defined on $[0, \infty)$, such that $f$ is continuous on $[0, \infty)$ and Lipschitz continuous on $(0, \infty)$, with $f(0) = 0$. We consider the boundary value problem

$$\Delta u + f(u) = 0, \ x \in \mathbb{R}^N \ (N \geq 2); \ \lim_{|x| \to \infty} u(x) = 0. \tag{1}$$

A *ground state solution* of (1) is a nontrivial solution that does not change sign. (We assume that it is nonnegative everywhere.) We prove the following result.

**Theorem 1.** Let
$$F(u) = \int_0^u f(v)dv, \ u \geq 0, \tag{2}$$

and
$$\beta = \inf\{u > 0 : F(u) > 0\}. \tag{3}$$

If $\beta > 0$ and $u \mapsto f(u)/(u - \beta)$ is monotone nonincreasing on $(\beta, \infty)$, then (1) admits at most one ground state solution.

This result generalizes and extends earlier results of McLeod and Serrin [1], Peletier and Serrin [2], and Peletier and Serrin [3].

---

This work was supported by the Applied Mathematical Sciences subprogram of the Office of Energy Research, U.S. Department of Energy, under contract W-31-109-Eng-38.

In Section 2 we present a brief outline of the proof of the theorem; details can be found in our article [4]. A further generalization for quasilinear diffusion equations of the type $\nabla \cdot (a(|\nabla u|)\nabla u + f(u) = 0$ can be found in our article [6].

The proof of the theorem generalizes to the case of a bounded radially symmetric domain in the following sense. If $f$ satisfies the conditions of the theorem, then the boundary value problem

$$\Delta u + f(u) = 0, \; x \in B_R; \; u(x) = 0, \; \mathbf{n} \cdot \nabla u(x) = 0, \; x \in \partial B_R, \qquad (4)$$

admits at most one ground state solution. Here, $B_R$ is the ball of finite radius $R$ centered at the origin in $\mathbb{R}^N$; $\partial B_R$ is its boundary, and $\mathbf{n}$ is the outward unit normal at the boundary. Notice that the zero normal gradient condition at the boundary must be specified in this case. (It is satisfied automatically in the unbounded case.)

## 2. Proof of Theorem 1

### 2.1. Preliminaries.

Any ground state solution of (1) is radially symmetric (see [5]), so $u$ depends only on $r = |x|$. If $'$ denotes differentiation with respect to $r$, then $u$ satisfies

$$u'' + \frac{N-1}{r} u' + f(u) = 0, \; r > 0; \qquad (5)$$

$$u'(0) = 0; \; \lim_{r \to \infty} u(r) = 0. \qquad (6)$$

Two identities play a crucial role in the following analysis. They are obtained by multiplying (5) by $u'$ and $r^{2(N-1)} u'$, respectively, and integrating over $(r_1, r_2)$,

$$\left[ \frac{1}{2}(u'(r))^2 + F(u(r)) \right]_{r_1}^{r_2} = -(N-1) \int_{r_1}^{r_2} \frac{1}{s} (u'(s))^2 \, ds, \qquad (7)$$

$$\left[ r^{2(N-1)} \left( \frac{1}{2}(u'(r))^2 + F(u(r)) \right) \right]_{r_1}^{r_2} = 2(N-1) \int_{r_1}^{r_2} s^{2N-3} F((u(s)) \, ds. \qquad (8)$$

One can show, using (7), that $\lim_{r \to \infty} u'(r)$ exists and

$$\lim_{r \to \infty} u'(r) = 0; \qquad (9)$$

and similarly, using (8), that $\lim_{r \to \infty} r^{2(N-1)} (\frac{1}{2}(u'(r))^2 + F(u(r)))$ exists and

$$\lim_{r \to \infty} r^{2(N-1)} \left( \frac{1}{2}(u'(r))^2 + F(u(r)) \right) = K, \qquad (10)$$

where $K = 0$ if $N = 2$; if $N > 2$, then $\lim_{r \to \infty} r^{N-2} u(r) = (N-2)^{-1} \sqrt{2K}$. Letting $r_2 \to \infty$ in (7), we obtain the identity

$$\frac{1}{2}(u'(r))^2 + F(u(r)) = (N-1) \int_r^{\infty} \frac{1}{s} (u'(s))^2 \, ds, \; r \geq 0, \qquad (11)$$

and similarly, from (8),

$$r^{2(N-1)}\left(\frac{1}{2}(u'(r))^2 + F(u(r))\right) = K - 2(N-1)\int_r^\infty s^{2N-3}F((u(s))\,ds, \; r \geq 0. \tag{12}$$

We observe that, whereas (9) follows from the definition of a ground state solution, the analogous condition $u'(R) = 0$ must be *imposed* if the domain under consideration is the ball $B_R$; cf. (4).

**Lemma 1.** *If $u$ is a ground state solution of (1), then $u(0) > \beta$.*

**Proof.** Taking $r = 0$ in (11), we find that $F(u(0)) > 0$; hence, $u(0) > \beta$. $\square$

**Lemma 2.** *Any ground state solution of (1) is monotonically decreasing on its support.*

**Proof.** Let $u$ be a ground state solution of (1) and let $R$ be the lowest upper bound (possibly $\infty$) of the support of $u$.

Let $a = \inf\{r \in [0, R) : u'(s) < 0 \text{ for all } s \in (r, R)\}$. We have $u(a) > 0$ and $u'(a) = 0$. Suppose $a > 0$ and $u$ has a local maximum at $a$. Then there exists a point $b \in [0, a)$, such that $u'(b) = 0$ and $u' > 0$ on $(b, a)$. Because $\lim_{r \to R} u(r) = 0$, there must be a point $c \in (a, R)$ where $u(c) = u(b)$.

Taking $r_1 = b$ and $r_2 = c$ in (7), we arrive at a contradiction. We must therefore conclude that either $a = 0$ or, if $a > 0$, then $u''(a) = 0$. The latter configuration is impossible, because $f$ is Lipschitz at $u(a)$, so $u(r) \equiv u(a)$ is the (unique) solution of (5) that starts at $u(a)$ with zero slope. It must therefore be the case that $a = 0$. $\square$

### 2.2. Distinct solutions do not intersect.

We assume that $u_1$ and $u_2$ are two ground state solutions of (1) and show that, if the graphs of $u_1$ and $u_2$ intersect, then $u_1$ and $u_2$ are identical.

**Lemma 3.** *If $u_1(r) = u_2(r) > \beta$ for some $r \geq 0$, then $u_1$ and $u_2$ are identical.*

**Proof.** The lemma follows from the sublinearity of $f$. Suppose that $u_1(a) = u_2(a) = \tau$ for some $a \geq 0$, where $\tau > \beta$, and that $u_1 > u_2$ on $[0, a)$. The equality $u_1'(a) = u_2'(a)$ is ruled out, because $f$ is Lipschitz at $u_1(a)$, so it must be the case that $u_1'(a) < u_2'(a)$. We have $(u_2 - \beta)f(u_1) - (u_1 - \beta)f(u_2) \leq 0$ on $[0, a]$. Since $u_1$ and $u_2$ satisfy (5), it follows that

$$(u_2 - \beta)(u_1'' + ((N-1)/r)u_1') \geq (u_1 - \beta)(u_2'' + ((N-1)/r)u_2'), \tag{13}$$

and therefore

$$((u_2 - \beta)r^{N-1}u_1')' \geq ((u_1 - \beta)r^{N-1}u_2')', \tag{14}$$

on $[0, a]$. Upon integration over $[0, a]$, we find that $u_1'(a) \geq u_2'(a)$, a contradiction. $\square$

**Lemma 4.** If $0 < u_1(r) = u_2(r) \leq \beta$ for some $r \geq 0$, then $u_1$ and $u_2$ are identical.

**Proof.** We prove the lemma in two steps. In the first step we rule out the possibility that the graphs of $u_1$ and $u_2$ have more than one point in common, once they are at or below the horizontal line $u = \beta$. In the second step we show that they cannot even have a single point in common.

Suppose that there are two distinct points $a$ and $b$ ($a < b$) such that $\beta \geq u_1(a) = u_2(a) > u_1(b) = u_2(b) > 0$. Without loss of generality, we may assume that $u_1 > u_2$ on $(a, b)$. By continuity, there exists a pair of points $(c, d)$, with $a < c < d < b$, such that $u_1(d) = u_2(c)$ and $u_1'(d) = u_1'(c)$. Applying (8) to $u_1$ on $[a, d]$ and to $u_2$ on $[a, c]$ and subtracting the two expressions, we arrive at the identity

$$(d^{2(N-1)} - c^{2(N-1)})\left(\frac{1}{2}(u_1'(d))^2 + F(u_1(d))\right) - \frac{1}{2}a^{2(N-1)}[(u_1'(a))^2 - (u_2'(a))^2]$$
$$= 2(N-1)\int_{u_1(d)}^{u_1(a)} \left(\frac{(r_1(u))^{2N-3}}{|u_1'(r_1(u))|} - \frac{(r_2(u))^{2N-3}}{|u_2'(r_2(u))|}\right) F(u)\,du. \tag{15}$$

Here, $r_1$ and $r_2$ are the inverse functions for $u_1$ and $u_2$, respectively (i.e., $u_j(r_j(u)) = u$ for $0 \leq u \leq u_j(0)$, $j = 1, 2$). The expression in the left member is positive, while the right member is negative, a contradiction. The possibility of two points of intersection is thus ruled out.

Suppose that there is a single point $a > 0$ where $0 < u_1(a) = u_2(a) \leq \beta$. Without loss of generality we may assume that $u_1(r) > u_2(r)$ for $r > a$.

Let

$$K_j = \lim_{r \to \infty} r^{2(N-1)}\left(\frac{1}{2}(u_j'(r))^2 + F(u(r))\right), \quad j = 1, 2. \tag{16}$$

Applying (8) to $u_1$ and $u_2$ on $[a, r]$ and subtracting the resulting equations, we obtain

$$\frac{1}{2}a^{2(N-1)}[(u_1'(a))^2 - (u_2'(a))^2]$$
$$= K_1 - K_2 - 2(N-1)\int_{u(r)}^{u(a)} \left(\frac{(r_1(u))^{2N-3}}{|u_1'(r_1(u))|} - \frac{(r_2(u))^{2N-3}}{|u_2'(r_1(u))|}\right) F(u)\,du. \tag{17}$$

The expression in the left member is negative. Under the integral sign, the expression inside the parentheses is positive, while $F(u)$ is zero or negative, so the integral is certainly negative. Hence, if $K_1 \geq K_2$ (which is certainly true if $N = 2$), the expression in the right member is positive, and we have a contradiction.

It remains to investigate those cases where $N > 2$ and $K_1 < K_2$. Take $\epsilon < \frac{1}{8}(K_2 - K_1)$ and choose $r$ sufficiently large that

$$r^{2(N-1)}\left(\frac{1}{2}(u_1'(r))^2 + F(u_1(r))\right) < K_1 + \epsilon, \tag{18}$$

$$r^{2(N-1)} \left( \frac{1}{2}(u_2'(r))^2 + F(u_2(r)) \right) > K_2 - \epsilon. \tag{19}$$

From (18) we obtain
$$u_1(r) < \frac{\sqrt{2(K_1 + \epsilon)} + \epsilon}{(N-2)r^{N-2}}. \tag{20}$$

By reducing $\epsilon$ if necessary, we can certainly achieve that $\sqrt{2(K_1 + \epsilon)} + \epsilon < \sqrt{2(K_2 - \epsilon)}$. Thus,
$$u_1(r) < \frac{\sqrt{2(K_2 - \epsilon)}}{(N-2)r^{N-2}}. \tag{21}$$

On the other hand, it follows from (19) that
$$u_2(r) > \frac{\sqrt{2(K_2 - \epsilon)}}{(N-2)r^{N-2}}. \tag{22}$$

These results imply that $u_1(r) < u_2(r)$ for $r$ sufficiently large. But this conclusion contradicts the earlier assumption that $u_1(r) > u_2(r)$ for all $r > a$. Thus, the possibility that the graphs of $u_1$ and $u_2$ intersect is ruled out. □

On the basis of Lemmas 3 and 4 we conclude that distinct ground state solutions of (1) do not intersect.

### 2.3. Distinct solutions must intersect.

According to Lemma 2, any ground state solution of (1) is (strictly) decreasing on its support. Thus, if $r \mapsto u(r)$ is a ground state solution, the inverse $u \mapsto r(u)$ is well defined on $[0, u(0)]$ by the identity $u(r(u)) = u$. Let $v$ be defined by the expression
$$v(u) = \frac{1}{2}(u'(r(u)))^2, \ 0 \leq u \leq u(0). \tag{23}$$

Thus,
$$u'(r) = -\sqrt{2v(u(r))}, \ r \geq 0. \tag{24}$$

We now use the pair $(u, v)$ as the coordinates for a phase plane analysis.

From (23) we obtain $dv/du = u''(r(u))$. As $u$ satisfies (5), it follows that
$$\frac{dv}{du} = \frac{N-1}{r(u)} \sqrt{2v} - f(u), \ 0 < u < u(0). \tag{25}$$

Furthermore,
$$v(0) = 0, \ v(u(0)) = 0. \tag{26}$$

We prove the following lemma.

**Lemma 5.** If $u_1$ and $u_2$ are two distinct ground state solutions of (1), then $u_1(r) = u_2(r)$ for at least one value $r > 0$.

**Proof.** Let $u_1$ and $u_2$ denote two *distinct* ground state solutions of (1). The graphs of $u_1$ and $u_2$ do not intersect; without loss of generality we assume that $u_1(r) > u_2(r)$ for all $r > 0$. Denoting the inverse functions for $u_1$ and $u_2$ by $r_1$ and $r_2$, we then have $r_1(u) > r_2(u)$ for all $u \in (0, u_2(0))$.

We now analyze the trajectories of the two solutions in the $(u, v)$-phase plane, distinguishing them by their respective indices.

Because $r_1(u) > r_2(u)$ near 0, $v_1$ and $v_2$ satisfy

$$\frac{dv_1}{du} \leq \frac{N-1}{r_2(u)}\sqrt{2v_1} - f(u), \ u > 0; \ v_1(0) = 0; \tag{27}$$

and

$$\frac{dv_2}{du} = \frac{N-1}{r_2(u)}\sqrt{2v_2} - f(u), \ u > 0; \ v_2(0) = 0. \tag{28}$$

Notice that the right hand sides of the differential equations are not Lipschitz. Hence, it is only possible to compare the *maximal* solutions of these initial value problems, unless we can somehow guarantee that there are no other solutions. The condition $\beta > 0$ serves this purpose.

We refer to our article [7], where we investigated initial value problems of the type

$$x' = p(t)x^\alpha + q(t), \ t > 0; \ x(0) = 0, \tag{29}$$

where $0 < \alpha < 1$ and $p$ and $q$ are integrable near 0. We showed that (29) has at most one nontrivial nonnegative solution if (i) $p$ and the first integral $Q$ of $q$ are nonnegative near 0; and (ii) for every $t > 0$, there is a point $\tau \in (0, t)$, where $Q(\tau) > 0$.

In the case of (27) and (28), where $\alpha = \frac{1}{2}$, $p(t) = (N-1)\sqrt{2}/r(t)$, and $q(t) = -f(t)$, the condition (i) is satisfied, and (ii) is satisfied if $\beta > 0$, unless $f$ vanishes identically near 0. If $f$ vanishes identically near 0, a trivial modification suffices to establish uniqueness, again provided that $\beta > 0$.

The direct comparison yields the inequality

$$v_1(u) \leq v_2(u), \ u \in [0, u_2(0)]. \tag{30}$$

If $u_1(0) > u_2(0)$, then $v_1(u_2(0)) > 0$, while $v_2(u_2(0)) = 0$. This would clearly contradict (30), so at this point we must conclude that $u_1(0) = u_2(0)$.

The inequality (30) implies that $|u_1'(r_1(u))| \leq |u_2'(r_2(u))|$. Furthermore, $r_1(u) \geq r_2(u)$, so

$$\frac{|u_1'(r_1(u))|}{r_1(u)} \leq \frac{|u_2'(r_2(u))|}{r_2(u)}, \ u \in [0, u_2(0)]. \tag{31}$$

Next, we apply (11) to $u_1$ and $u_2$ at $r = 0$ and subtract the resulting equations. We find

$$\int_0^\infty \frac{1}{r}(u_1'(r))^2 \, dr = \int_0^\infty \frac{1}{r}(u_2'(r))^2 \, dr, \qquad (32)$$

or, after a transformation of variables,

$$\int_0^{u_1(0)} \frac{|u_1'(r_1(u))|}{r_1(u)} du = \int_0^{u_2(0)} \frac{|u_2'(r_2(u))|}{r_2(u)} du. \qquad (33)$$

We recall that $u_1(0) = u_2(0)$ and conclude that the inequality (31) is compatible with the identity (33) if and only if $u_1'(r_1(u)) = u_2'(r_2(u))$ for all $u \in [0, u_1(0)]$. This equality, in turn, implies that $u_1$ and $u_2$ coincide everywhere. But here we have arrived at a contradiction, since we had assumed that $u_1$ and $u_2$ were distinct. Hence, if $u_1$ and $u_2$ are distinct, their graphs must intersect at some point $r > 0$. $\square$

The Monotone Separation Lemma of Peletier and Serrin [2, Lemma 9] is an immediate consequence of Lemma 5. We formulate it as a corollary.

**Corollary 1.** *If $u_1$ and $u_2$ are two distinct ground state solutions of (1) and $u_1(r) = u_2(r) = \tau$ for some $r > 0$, then $u \mapsto r_1(u) - r_2(u)$ is monotone nonincreasing on $[0, \tau]$.*

### 2.4. Completion of the proof of Theorem 1.

In Section 2.2 we found that if the graphs of two ground state solutions of (1) intersect at some point, then they coincide everywhere. On the other hand, according to Lemma 5, the graphs of two distinct ground state solutions must intersect at some point. Clearly, we have a contradiction, unless (1) admits no more than one ground state solution, as asserted.

## References

[1] K. McLeod and J. Serrin, *Uniqueness of the ground state solution for $\Delta u + f(u) = 0$*, Proc. Nat. Acad. Sci. USA **78** (1981), 6592–6595.

[2] L. A. Peletier and J. Serrin, *Uniqueness of positive solutions of of semilinear equations in $\mathbb{R}^n$.*, Arch. Rat. Mech. Anal. **81** (1983), 181–197.

[3] L. A. Peletier and J. Serrin, *Uniqueness of non-negative solutions of semilinear equations in $\mathbb{R}^n$.*, J. Diff. Eq. **61** (1986), 380–397.

[4] H. G. Kaper and Man Kam Kwong, *Uniqueness of Non-Negative Solutions of a Class of Semilinear Elliptic Equations*, Nonlinear Diffusion Equations and Their Equilibrium States Vol. 2, W.-M. Ni, L. A. Peletier, and J. Serrin (Eds.), Springer-Verlag, NY, 1988, pp. 1–17.

[5] B. Gidas, Wei-Ming Ni and L. Nirenberg, *Symmetry and related properties via the maximum principle*, Commun. Math. Phys. **68** (1979), 209–243.

[6] H. G. Kaper and Man Kam Kwong, *Uniqueness Results for Some Nonlinear Initial and Boundary Value Problems*, Arch. Rat. Mech. Anal. **102** (1988), 45–56.

[7] H. G. Kaper and Man Kam Kwong, *Uniqueness for a class of non-linear initial value problems*, J. Math. Anal. Applic. **130** (1988), 467–473.


Hans G. Kaper
Man Kam Kwong
Mathematics and Computer Science Division
Argonne National Laboratory
Argonne, IL 60439-4844, USA

# Uniform Energy Decay of a Class of Cantilevered Nonlinear Beams with Nonlinear Dissipation at the Free End

J. LAGNESE*
G. LEUGERING**

Department of Mathematics, Georgetown University

## 1. Problem formulation and statement of result

We begin by briefly describing a derivation of a thin, uniform nonlinear, dynamic beam model which is a prototype of the nonlinear equations to be considered in this paper. The derivation is adapted from ideas appearing in Rogers and Russell [3] and in Kane, Ryan and Banerjee [2].

Consider an elastic body whose points, in its reference state, are described in rectangular coordinates by

$$\{(x,y,z)|\, 0 \le x \le L,\ -1 \le y \le 1,\ -h/2 \le z \le h/2\}.$$

The line segment $0 \le x \le L$, $y = z = 0$ is called the *centerline* of the body, and the sets

$$A(x) = \{(x,y,z)|\, x = x,\ -1 \le y \le 1,\ -h/2 \le z \le h/2\}$$

are its *cross sections*. We shall henceforth refer to this body as a *beam* since the assumptions that will be made below regarding the admissible motions of the body characterize a class of motions which depend on $x$ and $t$ only, $t$ denoting the time variable.

Let $\mathbf{r}(x,t)$ denote the position vector at time $t$ of the particle which occupies position $(x,0,0)$ on the centerline in the reference configuration (so that $\mathbf{r}(x,t) - \{x,0,0\}$ is the displacement vector of the particle).

---

*Research supported by the Air Force Office of Scientific Research through grant AFOSR 88-0337. **Research supported by the Deutsche Forschungsgemeinschaft.

**Assumption 1.** The cross-sections move rigidly, i.e., if $\mathbf{p}(x, y, z, t)$ is the vector describing the displacement of the point $(x, y, z)$, then $\mathbf{p}$ is determined by $\mathbf{r}(x, t)$ and two orthonormal vectors $\mathbf{d}_2(x, t)$ and $\mathbf{d}_3(x, t)$ through the formula

$$\mathbf{p}(x, y, z, t) = [\mathbf{r}(x, t) - \{x, 0, 0\}] + y\mathbf{d}_2(x, t) + z\mathbf{d}_3(x, t).$$

We set $\mathbf{d}_1 = \mathbf{d}_3 \times \mathbf{d}_2$. The orthonormal system $(\mathbf{d}_1, \mathbf{d}_2, \mathbf{d}_3)$ may be visualized as a moving coordinate system with $\mathbf{d}_2(x, t)$ and $\mathbf{d}_3(x, t)$ in the plane of the deformed cross section $A(x)$; one has $\mathbf{d}_i = \mathbf{e}_i$ in the reference configuration, where $(\mathbf{e}_1, \mathbf{e}_2, \mathbf{e}_3)$ is the natural basis for $\mathbb{R}^3$.

**Assumption 2.** The centerline is constrained to move in the $xz$-plane, i.e.,

$$\mathbf{r}(x, t) = [u(x, t) + x]\mathbf{e}_1 + w(x, t)\mathbf{e}_3.$$

The quantities $u$ and $w$ represent, respectively, longitudinal and vertical displacement of the point $(x, 0, 0)$

The *strains* of the beam consist of six quantities. The first three are the components $v_i$ of $\partial \mathbf{r}$ in the $\mathbf{d}_i$ basis (where $\partial = \partial/\partial x$), that is

$$v_i = \partial \mathbf{r} \cdot \mathbf{d}_i.$$

**Assumption 3.** There is no shearing of cross sections, i.e., $v_2 = v_3 = 0$.

The remaining three components of strain are related to bending and twisting motions and are defined as follows. Introduce the vector $\mathbf{q}$ by

$$\partial \mathbf{d}_k = \mathbf{q} \times \mathbf{d}_k.$$

$\mathbf{q}$ exists and is unique since the $\mathbf{d}_i$'s form an orthonormal basis. The final three components of strain are the components of $\mathbf{q}$ in the $\mathbf{d}_i$ basis:

$$q_i = \mathbf{q} \cdot \mathbf{d}_i.$$

Components $q_2$ and $q_3$ measure the amount of bending about $\mathbf{d}_2$ and $\mathbf{d}_3$, respectively, while $q_1$ describes the amount of twist about $\mathbf{d}_1$. Assumption 2 implies that $q_3 = 0$.

**Assumption 4.** There is no twisting about $\mathbf{d}_1$: $q_1 = 0$.

We introduce the stretch $s(x, t)$ at the point $x$ of the centerline, induced by the deformation, through

$$s(x, t) = \int_0^{x+u(x,t)} [1 + (\partial w(\xi, t))^2]^{1/2} d\xi - x. \tag{1.1}$$

Any set of forces acting on the particular cross-section located at $x$ in the undeformed state can be replaced by a couple of torque $T$ and a resultant force $R$ such that

$$\mathbf{T} = T_1\mathbf{d}_1 + M_2\mathbf{d}_2 + M_3\mathbf{d}_3,$$

$$\mathbf{R} = P_1 \mathbf{d}_1 + V_2 \mathbf{d}_2 + V_3 \mathbf{d}_3.$$

Here $T_1$ is an axial torque, $M_i$ is a bending moment about $\mathbf{d}_i$, and $V_i$ are the shear components of $R$. Assumption 3 requires that $V_2 = V_3 = 0$, while assumptions 2 and 4 require $T_1 = M_3 = 0$. Following [2] we have

$$P_1 = EA \frac{\partial s}{\partial x}(x,t),$$

$$M_2 = -EI \frac{\partial^2 w}{\partial x^2}(x,t),$$

where the physical constants are $A$, the area of a cross section, $I$ its moment of inertia with respect to the $z$-axis, and Young's modulus $E$. (These are assumed to be *constants* only to simplify some of the computations below. This assumption is inessential.) Therefore, the *strain energy*

$$U = \int_0^L \frac{(P_1)^2}{2EA} dx + \int_0^L \frac{(M_2)^2}{2EI} dx$$

can be expressed as

$$U = \frac{1}{2} \int_0^L EA(\partial s)^2 dx + \frac{1}{2} \int_0^L EI(\partial^2 w)^2 dx.$$

From (1.1) we have

$$\partial s(x,t) = [1 + (\partial w(x + u(x,t),t))^2]^{1/2}[1 + \partial u(x,t)] - 1.$$

**Assumption 5.** The longitudinal deformation $u(x,t)$ is small compared with $x$ and $\partial u(x,t)$ is small compared to 1, i.e., $\partial s(x,t)$ is well-approximated by

$$\partial s(x,t) = [1 + (\partial w(x,t))^2]^{1/2} - 1.$$

With assumption 5, the strain energy of the deformation is given by

$$U = \frac{1}{2} \int_0^L \{EA[(1 + (\partial w)^2]^{1/2} - 1]^2 + EI(\partial^2 w)^2\} dx,$$

The *kinetic energy* is given by

$$K = \frac{1}{2} \int_0^L \rho A \dot{w}^2 dx,$$

where $\dot{} = \partial/\partial t$, $\rho$ (assumed to be a constant) is the mass density per unit volume of the beam in the reference configuration and where we have neglected the rotational inertia of the cross sections (which would introduce the additional term $\rho AI(\partial \dot{w})^2$ into $K$). The *total energy* is then

$$\mathcal{E} = U + K.$$

We may derive the equations of motion together with appropriate boundary conditions by requiring conservation of energy. The Lagrangian approach, of course, works equally well. We assume the beam to be fixed at the end $x = 0$ and, if uncontrolled, to be free at $x = L$. Calculation of $\dot{\mathcal{E}}(t) = 0$ leads, after routine integration by parts and variation arguments, to the nonlinear partial differential equation

$$\rho A \ddot{w} + EI\partial^4 w - \frac{EA}{2}\partial g(\partial w) = 0, \quad (0 < x < L), \tag{1.2}$$

where

$$g(\xi) = 2\xi\left(1 - \frac{1}{\sqrt{1+\xi^2}}\right), \tag{1.3}$$

and to the boundary conditions

$$w(0,t) = \partial w(0,t) = 0, \tag{1.4}$$

$$EI\partial^2 w(L,t) = 0, \quad EI\partial^3 w(L,t) - \frac{EA}{2}g(\partial w(L,t)) = 0. \tag{1.5}$$

In fact, the specific form of the function $g$ will not be important in what follows, and we only assume (for the moment) that $g : \mathbb{R} \to \mathbb{R}$ satisfies

$$g \text{ is continuous, nondecreasing and } g(0) = 0. \tag{1.6}$$

We define a strain energy functional for this $g$ through

$$U = \frac{1}{2}\int_0^L [EAG(\partial w) + EI(\partial^2 w)^2]dx,$$

where

$$G(s) = \int_0^s g(\xi)d\xi.$$

Note that

$$sg(s) \geq G(s) \geq 0, \quad \forall s \in \mathbb{R}.$$

Setting $\mathcal{E} = U + K$ as before, the requirement that $\dot{\mathcal{E}}(t) = 0$ leads to the system (1.2), (1.4) and (1.5). In (1.2), $\partial g(\partial w)$ may be interpreted in the sense of distributions on $\mathbb{R}$ if, say $w \in C^1(0,L)$ in the $x$ variable.

We now introduce a feedback control into the system (1.2), (1.4), (1.5) through the boundary condition

$$EI\partial^3 w(L,t) - EAg(\partial w(L,t)) = f(\dot{w}(L,t)). \tag{1.7}$$

Physically, $f$ represents a shear force (nonlinear friction) in the $\mathbf{d}_3$ direction applied at the free end of the beam. If $\dot{\mathcal{E}}(t)$ is calculated with the boundary condition (1.7) in effect, it is found that

$$\dot{\mathcal{E}}(t) = -\dot{w}(L,t)f(\dot{w}(L,t)).$$

Thus (1.7) will be a dissipative boundary condition if $f$ satisfies

$$sf(s) \geq 0, \quad \forall s \in \mathbb{R}.$$

If we rescale the time variable by making the change $t \to t\sqrt{\rho A/EI}$, the closed-loop system takes the form

$$\ddot{w} + \partial^4 w - \gamma^2 \partial g(\partial w) = 0, \quad \gamma^2 = A/2I, \quad (0 < x < L), \tag{1.8}$$

$$w(0,t) = \partial w(0,t) = 0, \tag{1.9}$$

$$\partial^2 w(L,t) = 0, \quad \partial^3 w(L,t) - \gamma^2 g(\partial w(L,t)) = f(\dot{w}(L,t)), \tag{1.10}$$

in which the notational change $f \to EIf$ has been made. To complete the description of the system, initial data

$$w(0) = w^0, \quad \dot{w}(0) = w^1, \quad (0 < x < L), \tag{1.11}$$

should be prescribed. The energy for the system (1.8)–(1.10) is

$$\mathcal{E}(t) = \frac{1}{2} \int_0^L [\dot{w}^2 + (\partial^2 w)^2 + 2\gamma^2 G(\partial w)]dx.$$

We now pose the problem to be considered in this paper.

**Uniform Stabilization Problem.** Determine those feedback laws

$$f = f(\dot{w}(L,t))$$

for which $\mathcal{E}(t) \to 0$ as $t \to \infty$, *uniformly* on each bounded set $\mathcal{E}(0) \leq M$ of initial data.

**Remark 1.1.** In the uniform stabilization problem, it is not only the decay of energy that is to be established but also the *rate of decay* of $\mathcal{E}(t)$ in terms of elastic parameters and parameters associated with the particular feedback law selected.

The result to be proved is as follows.

**Theorem.** Assume that $f$ and $g$ are continuous, nondecreasing functions with $f(0) = g(0) = 0$. In addition, suppose that $g$ satisfies

$$sg(s) \geq (1+\alpha)G(s) \tag{1.12}$$

for some $\alpha > 0$ and that $f$ has the following growth properties:

$$\begin{cases} c_0|s|^p \leq |f(s)| \leq C_0|s|^\lambda & \text{if } |s| \leq 1, \\ c_0|\xi| \leq |f(s)| \leq C_0|s|^q & \text{if } |s| > 1, \end{cases} \tag{1.13}$$

where $\lambda \in (0,1]$, $p \geq \lambda$ and $q \geq 1$. Let $w$ be a classical solution of (1.8)–(1.10) and let $M > 0$. There are constants $C > 0$ and $\omega = \omega(M) > 0$ such that the following estimates hold provided $\mathcal{E}(0) \leq M$: (i) if $p = \lambda = q = 1$,

$$\mathcal{E}(t) \leq Ce^{-\omega t}\mathcal{E}(0);$$

(ii) if $p + 1 > 2\lambda$, then

$$\mathcal{E}(t) \leq C[1 + \omega t(\mathcal{E}(0))^{(p+1-2\lambda)/2\lambda}]^{-2\lambda/(p+1-2\lambda)}\mathcal{E}(0).$$

**Remark 1.2.** The constant $C$ may be chosen independent of $\mathcal{E}(0)$ but $\omega \to 0$ as $M \to \infty$ at least as fast as $M^{(p+1-2\lambda)/2\lambda}$. When $p = q = \lambda = 1$, $\omega$ is independent of $\mathcal{E}(0)$.

**Remark 1.3.** It is easy to see that the function $g$ defined in (1.3) satisfies (1.12) with $\alpha = 1$. (1.12) is also satisfied if, for example, $g(s) = s|s|^{\alpha-1}$ with $\alpha > 0$.

**Remark 1.4.** In the case $p = \lambda < 1$, Theorem 1 gives a decay rate $\mathcal{E}(t) \sim t^{-2p/(1-p)}$. This is in agreement with asymptotic energy estimates obtained by Conrad, Leblond and Marmorat [1] for solutions of the Bernoulli-Euler beam equation (i.e., $g = 0$) subject to the boundary conditions (1.9), (1.10), where $f$ satisfies (1.13) with $p \in [\lambda, 1]$ and $q = 1$.

**Remark 1.5.** Existence of global weak solutions of (1.8)–(1.11) may be proved by the Faedo-Galerkin approximation method if, for example, $f$ satisfies the conditions of Theorem 1 and $g$ satisfies (1.6) and is locally Lipschitzian. However, existence of classical solutions remains an open question.

## 2. Proof of Theorem 1

Let $\varepsilon > 0$, $\beta \geq 0$ and define

$$F_\varepsilon(t) = \mathcal{E}(t) + \varepsilon \rho(t)(\mathcal{E}(t))^\beta \tag{2.1}$$

where

$$\rho(t) = \int_0^L x\dot{w}\partial w\, dx. \tag{2.2}$$

We are going to prove that for $\varepsilon$ sufficiently small and for the choice

$$\beta = \frac{p+1-2\lambda}{2\lambda},$$

the function $F_\varepsilon(t)$ is a Lyapunov functional for the system (1.5)–(1.7), specifically that

$$\dot{F}_\varepsilon(t) \leq -\varepsilon \frac{\min(1,\alpha)}{2}(\mathcal{E}(t))^{(p+1)/2\lambda}. \tag{2.3}$$

The choice of the particular functional $F_\varepsilon$ as an appropriate Lyapunov functional, and the calculation of $\dot{F}_\varepsilon$, is based on the following **energy identity**, valid for classical solutions of (1.5)–(1.7) and arbitrary $T > 0$:

$$0 = \rho(T) - \rho(0) + \frac{1}{2}\int_0^T\int_0^L \dot{w}^2 dx dt + \frac{3}{2}\int_0^T\int_0^L (\partial^2 w)^2 dx dt$$
$$+ \gamma^2\int_0^T\int_0^L [\partial w g(\partial w) - G(\partial w)]dx dt + L\int_0^T f(\dot{w}(L,t))\partial w(L,t)dt \tag{2.4}$$
$$- \frac{L}{2}\int_0^T \dot{w}^2(L,t)dt + \gamma^2 L\int_0^T G(\partial w(L,t))dt.$$

Identity (2.4) will be proved at the end of this section. Let us see why it implies (2.3).

From (2.1),

$$\dot{F}_\varepsilon(t) = \dot{\mathcal{E}}(t) + \varepsilon\beta\rho(t)(\mathcal{E}(t))^{\beta-1}\dot{\mathcal{E}}(t) + \varepsilon\dot{\rho}(t)(\mathcal{E}(t))^\beta.$$

From the definition of $\rho(t)$, it is seen that

$$|\rho(t)| \leq L^2\mathcal{E}(t).$$

Therefore (since $\dot{\mathcal{E}}(t) \leq 0$)

$$\dot{F}_\varepsilon(t) \leq [1 - \varepsilon\beta L^2(\mathcal{E}(0))^\beta]\dot{\mathcal{E}}(t) + \varepsilon\dot{\rho}(t)(\mathcal{E}(t))^\beta. \tag{2.5}$$

The quantity $\dot{\rho}(t)$ may be calculated from (2.4). One obtains

$$\dot{\rho}(t) = -\int_0^L \left[\frac{1}{2}\dot{w}^2 + \frac{3}{2}(\partial^2 w)^2 + \gamma^2[\partial w g(\partial w) - G(\partial w)]\right] dx$$
$$+ \frac{L}{2}\dot{w}^2(L,t) - \gamma^2 L G(\partial w(L,t)) - Lf(\dot{w}(L,t))\partial w(L,t)$$
$$\leq -\int_0^L \left[\frac{1}{2}\dot{w}^2 + \frac{3}{2}(\partial^2 w)^2 + \alpha\gamma^2 G(\partial w)\right] dx \tag{2.6}$$
$$+ \frac{L}{2}\dot{w}^2(L,t) - \gamma^2 L G(\partial w(L,t)) - Lf(\dot{w}(L,t))\partial w(L,t)$$
$$\leq -\min(1,\alpha)\mathcal{E}(t) + \frac{L}{2}\dot{w}^2(L,t) - Lf(\dot{w}(L,t))\partial w(L,t).$$

The last term on the right side of (2.6) will be estimated next. To abbreviate the notation, the argument $(L,t)$ will be omitted.

If $|\dot{w}(L,t)| \leq 1$ we have

$$\begin{aligned}
|f(\dot{w})\partial w| &\leq C_0|\dot{w}|^\lambda|\partial w| \\
&\leq \frac{C_0}{4\delta}|\dot{w}|^{2\lambda} + \delta|\partial w|^2 \\
&\leq \frac{C_0}{4\delta}|\dot{w}|^{2\lambda} + \delta L \int_0^L (\partial^2 w)^2 dx \\
&\leq \frac{C_0}{4\delta}|\dot{w}|^{2\lambda} + 2\delta L \mathcal{E}(t)
\end{aligned} \qquad (2.7)$$

where $\delta > 0$ is arbitrary.

If, on the other hand, $|\dot{w}(L,t)| > 1$,

$$|f(\dot{w})\partial w| = \left|\frac{f(\dot{w})}{\dot{w}}\dot{w}\partial w\right| \leq \frac{1}{4\delta}\dot{w}f(\dot{w}) + \delta\frac{f(\dot{w})}{\dot{w}}(\partial w)^2. \qquad (2.8)$$

We assume that $q > 1$ and leave the simpler case $q = 1$ to the reader. The last term in (2.8) is then bounded above by

$$\begin{aligned}
&\frac{\delta(q-1)}{q+1}\left(\frac{f(\dot{w})}{\dot{w}}\right)^{(q+1)/(q-1)} + \frac{2\delta}{q+1}(\partial w)^{q+1} \\
&= \frac{\delta(q-1)}{q+1}\dot{w}f(\dot{w})\left(\frac{f(\dot{w})}{|\dot{w}|^q}\right)^{2/(q-1)} + \frac{2\delta}{q+1}(\partial w)^{q+1} \\
&\leq \frac{\delta(q-1)}{q+1}C_0^{2/(q-1)}\dot{w}f(\dot{w}) + \frac{2\delta}{q+1}(2L)^{(q+1)/2}(\mathcal{E}(t))^{(q+1)/2} \\
&\leq \frac{\delta(q-1)}{q+1}C_0^{2/(q-1)}\dot{w}f(\dot{w}) + \frac{2\delta}{q+1}(2L)^{(q+1)/2}(\mathcal{E}(0))^{(q-1)/2}\mathcal{E}(t).
\end{aligned}$$

Use of this last estimate in (2.8) yields, for $|\dot{w}(L,t)| > 1$ and $\delta < 1$,

$$|f(\dot{w})\partial w| \leq \frac{c_1}{\delta}\dot{w}f(\dot{w}) + \delta c_2(\mathcal{E}(0))^{(q-1)/2}\mathcal{E}(t) \qquad (2.9)$$

for suitable constants $c_1$ and $c_2$ which are independent of $\delta$ and $\mathcal{E}(0)$. The combination of (2.7) and (2.9) gives the estimate

$$\begin{aligned}
|f(\dot{w}(L,t))\partial w(L,t)| &\leq \frac{C_0}{4\delta}\chi(|\dot{w}(L,t)|)|\dot{w}(L,t)|^{2\lambda} \\
&+ \frac{c_1}{\delta}[1 - \chi(|\dot{w}(L,t)|)]\dot{w}(L,t)f(\dot{w}(L,t)) \\
&+ \delta[2L + c_2(\mathcal{E}(0))^{(q-1)/2}]\mathcal{E}(t),
\end{aligned} \qquad (2.10)$$

where $\chi$ denotes the characteristic function of the interval $[0,1]$. If (2.10) is substituted into (2.6), the following estimate for $\dot{\rho}(t)$ is obtained:

$$\begin{aligned}
\dot{\rho}(t) \leq &-\{\min(1,\alpha) - L\delta[2L + c_2(\mathcal{E}(0))^{(q-1)/2}]\}\mathcal{E}(t) \\
&+ \frac{LC_0}{4\delta}\chi(|\dot{w}(L,t)|)|\dot{w}(L,t)|^{2\lambda} \\
&+ \frac{Lc_1}{\delta}[1 - \chi(|\dot{w}(L,t)|)]\dot{w}(L,t)f(\dot{w}(L,t)) \\
&+ \frac{L}{2}\dot{w}^2(L,t).
\end{aligned} \quad (2.11)$$

The inequality (2.11) is now used in (2.5), leading to the following estimate of $\dot{F}_\varepsilon(t)$ [recall that $\dot{\mathcal{E}}(t) = -\dot{w}(L,t)f(\dot{w}(L,t))$]:

$$\begin{aligned}
\dot{F}_\varepsilon(t) \leq &\left[1 - \varepsilon\beta L^2(\mathcal{E}(0))^\beta - \frac{\varepsilon Lc_1}{\delta}(\mathcal{E}(0))^\beta\right]\dot{\mathcal{E}}(t) \\
&- \varepsilon\{\min(1,\alpha) - L\delta[2L + c_2(\mathcal{E}(0))^{(q-1)/2}]\}\mathcal{E}(t)^{\beta+1} \\
&+ \frac{\varepsilon LC_0}{4\delta}\chi(|\dot{w}(L,t)|)|\dot{w}(L,t)|^{2\lambda}(\mathcal{E}(t))^\beta \\
&+ \frac{\varepsilon L}{2}\dot{w}^2(L,t)(\mathcal{E}(t))^\beta.
\end{aligned} \quad (2.12)$$

The last two terms in (2.12) still have to be estimated.

We shall assume that $p + 1 > 2\lambda$ and leave the other possibility $p = \lambda = 1$ to the reader. The inequality

$$|ab| \leq \frac{1}{\delta r}|a|^r + \frac{\delta^{s-1}}{s}|b|^s \quad (2.13)$$

($\delta > 0$, $r > 1$, $s > 1$, $1/r + 1/s = 1$) will be needed.

We apply (2.13) with $r = (p+1)/2\lambda$ and $s = (p+1)/(p+1-2\lambda)$ in order to obtain, for every $\eta > 0$,

$$\begin{aligned}
\chi(|\dot{w}|)&|\dot{w}|^{2\lambda}(\mathcal{E}(t))^\beta \\
&\leq \frac{2\lambda}{\eta(p+1)}|\dot{w}|^{p+1} + \frac{p+1-2\lambda}{p+1}\eta^{2\lambda/(p+1-2\lambda)}(\mathcal{E}(t))^{\beta(p+1)/(p+1-2\lambda)} \\
&\leq \frac{2\lambda}{\eta(p+1)}\dot{w}f(\dot{w}) + \frac{p+1-2\lambda}{p+1}\eta^{2\lambda/(p+1-2\lambda)}(\mathcal{E}(t))^{\beta(p+1)/(p+1-2\lambda)}.
\end{aligned} \quad (2.14)$$

To estimate the last term in (2.12), the two possibilities $\lambda < p \leq 1$ and $p > 1$ will be considered separately.

If $\lambda < p \leq 1$ we simply use the estimate

$$\dot{w}^2(L,t)(\mathcal{E}(t))^\beta \leq \frac{1}{c_0}\dot{w}(L,t)f(\dot{w}(L,t))(\mathcal{E}(0))^\beta, \quad (2.15)$$

which follows from our assumptions, since $|\xi|^2 \leq |\xi|^{p+1}$ for $p \leq 1$ and $|\xi| \leq 1$.

Suppose that $p > 1$. If $|\dot{w}| > 1$ then (2.15) continues to hold. If, on the other hand, $|\dot{w}| \leq 1$ we apply (2.13) with $r = (p+1)/2$, $s = (p+1)/(p-1)$ and obtain

$$\dot{w}^2(\mathcal{E}(t))^\beta \leq \frac{2}{\delta(p+1)}|\dot{w}|^{p+1} + \frac{p-1}{p+1}\delta^{2/(p-1)}(\mathcal{E}(t))^{\beta(p+1)/(p-1)}$$
$$\leq \frac{2}{\delta(p+1)c_0}\dot{w}f(\dot{w}) + \frac{p-1}{p+1}\delta^{2/(p-1)}(\mathcal{E}(t))^{\beta(p+1)/(p-1)}. \quad (2.16)$$

Substitute (2.14)–(2.16) into (2.12) to obtain

$$\dot{F}_\varepsilon(t) \leq \left[1 - \varepsilon\beta L^2(\mathcal{E}(0))^\beta - \frac{\varepsilon L c_1}{\delta}(\mathcal{E}(0))^\beta \right.$$
$$\left. - \frac{\varepsilon L}{2c_0}(\mathcal{E}(0))^\beta - \frac{\varepsilon\lambda L C_0}{2\delta\eta(p+1)} - \frac{\varepsilon L}{\delta(p+1)c_0}\right]\dot{\mathcal{E}}(t)$$
$$- \varepsilon\left\{\min(1,\alpha) - L\delta[2L + c_2(\mathcal{E}(0))^{(q-1)/2}]\right\}\mathcal{E}(t)^{\beta+1} \quad (2.17)$$
$$+ \frac{\varepsilon L C_0(p+1-2\lambda)}{4\delta(p+1)}\eta^{2\lambda/(p+1-2\lambda)}(\mathcal{E}(t))^{\beta(p+1)/(p+1-2\lambda)}$$
$$+ \frac{\varepsilon L(p-1)}{p+1}\delta^{2/(p-1)}(\mathcal{E}(t))^{\beta(p+1)/(p-1)}$$

with the understanding that the last term on the right, and the last term in the bracket which multiplies $\dot{\mathcal{E}}(t)$, are to be omitted if $p \leq 1$. To complete the proof of (2.3), we suppose that $p > 1$ and leave it to the reader to make the minor changes needed for the opposite case.

The inequality (2.17) is valid for every $\beta > 0$. We now choose $\beta$ so that

$$\frac{\beta(p+1)}{p+1-2\lambda} = \beta + 1,$$

that is, $\beta = (p+1-2\lambda)/2\lambda$. Then $\beta + 1 = (p+1)/2\lambda$ and

$$(\mathcal{E}(t))^{\beta(p+1)/(p-1)} \leq (\mathcal{E}(t))^{\beta+1}(\mathcal{E}(0))^{(1-\lambda)(p+1)/\lambda(p-1)}.$$

The last three terms in (2.17) may therefore be combined to yield the estimate

$$\dot{F}_\varepsilon(t) \leq \left[1 - \varepsilon\beta L^2(\mathcal{E}(0))^\beta - \frac{\varepsilon L c_1}{\delta}(\mathcal{E}(0))^\beta \right.$$
$$\left. - \frac{\varepsilon L}{2c_0}(\mathcal{E}(0))^\beta - \frac{\varepsilon L}{\delta(p+1)c_0} - \frac{\varepsilon\lambda L C_0}{2c_0\delta\eta(p+1)}\right]\dot{\mathcal{E}}(t)$$
$$- \varepsilon\left\{\min(1,\alpha) - L\delta[2L + c_2(\mathcal{E}(0))^{(q-1)/2}] \right. \quad (2.18)$$
$$- \frac{L(p-1)}{p+1}\delta^{2/(p-1)}(\mathcal{E}(0))^{(1-\lambda)(p+1)/\lambda(p-1)}$$
$$\left. - \frac{LC_0(p+1-2\lambda)}{4\delta(p+1)}\eta^{2\lambda/(p+1-2\lambda)}\right\}\mathcal{E}(t)^{\beta+1}.$$

Suppose that $\mathcal{E}(0) \leq M$. We choose $\delta$ as follows:

$$\delta L[2L + c_2 M^{(q-1)/2}] + \frac{L(p-1)}{p+1}\delta^{2/(p-1)}M^{(1-\lambda)(p+1)/\lambda(p-1)} \leq \frac{1}{4}\min(1,\alpha).$$

With $\delta$ selected, choose $\eta$ so that

$$\frac{LC_0(p+1-2\lambda)}{4\delta(p+1)}\eta^{2\lambda/(p+1-2\lambda)} \leq \frac{1}{4}\min(1,\alpha).$$

Having selected $\delta$ and $\eta$, choose $\varepsilon$ so that

$$\varepsilon\left[\left(\beta L^2 + \frac{Lc_1}{\delta} + \frac{L}{2c_0}\right)M^\beta + \frac{L}{\delta(p+1)c_0} + \frac{\lambda LC_0}{2c_0\delta\eta(p+1)}\right] \leq 1.$$

Then we obtain from (2.18)

$$\dot{F}_\varepsilon(t) \leq -\varepsilon\sigma(\mathcal{E}(t))^{\beta+1}, \quad \sigma = \frac{\min(1,\alpha)}{2}, \quad \beta+1 = \frac{p+1}{2\lambda}. \tag{2.19}$$

Since $|\rho(t)| \leq L^2\mathcal{E}(t)$ we have

$$|F_\varepsilon(t) - \mathcal{E}(t)| \leq L^2(\mathcal{E}(t))^{\beta+1}$$
$$\leq L^2 M^\beta \mathcal{E}(t).$$

Therefore

$$(1-\varepsilon L^2 M^\beta)\mathcal{E}(t) \leq F_\varepsilon(t) \leq (1+\varepsilon L^2 M^\beta)\mathcal{E}(t).$$

Consequently,

$$\dot{F}_\varepsilon(t) \leq -\varepsilon\sigma(1+\varepsilon L^2 M^\beta)^{-\beta-1}(F_\varepsilon(t))^{\beta+1}$$
$$=: -K(F_\varepsilon(t))^{\beta+1}.$$

This yields, upon integration,

$$(F_\varepsilon(t))^\beta \leq \frac{(F_\varepsilon(0))^\beta}{1+(F_\varepsilon(0))^\beta K\beta t}. \tag{2.20}$$

Since

$$0 < F_\varepsilon(0) \leq (1+\varepsilon L^2 M^\beta)\mathcal{E}(0),$$

and since the function $\xi \to \xi/(1+k^2\xi)$ is strictly increasing on $\xi > 0$, it follows from (2.20) that

$$(F_\varepsilon(t))^\beta \leq \frac{(1+\varepsilon L^2 M^\beta)^\beta(\mathcal{E}(0))^\beta}{1+(1+\varepsilon L^2 M^\beta)^\beta(\mathcal{E}(0))^\beta K\beta t}$$
$$= \frac{(1+\varepsilon L^2 M^\beta)^\beta(\mathcal{E}(0))^\beta}{1+\varepsilon\sigma(1+\varepsilon L^2 M^\beta)^{-1}(\mathcal{E}(0))^\beta \beta t}.$$

Therefore
$$F_\varepsilon(t) \le \frac{(1+\varepsilon L^2 M^\beta)\mathcal{E}(0)}{[1+\omega t(\mathcal{E}(0))^\beta]^{1/\beta}}, \quad \omega = \frac{\varepsilon\sigma\beta}{1+\varepsilon L^2 M^\beta},$$

hence
$$\mathcal{E}(t) \le C[1+\omega t(\mathcal{E}(0))^\beta]^{-1/\beta}\mathcal{E}(0),$$

if $\varepsilon < L^{-2}M^{-\beta}$, where
$$C = \frac{1+\varepsilon L^2 M^\beta}{1-\varepsilon L^2 M^\beta}, \quad \beta = \frac{p+1-2\lambda}{2\lambda}.$$

**Remark 2.1.** Since $\varepsilon < L^{-2}M^{-\beta}$, it is seen that $\omega \to 0$ as $M \to \infty$ as least as fast as $M^{-\beta}$. On the other hand, by insisting that $\varepsilon \le (1/2)L^{-2}M^{-\beta}$ (for example), we may bound $C$ independently of $M$.

**Proof of (2.4).** Multiply (1.5) by $x\partial w$ and integrate the product over $(0, L) \times (0, T)$:
$$\int_0^T \int_0^L x\partial w[\ddot{w} + \partial^4 w - \gamma^2 \partial g(\partial w)]\,dx\,dt = 0. \tag{2.21}$$

We have
$$\int_0^T \int_0^L x\partial w \ddot{w}\,dx\,dt = \rho(T) - \rho(0) - \int_0^T \int_0^L x\dot{w}\partial\dot{w}\,dx\,dt$$
$$= \rho(T) - \rho(0) - \frac{1}{2}\int_0^T \int_0^L \partial(x\dot{w}^2)\,dx\,dt + \frac{1}{2}\int_0^T \int_0^L \dot{w}^2\,dx\,dt \tag{2.22}$$
$$= \rho(T) - \rho(0) + \frac{1}{2}\int_0^T \int_0^L \dot{w}^2\,dx\,dt - \frac{L}{2}\int_0^T \dot{w}^2(L,t)\,dt,$$

and
$$\int_0^T \int_0^L x\partial w \partial^4 w\,dx\,dt = L\int_0^T \partial w(L,t)\partial^3 w(L,t)\,dt + \int_0^T \int_0^L \partial^2(x\partial w)\partial^2 w\,dx\,dt$$
$$= L\int_0^T \partial w(L,t)\partial^3 w(L,t)\,dt + \frac{3}{2}\int_0^T \int_0^L (\partial^2 w)^2\,dx\,dt \tag{2.23}$$
$$+ \frac{1}{2}\int_0^T \int_0^L \partial[x(\partial^2 w)^2]\,dx\,dt$$
$$= L\int_0^T \partial w(L,t)\partial^3 w(L,t)\,dt + \frac{3}{2}\int_0^T \int_0^L (\partial^2 w)^2\,dx\,dt.$$

Finally,

$$\begin{aligned}
-\int_0^T\int_0^L x\partial w \partial g(\partial w) dx dt &= -L\int_0^T \partial w(L,t)g(\partial w(L,t))dt \\
&\quad + \int_0^T\int_0^L g(\partial w)(x\partial^2 w + \partial w)dx dt \\
&= -L\int_0^T \partial w(L,t)g(\partial w(L,t))dt \\
&\quad + \int_0^T\int_0^L \partial[xG(\partial w)]dx dt \qquad (2.24)\\
&\quad + \int_0^T\int_0^L [\partial w g(\partial w) - G(\partial w)]dx dt \\
&= L\int_0^T [G(\partial w(L,t)) - \partial w(L,t)g(\partial w(L,t))]dt \\
&\quad + \int_0^T\int_0^L [\partial w g(\partial w) - G(\partial w)]dx dt.
\end{aligned}$$

Identity (2.4) follows upon substitution of (2.22)–(2.24) into (2.21) and use of the boundary condition (1.10).

## References

[1] Conrad, F., Leblond, J. and Marmorat, J.-P., *Energy decay estimates for a beam with nonlinear boundary feedback*, Proc. COMCON Workshop on Stabilization of Flexible Structures, Montpellier, France, January, 1989.

[2] Kane, T. R., Ryan, R. R. and Banerjee, A. K., *Dynamics of a beam attached to a moving base*, J. Guidance, Control and Dynamics **10** (1987).

[3] Rogers, R. C. and Russell, D. L., *Derivation of linear beam equations using nonlinear continuum mechanics* (to appear).

J.Lagnese
G.Leugering
Department of Mathematics
Georgetown University
Washington, DC 20057, USA

# Neumann Boundary Stabilization of Structurally Damped Time Periodic Wave and Plate Equations

ALESSANDRA LUNARDI[*]

Dipartimento di Matematica, Università di Cagliari

## 1. Introduction

Let $\Omega \subset \mathbb{R}^n$ be a bounded open set with $C^2$ boundary $\partial\Omega$ and exterior unit normal vector $\nu$. Let $\rho$ be a $T$-periodic positive $C^1$ function, and let $A : D(A) = \{\varphi \in H^2(\Omega) : \frac{\partial \varphi}{\partial \nu} = 0\} \to L^2(\Omega), A\varphi = \Delta\varphi$ be the realization of the Laplace operator in $L^2(\Omega)$ with homogeneous Neumann boundary condition. We shall study the stabilizability problem for the strongly damped wave equation and the proportionally damped wave equation in $(0, +\infty) \times \Omega$:

$$\begin{aligned} u_{tt}(t,x) &= \Delta u(t,x) + \rho(t)\Delta u_t(t,x),\ t>0, x \in \Omega, \\ u(0,x) &= u_0(x),\ u_t(0,x) = v_0(x),\ x \in \Omega, \\ \frac{\partial}{\partial \nu}u(t,x) &= (\Phi g_1(t))(x),\ t>0,\ x \in \partial\Omega; \end{aligned} \quad (1.1)$$

$$\begin{aligned} u_{tt}(t,x) &= \Delta u(t,x) + \rho(t)(-A)^{\frac{1}{2}}u_t(t,x),\ t>0, x \in \Omega, \\ u(0,x) &= u_0(x),\ u_t(0,x) = v_0(x),\ x \in \Omega, \\ \frac{\partial}{\partial \nu}u(t,x) &= (\Phi g_2(t))(x),\ t>0,\ x \in \partial\Omega; \end{aligned} \quad (1.2)$$

where $g_1, g_2 : [0,+\infty[ \to Z$ ($Z$ = Banach space of controls) are the controls, and $\Phi$ is a linear bounded operator from $Z$ into $L^2(\partial\Omega)$. Analogously, we consider

---

[*] The author is a member of G.N.A.F.A. of C.N.R., and was partially supported by the Italian National Project M.P.I. "Equazioni di Evoluzione e Applicazioni Fisico-Matematiche".

two damped plate equations:

$$\begin{aligned}&u_{tt}(t,x) = -\Delta^2 u(t,x) - \rho(t)\Delta^2 u_t(t,x), \ t > 0, x \in \Omega \\ &u(0,x) = u_0(x), \ u_t(0,x) = v_0(x), \ x \in \Omega, \\ &\frac{\partial}{\partial \nu}u(t,x) = 0, \ \frac{\partial}{\partial \nu}\Delta u(t,x) = (\Phi g_3(t))(x), \ t > 0, \ x \in \partial\Omega;\end{aligned} \quad (1.3)$$

$$\begin{aligned}&u_{tt}(t,x) = (-\Delta)^2 u(t,x) + \rho(t)(-\Delta)^2 u_t(t,x), \ t > 0, x \in \Omega \\ &u(0,x) = u_0(x), \ u_t(0,x) = v_0(x), \ x \in \Omega, \\ &\frac{\partial}{\partial \nu}u(t,x) = 0, \ \frac{\partial}{\partial \nu}\Delta u(t,x) = (\Phi g_4(t))(x), \ t > 0, \ x \in \partial\Omega;\end{aligned} \quad (1.4)$$

In the case $F = 0$, systems (1.1),...,(1.4) are not stable, since they admit solutions of the form $u(t,x) = a + bt$. Interior stabilization of system (1.1) was considered in [9] in the autonomous case $\rho(t) = \rho$, and in [4] in the nonautonomous case. Several regularity results concerning equations (1.3) and (1.4) (also with other boundary conditions) were given in [10]. Here we prove that systems (1.1), (1.2) (resp. (1.3), (1.4)) are exponentially stabilizable in the $H^1 \times L^2$ (resp. $H^2 \times L^2$) norm if and only if

$$\exists z \in Z \text{ such that } \int_{\partial\Omega} (\Phi z)(x) d\sigma_x \neq 0 \quad (1.5)$$

The idea of the proof is the following. Setting $u_t = v$, we reduce systems (1.1) and (1.2) to first order systems in the product space $\mathcal{X} = H^1(\Omega) \times L^2(\Omega)$ :

$$\begin{aligned}&\begin{bmatrix} u' \\ v' \end{bmatrix} = \begin{bmatrix} 0 & 1 \\ \Delta & \rho(t)\Delta \end{bmatrix} \begin{bmatrix} u \\ v \end{bmatrix}, \ t > 0 \\ &\begin{bmatrix} u(0) \\ v(0) \end{bmatrix} = \begin{bmatrix} u_0 \\ v_0 \end{bmatrix}, \\ &\frac{\partial u}{\partial \nu} = (\Phi g_1(t)), \ t > 0;\end{aligned} \quad (1.6)$$

$$\begin{aligned}&\begin{bmatrix} u' \\ v' \end{bmatrix} = \begin{bmatrix} 0 & 1 \\ \Delta & \rho(t)(-A)^{\frac{1}{2}} \end{bmatrix} \begin{bmatrix} u \\ v \end{bmatrix}, \ t > 0, \\ &\begin{bmatrix} u(0) \\ v(0) \end{bmatrix} = \begin{bmatrix} u_0 \\ v_0 \end{bmatrix}, \\ &\frac{\partial u}{\partial \nu} = (\Phi g_2(t)), \ t > 0;\end{aligned} \quad (1.7)$$

and we study the evolution operators $G(t,s)$ generated in $\mathcal{X}$ by the families $\mathcal{A}(t) = \begin{bmatrix} 0 & 1 \\ \Delta & \rho(t)\Delta \end{bmatrix}$ and $\mathcal{B}(t) = \begin{bmatrix} 0 & 1 \\ \Delta & \rho(t)(-A)^{\frac{1}{2}} \end{bmatrix}$, with domains $D(\mathcal{A}(t)) = \left\{ \begin{bmatrix} u \\ v \end{bmatrix} \in H^1(\Omega) \times H^1(\Omega) : u + \rho(t)v \in D(A) \right\}$, and $D(\mathcal{B}(t)) = D(A) \times H^1(\Omega)$

respectively. Analogously, we reduce systems (1.3) and (1.4) to first order systems in the product space $\mathcal{Y} = D(A) \times L^2(\Omega)$ :

$$\begin{bmatrix} u' \\ v' \end{bmatrix} = \begin{bmatrix} 0 & 1 \\ -\Delta^2 & -\rho(t)\Delta^2 \end{bmatrix} \begin{bmatrix} u \\ v \end{bmatrix}, \; t > 0$$
$$\begin{bmatrix} u(0) \\ v(0) \end{bmatrix} = \begin{bmatrix} u_0 \\ v_0 \end{bmatrix}, \quad (1.8)$$
$$\frac{\partial u}{\partial \nu} = 0, \; \frac{\partial \Delta u}{\partial \nu} = (\Phi g_3(t)), \; t > 0$$

$$\begin{bmatrix} u' \\ v' \end{bmatrix} = \begin{bmatrix} 0 & 1 \\ -\Delta^2 & \rho(t)\Delta \end{bmatrix} \begin{bmatrix} u \\ v \end{bmatrix}, \; t > 0$$
$$\begin{bmatrix} u(0) \\ v(0) \end{bmatrix} = \begin{bmatrix} u_0 \\ v_0 \end{bmatrix} \quad (1.9)$$
$$\frac{\partial u}{\partial \nu} = 0, \; \frac{\partial \Delta u}{\partial \nu} = (\Phi g_4(t)), \; t > 0$$

and we study the evolution operators $G(t,s)$ generated in $\mathcal{Y}$ by the families $\mathcal{L}(t) = \begin{bmatrix} 0 & 1 \\ -A^2 & -\rho(t)A^2 \end{bmatrix}$ and $\mathcal{M}(t) = \begin{bmatrix} 0 & 1 \\ -A^2 & \rho(t)A \end{bmatrix}$, with domains $D(\mathcal{L}(t)) = \left\{ \begin{bmatrix} u \\ v \end{bmatrix} \in H^2(\Omega) \times H^2(\Omega) : \frac{\partial u}{\partial \nu} = 0, \frac{\partial v}{\partial \nu} = 0, u + \rho(t)v \in D(A^2) \right\}$, and $D(\mathcal{M}(t)) = D(A^2) \times D(A)$ respectively.

The theories developed in [9]–[10] and in [1]–[2] are applicable to our systems, provided we reduce them to homogeneous boundary problems. This is done in the usual way, by using the Neumann mapping $N$ (defined by $N\xi = z$, where $z$ is the solution of the elliptic problem $\Delta z = z$ in $\Omega$, $\frac{\partial z}{\partial \nu} = \xi$ in $\partial\Omega$), and introducing the new unknown $w(t,x) = u(t,x) - (N\Phi g_i(t))(x)$ (i=1,2) in problems (1.1),(1.2), and the new unknown $z(t,x) = u(t,x) - ((I-A)^{-1}N\Phi g_i(t))(x)$ (i=3,4) in problems (1.3),(1.4). Therefore, the couples $\begin{bmatrix} w \\ w_t \end{bmatrix} \begin{bmatrix} z \\ z_t \end{bmatrix}$ satisfy homogeneous systems, and they can be represented by the variation of constants formula, provided that the $g_i$'s are twice differentiable. Once a function $W$ has a representation of the form $W(t) = \int_0^t G(t,s)F(s)ds + G(t,0)W_0$, ( $F$ being an exponentially decaying function), a necessary and sufficient compatibility condition between $W_0$ and $F$ is known ( [8], [5] ) in order that $W$ decays exponentially as $t \to \infty$. After some computation, we find that such a condition is satisfied if and only if conditions

$$Pu_0 = \int_0^\infty [s - \frac{\partial(s\rho(s))}{\partial s}]P'\Phi g_i(s)ds,$$
$$Pv_0 = \int_0^\infty [\rho'(s) - 1]P'\Phi g_i(s)ds + \rho(0)P'\Phi g_i(0) \quad (1.10)$$

($i = 1, 3$) hold in the case of problems (1.1),(1.3), and conditions

$$Pu_0 = \int_0^\infty sP'\Phi g_i(s)ds,$$
$$Pv_0 = -\int_0^\infty P'\Phi g_i(s)ds \qquad (1.11)$$

($i = 2, 4$) hold in the case of problems (1.2),(1.4). Here we have set

$$P\xi = (\text{ meas } \Omega)^{-1} \int_\Omega \xi(x)dx, \xi \in L^2(\Omega);$$
$$P'\eta = (\text{ meas } \partial\Omega)^{-1} \int_{\partial\Omega} \eta(x)d\sigma_x, \eta \in L^2(\partial\Omega). \qquad (1.12)$$

Therefore, the stabilization problem for systems (1.1) and (1.3) (resp. (1.2) and (1.4)) is reduced to the problem of finding a function $g_i$ satisfying (1.10) (resp. (1.11)). Of course, if either (1.10) or (1.11) are solvable for arbitrary initial values $(u_0, v_0)$, then (1.5) is necessarily satisfied; it is not difficult to see that the converse is also true.

## 2. The abstract setting

Let $\Omega \subset \mathbb{R}^n$ be a bounded open set with $C^2$ boundary $\partial\Omega$ and exterior unit normal vector $\nu$, and set $D(A) = \{\varphi \in H^2(\Omega) : \frac{\partial \varphi}{\partial \nu} = 0\}$, $A : D(A) \to L^2(\Omega)$, $A\varphi = \Delta\varphi$. Then the spectrum of $A$ consists of a sequence of eigenvalues:

$$\sigma(A) = \{-\lambda_k\}_{k \in \mathbb{N}}; \lambda_0 = 0, \lim_{k \to \infty} \lambda_k = +\infty \qquad (2.1)$$

so that the spectrum of $-A^2 : D(A^2) = \{\varphi \in H^4(\Omega) : \frac{\partial \varphi}{\partial \nu} = \frac{\partial \Delta \varphi}{\partial \nu} = 0\}$ consists of the sequence of eigenvalues

$$\sigma(-A^2) = \{-\lambda_k^2\}_{k \in \mathbb{N}}. \qquad (2.2)$$

Let $\{e_{kh}\}_{k \in \mathbb{N}, h=1,..,m(k)}$, be a complete orthonormal system spanning $L^2(\Omega)$ and consisting of eigenvectors of $A$ : $Ae_{kh} = -\lambda_k e_{kh}$ for each $k \in \mathbb{N}$ and $h = 0,..,m(k)$. We shall represent every function $\varphi \in L^2(\Omega)$ in the form $\varphi = \sum_{k=0}^\infty \sum_{h=1}^{m(k)} \varphi_{kh} e_{kh}$. The fractional power $(-A)^{\frac{1}{2}} : D((-A)^{\frac{1}{2}}) = H^1(\Omega) \to L^2(\Omega)$ is well defined by

$$(-A)^{\frac{1}{2}}\varphi = \sum_{k=0}^\infty \sum_{h=1}^{m(k)} (\lambda_k)^{\frac{1}{2}} \varphi_{kh} e_{kh} \ \forall \varphi \in H^1(\Omega). \qquad (2.3)$$

Set moreover

$$\mathcal{X} = H^1(\Omega) \times L^2(\Omega), \mathcal{Y} = D(A) \times L^2(\Omega). \qquad (2.4)$$

If $\rho : \mathbb{R} \to \mathbb{R}$ is a positive T- periodic $C^1$ function, define

$$D(\mathcal{A}(t)) = \left\{ \begin{bmatrix} u \\ v \end{bmatrix} \in H^1(\Omega) \times H^1(\Omega) : u + \rho(t)v \in D(A) \right\} \tag{2.5}$$

$$\mathcal{A}(t) : D(\mathcal{A}(t)) \to \mathcal{X}, \mathcal{A}(t) = \begin{bmatrix} 0 & 1 \\ A & \rho(t)A \end{bmatrix}.$$

$$D(\mathcal{B}(t)) = D(A) \times H^1(\Omega), \; \mathcal{B}(t) : D(\mathcal{B}(t)) \to \mathcal{X},$$
$$\mathcal{B}(t) = \begin{bmatrix} 0 & 1 \\ A & \rho(t)(-A)^{\frac{1}{2}} \end{bmatrix}. \tag{2.6}$$

and

$$D(\mathcal{L}(t)) = \left\{ \begin{bmatrix} u \\ v \end{bmatrix} \in D(A) \times D(A) : u + \rho(t)v \in D(A^2) \right\} \tag{2.7}$$

$$\mathcal{L}(t) : D(\mathcal{L}(t)) \to \mathcal{Y}, \; \mathcal{L}(t) = \begin{bmatrix} 0 & 1 \\ -A^2 & -\rho(t)A^2 \end{bmatrix}.$$

$$D(\mathcal{M}(t)) = D(A^2) \times D(A) \; \mathcal{M}(t) : D(\mathcal{M}(t)) \to \mathcal{Y},$$
$$\mathcal{M}(t) = \begin{bmatrix} 0 & 1 \\ -A^2 & \rho(t)A \end{bmatrix}. \tag{2.8}$$

It is well known that both $\mathcal{A}(t) : D(\mathcal{A}(t)) \to \mathcal{X}, \mathcal{B}(t) : D(\mathcal{B}(t)) \to \mathcal{X}$ generate analytic semigroups in $\mathcal{X}$, and both $\mathcal{L}(t) : D(\mathcal{L}(t)) \to \mathcal{Y}, \mathcal{M}(t) : D(\mathcal{M}(t)) \to \mathcal{Y}$ generate analytic semigroups in $\mathcal{Y}$ (see e.g. [3] ) for every $t \in \mathbb{R}$. Since the operators $\mathcal{B}(t)$ and $\mathcal{M}(t)$ have respective domains $\mathcal{D} = D(A) \times H^1(\Omega)$ and $\mathcal{E} = D(A^2) \times D(A)$ independent of $t$, and (as easily seen) $t \to \mathcal{B}(t)$ belongs to $C^1(\mathbb{R}, L(\mathcal{D}, \mathcal{X})), t \to \mathcal{M}(t)$ belongs to $C^1(\mathbb{R}, L(\mathcal{E}, \mathcal{Y}))$, then the family $\{\mathcal{B}(t)\}$ generates an evolution operator $G_\mathcal{B}(t, s)$ in $\mathcal{X}$, and the family $\{\mathcal{M}(t)\}$ generates an evolution operator $G_\mathcal{M}(t, s)$ in $\mathcal{Y}$ (see [9], [10] ). On the contrary, the domains of $\mathcal{A}(t)$ and $\mathcal{L}(t)$ depend on time, but it is easy to check that also the families $\{\mathcal{A}(t)\}$ and $\{\mathcal{L}(t)\}$ generate evolution operators in $\mathcal{X}$ and in $\mathcal{Y}$ respectively, by applying either the result of [7] or the following one (a simplified version of the main theorem of [1] ):

**Proposition 2.1.** *Let $X$ be a Banach space, and let $A(t) : D(A(t)) \subset X \to X (t \in \mathbb{R})$ be linear operators such that there are $\omega \in \mathbb{R}, \theta > \frac{\pi}{2}, c, \varepsilon > 0$ with*

(a) $\rho(A(t)) \supset \{\lambda \in \mathbb{C} : \; \mathrm{Re}\; \lambda \geq \omega\}, \|\lambda(\lambda - A(t)^{-1}\|_{L(X)} \leq M \; \forall t \in \mathbb{R},$

(b) $\|A(t)(\lambda - A(t))^{-1}[(A(t) - \omega)^{-1} - (A(s) - \omega)^{-1}\|_{L(X)} \leq |t - s||\lambda|^{-\varepsilon}$, $\forall \, t, s \in \mathbb{R}.$

*Then the family $\{A(t)\}$ generates an evolution operator $U_A(t, s)$ in $X$.*

By using the representation formula

$$(\lambda - \mathcal{A}(t))^{-1} = \frac{1}{\lambda \rho(t) + 1} \begin{bmatrix} (\lambda - \rho(t)A) & 1 \\ A & \lambda \end{bmatrix} (\frac{\lambda^2}{\lambda \rho(t) + 1} - A)^{-1}$$

which holds for every $\lambda \neq 0$, $\lambda \neq -\frac{1}{\rho(t)}$, $\lambda \neq -\frac{1}{2}[\lambda_k \rho(t) \pm ((\lambda_k \rho(t))^2 - 4\lambda_k)^{\frac{1}{2}}]$, and

$$(\lambda - \mathcal{L}(t))^{-1} = \frac{1}{\lambda \rho(t) + 1} \begin{bmatrix} (\lambda + \rho(t)A^2) & 1 \\ -A^2 & \lambda \end{bmatrix} (\frac{\lambda^2}{\lambda \rho(t) + 1} + A^2)^{-1},$$

one can check easily that the assumptions of Proposition 2.1 are satisfied by both families $\{\mathcal{A}(t)\}$ (with $X = \mathcal{X}$) and $\{\mathcal{L}(t)\}$ (with $X = \mathcal{Y}$). We denote by $U_\mathcal{B}(t,s)$ (resp. $U_\mathcal{L}(t,s)$) the evolution operator generated by the family $\{\mathcal{A}(t)\}$ (resp. $\{\mathcal{L}(t)\}$) in $\mathcal{X}$ (resp. in $\mathcal{Y}$). The asymptotic behavior of any evolution operator $H(t,s)$ (under the assumptions of either [9]–[10] or [1]–[2]) is determined by the spectral properties of the families of operators $\{H(s+T,s) : s \in \mathbb{R}\}$ (see [8] and [5]). The following proposition holds:

**Proposition 2.2.** *The spectra of $G_\mathcal{B}(s+T,s)$, $G_\mathcal{M}(s+T,s)$ and of $U_\mathcal{A}(s+T,s)$, $U_\mathcal{L}(s+T,s)$ are independent of $s \in \mathbb{R}$. Moreover:*

(i) *If $G$ denotes either $G_\mathcal{B}$ or $G_\mathcal{M}$, then $\sigma(G(T,0)) = \{0\} \cup \{\eta_k\}_{k \in \mathbb{N}}$, where $\eta_0 = 1, |\eta_k| < 1$ for $k \geq 1$, $\lim_{k \to \infty} \eta_k = 0$, and $\eta_k$ is an eigenvalue for each $k$;*

(ii) *If $U$ denotes either $U_\mathcal{A}$ or $U_\mathcal{L}$, then $\sigma(U(T,0)) = \{0, e^{-\int_0^T \frac{1}{\rho(s)} ds}\} \cup \{\chi_\pm(k)\}_{k \in \mathbb{N}}$, where $\chi_\pm(k)$ are eigenvalues, $\lim_{k \to \infty} \chi_-(k) = 0$, $\lim_{k \to \infty} \chi_+(k) = e^{-\int_0^T \frac{1}{\rho(s)} ds}$, and $\chi_+(0) = 1, |\chi_-(0)| < 1, |\chi_\pm(0)| < 1$ for $k \geq 1$;*

(iii) *For every $s \in \mathbb{R}$ and for every sufficiently small circle $\gamma$ centered at 1, we have*

$$\frac{1}{2\pi i} \int_\gamma (z - G(s+T,s))^{-1} dz = \frac{1}{2\pi i} \int_\gamma (z - U(s+T,s))^{-1} dz = \begin{bmatrix} P & 0 \\ 0 & P \end{bmatrix}$$

*where $P$ (defined in (1.12)) is the projection on the space of the constant functions.*

**Proof.** Since $\mathcal{D}$ (resp. $\mathcal{E}$) is compactly embedded in $\mathcal{X}$ (resp. in $\mathcal{Y}$), then $G(s+T,s)$ is a compact operator for every $s$, so that its spectrum (except the point 0) consists of a sequence of eigenvalues converging to 0. It is independent of $s$ because the eigenvalues of $G(s+T,s)$ do not depend on $s$. To estimate the eigenvalues of $G(s+T,s)$, we recall that $G(s+T,s) \begin{bmatrix} u_0 \\ v_0 \end{bmatrix} = \begin{bmatrix} u(s+T,\cdot) \\ u_t(s+T,\cdot) \end{bmatrix}$, where $u$ is the solution of (1.2) (resp. (1.4)) with initial time $s$ and $\Phi = 0$, and $u(t,\cdot) = \sum_{k=0}^\infty \sum_{h=1}^{m(k)} u_{kh}(t) e_{kh}$, where $u_{kh}(t)$ is the solution of

$$\begin{aligned} u''_{kh}(t) &= -\lambda_k u_{kh}(t) - (\lambda_k)^{1/2} \rho(t) u'_{kh}(t), \\ u_{kh}(s) &= \langle u_0, e_{kh} \rangle, \ u'_{kh}(s) = \langle v_0, e_{kh} \rangle, \end{aligned} \tag{2.9a}$$

or, respectively,
$$u''_{kh}(t) = -\lambda_k^2 u_{kh}(t) - \lambda_k \rho(t) u'_{kh}(t), \qquad (2.9b)$$
$$u_{kh}(s) = \langle u_0, e_{kh}\rangle, \; u'_{kh}(s) = \langle v_0, e_{kh}\rangle.$$

From (2.9)(a)(b) we deduce easily that 1 is a double eigenvalue of $G(s+T,s)$, and the other eigenvalues have modulus smaller than 1. Therefore (i) is proved. (ii) is a consequence of Theorem 2.3 of [4]. Now we have

$$U(s+T,s)\begin{bmatrix} u_0 \\ v_0 \end{bmatrix} = \begin{bmatrix} u(s+T,\cdot) \\ u_t(s+T,\cdot) \end{bmatrix},$$

where $u$ is the solution of (1.1) (resp. (1.3)) with initial time $s$ and $\Phi = 0$, and $u(t,\cdot) = \sum_{k=0}^{\infty} \sum_{h=1}^{m(k)} u_{kh}(t) e_{kh}$, where $u_{kh}(t)$ is the solution of

$$u''_{kh}(t) = -\lambda_k u_{kh}(t) - \lambda_k \rho(t) u'_{kh}(t), \qquad (2.10a)$$
$$u_{kh}(s) = \langle u_0, e_{kh}\rangle, \; u'_{kh}(s) = \langle v_0, e_{kh}\rangle,$$

or, respectively,
$$u''_{kh}(t) = -\lambda_k^2 u_{kh}(t) - \lambda_k^2 \rho(t) u'_{kh}(t), \qquad (2.10b)$$
$$u_{kh}(s) = \langle u_0, e_{kh}\rangle, \; u'_{kh}(s) = \langle v_0, e_{kh}\rangle.$$

Let us show (iii). Set

$$G(s+T,s)\begin{bmatrix} u_0 \\ v_0 \end{bmatrix} = \sum_{k=0}^{\infty} \sum_{h=1}^{m(k)} P_{kh} G_{kh}(s+T,s) \begin{bmatrix} \langle u_0, e_{kh}\rangle \\ \langle v_0, e_{kh}\rangle \end{bmatrix},$$

and

$$U(s+T,s)\begin{bmatrix} u_0 \\ v_0 \end{bmatrix} = \sum_{k=0}^{\infty} \sum_{h=1}^{m(k)} P_{kh} U_{kh}(s+T,s) \begin{bmatrix} \langle u_0, e_{kh}\rangle \\ \langle v_0, e_{kh}\rangle \end{bmatrix},$$

where $P_{kh}\begin{bmatrix} a \\ b \end{bmatrix} = \begin{bmatrix} ae_{kh} \\ be_{kh} \end{bmatrix}$. Moreover, $G_{kh}(s+T,s)\begin{bmatrix} \langle u_0, e_{kh}\rangle \\ \langle v_0, e_{kh}\rangle \end{bmatrix} = \begin{bmatrix} u_{kh}(s+T) \\ u'_{kh}(s+T) \end{bmatrix}$, where $u_{kh}$ is the solution of (2.9)(a)(resp. (2.9)(b)), and

$$U_{kh}(s+T,s)\begin{bmatrix} \langle u_0, e_{kh}\rangle \\ \langle v_0, e_{kh}\rangle \end{bmatrix} = \begin{bmatrix} u_{kh}(s+T) \\ u'_{kh}(s+T) \end{bmatrix},$$

where $u_{kh}$ is the solution of (2.10)(a) (resp. (2.10)(b)). From (2.9) and (2.10) we find

$$G_{01}(t,s) = U_{01}(t,s) = \begin{bmatrix} 1 & t-s \\ 0 & 1 \end{bmatrix} \qquad (2.11)$$

Then we have

$$\frac{1}{2\pi i}\int_\gamma (z - G(s+T,s))^{-1} dz$$

$$= \frac{1}{2\pi i}\int_\gamma \sum_{k=0}^{\infty}\sum_{h=1}^{m(k)} P_{hk}(z - G_{kh}(s+T,s))^{-1}\begin{bmatrix}\langle u_0 e_{kh}\rangle\\ \langle v_0 e_{kh}\rangle\end{bmatrix} dz$$

$$= \frac{1}{2\pi i}\int_\gamma P_{01}\left(z - \begin{bmatrix}1 & T\\ 0 & 1\end{bmatrix}\right)^{-1}\begin{bmatrix}Pu_0\\ Pv_0\end{bmatrix} dz,$$

and, analogously,

$$\frac{1}{2\pi i}\int_\gamma (z - G(s+T,s))^{-1} dz = \frac{1}{2\pi i}\int_\gamma P_{01}\left(z - \begin{bmatrix}1 & T\\ 0 & 1\end{bmatrix}\right)^{-1}\begin{bmatrix}Pu_0\\ Pv_0\end{bmatrix} dz.$$

Statement (iii) follows now easily. □

For representing the solutions of (1.1),...,(1.4) by means of the evolution operators considered above, we introduce the Neumann mapping, defined by $N\xi = z$, where $z$ is the solution of

$$\begin{aligned}\Delta z &= z \quad \text{in } \Omega\\ \frac{\partial z}{\partial \nu} &= \xi \quad \text{in } \partial\Omega.\end{aligned} \qquad (2.12)$$

As well known, $N$ is a bounded operator from $L^2(\partial\Omega)$ to $H^{\frac{3}{2}}(\Omega)$ and from $H^{\frac{1}{2}}(\partial\Omega)$ to $H^2(\Omega)$. Moreover, $\Delta N = N$. This implies that, if $g_1 \in C^2([0,+\infty[, Z)$, $u_0 \in H^1(\Omega)$, $v_0 \in L^2(\Omega)$, and $u$ is a solution of (1.1), then, setting $w = u(t,x) - N\Phi g_1(t)(x)$, $w$ satisfies

$$\begin{aligned}w_{tt}(t,x) &= \Delta w(t,x) + \rho(t)\Delta w_t(t,x) + N\Phi g_1(t)(x)\\ &\quad + \rho(t)N\Phi g_1'(t)(x) - N\Phi g_1''(t)(x),\ t>0, x\in\Omega,\\ w(0,x) &= u_0(x) - N\Phi g_1(0)(x),\ x\in\Omega,\\ w_t(0,x) &= v_0(x) - N\Phi g_1'(0)(x),\ x\in\Omega,\\ \frac{\partial w(t,x)}{\partial \nu} &= 0,\ t>0, x\in\partial\Omega;\end{aligned} \qquad (2.13)$$

so that for every $t \geq 0$ we have

$$\begin{bmatrix}u\\ u_t\end{bmatrix} = U_{\mathcal{A}}(t,0)\begin{bmatrix}u_0 - N\Phi g_1(0)\\ v_0 - N\Phi g_1'(0)\end{bmatrix}\begin{bmatrix}N\Phi g_1(t)\\ N\Phi g_1'(t)\end{bmatrix}$$
$$+ \int_0^t U_{\mathcal{A}}(t,s)\begin{bmatrix}0\\ N\Phi g_1(s) + \rho(s)N\Phi g_1'(s) - N\Phi g_1''(s)\end{bmatrix} ds. \qquad (2.14)$$

Similarly, if $u$ is a solution to (1.1), then $w$ satisfies

$$w_{tt}(t,x) = \Delta w(t,x) + \rho(t)(-A)^{\frac{1}{2}} w_t(t,x) + N\Phi g_2(t)(x) +$$
$$+ \rho(t)(-A)^{\frac{1}{2}} N\Phi g_2'(t)(x) - N\Phi g_2''(t)(x), \ t > 0, x \in \Omega,$$
$$w(0,x) = u_0(x) - N\Phi g_2(0)(x), x \in \Omega, \ w_t(0,x) = v_0(x) - N\Phi g_2'(0)(x), x \in \Omega,$$
$$\frac{\partial w(t,x)}{\partial \nu} = 0, \ t > 0, x \in \partial\Omega,$$
(2.15)

so that for every $t \geq 0$ we have

$$\begin{bmatrix} u \\ u_t \end{bmatrix} = U_{\mathcal{B}}(t,0) \begin{bmatrix} u_0 - N\Phi g_2(0) \\ v_0 - N\Phi g_2'(0) \end{bmatrix} + \begin{bmatrix} N\Phi g_2(t) \\ N\Phi g_2'(t) \end{bmatrix}$$
$$+ \int_0^t U_{\mathcal{B}} \begin{bmatrix} 0 \\ N\Phi g_2(s) + \rho(s)(-A)^{\frac{1}{2}} N\Phi g_2'(s) - N\Phi g_2''(s) \end{bmatrix} ds.$$
(2.16)

The functions defined in formulas (2.14) and (2.16) are in fact solutions of problems (1.1) and (1.2), as the following proposition shows.

**Proposition 2.3.** Let $g_i \in C^2([0,+\infty[;Z), i = 1,2$, and let $u_0 \in H^1(\Omega)$, $v_0 \in L^2(\Omega)$. If $\Phi$ belongs to $L(Z, L^2(\partial\Omega))$, then the function $u$ given by (2.14) (resp. (2.16)) belongs to $C^2(]0,+\infty[;L^2(\Omega)) \cap C([0,+\infty[;H^1(\Omega))$, $u + \rho u'$ belongs to $C(]0,+\infty[;H^2(\Omega'))$ for every $\Omega' \subset\subset \Omega$ (resp. $u \in C^2(]0,+\infty[;L^2(\Omega)) \cap C([0,+\infty[;H^1(\Omega)) \cap C(]0,+\infty[;H^2(\Omega'))$ for every $\Omega' \subset\subset \Omega$), and $u$ is a solution of problem (1.1) (resp. (1.2)). If $\Phi$ belongs to $L(Z, H^{\frac{1}{2}}(\partial\Omega))$, then $u + \rho u'$ belongs to $C(]0,+\infty[;H^2(\Omega))$ (resp. $u \in C(]0,+\infty[;H^2(\Omega))$).

**Proof.** Let $\Phi$ belong to $L(Z, L^2(\partial\Omega))$. Then the function

$$t \to F(t) = \begin{bmatrix} 0 \\ N\Phi g_1(t) + \rho(t) N\Phi g_1'(t) - N\Phi g_1''(t) \end{bmatrix}$$

belongs to $C([0,+\infty[;\{0\} \times H^{\frac{3}{2}}(\Omega))$. Since $D(A) \times D(A)$ is contained in $D(\mathcal{A}(t))$ for every $t$, then $\{0\} \times H^{\frac{3}{2}}(\Omega)$ is contained in $D_{\mathcal{A}(t)}(\theta, 2)$ for every $\theta < \frac{3}{4}$. Hence, $\sup_{0 \leq t \leq a} \|F(t)\|_{D_{\mathcal{A}(t)}(\theta,\infty)} < +\infty$ for every $a > 0$ and $\theta < \frac{3}{4}$. Now, theorem 6.6 of [2] ensures that the function

$$t \to U_1(t) = \int_0^t U_{\mathcal{A}}(t,s) \begin{bmatrix} 0 \\ N\Phi g_1(s) + \rho(s) N\Phi g_1'(s) - N\Phi g_1''(s) \end{bmatrix}$$
$$+ U(t,0) \begin{bmatrix} u_0 - N\Phi g_1(0) \\ v_0 - N\Phi g_1'(0) \end{bmatrix}$$

belongs to $C([0,+\infty[;\mathcal{X}) \cap C^1(]0,+\infty[;\mathcal{X})$, $U_1$ belongs to the domain of $\mathcal{A}(t)$ for each $t > 0$, and $\mathcal{A}(t)U_1(t)$ belongs to $C(]0,+\infty[;\mathcal{X})$. This implies that, setting $U_1(t) = \begin{bmatrix} u_1(t) \\ v_1(t) \end{bmatrix}$, then $u_1 \in C^2(]0,+\infty[; L^2(\Omega)) \cap C([0,+\infty[; H^1(\Omega))$, $u_1 + \rho v_1 \in C(]0,+\infty[; H^2(\Omega))$, and

$$u_1'(t) = v_1(t),$$
$$v_1'(t) = u_1''(t) = \Delta u_1(t) + \rho(t)\Delta u_1'(t) + N\Phi g_1(t) + \rho(t)N\Phi g_1'(t) - N\Phi g_1''(t)$$
$$u_1(0) = u_0 - N\Phi g_1(0),\ v_1(0) = v_0 - N\Phi g_1'(0)$$
$$\frac{\partial u_1}{\partial \nu} = 0.$$

(2.17)

The function $U_2(t) = \begin{bmatrix} N\Phi g_1(t) \\ N\Phi g_1'(t) \end{bmatrix}$ belongs to

$$C^2([0,+\infty[; H^{\frac{3}{2}}(\Omega)) \times C^1(]0,+\infty[; H^{\frac{3}{2}}(\Omega)),$$

$u_2(t)$ and $v_2(t)$ belong to $C^\infty(\Omega')$ for every $\Omega' \subset \Omega$, and, setting $u_2(t) = N\Phi g_1(t)$ and $v_2(t) = N\Phi g_1'(t)$ we have

$$u_2'(t) = v_2(t),\ v_2'(t) = u_2''(t) = N\Phi g_1''(t)$$
$$u_2(0) = N\Phi g_1(0),\ v_2(0) = v_0 - N\Phi g_1'(0)$$
$$\frac{\partial u_2}{\partial \nu} = \Phi g_1(t)$$

(2.18)

Summing up, we find that $u$ belongs to $C^2(]0,+\infty[; L^2(\Omega)) \cap C([0,+\infty[; H^1(\Omega))$, $u + \rho u' \in C(]0,+\infty[; H^2(\Omega'))$, for every $\Omega' \subset\subset \Omega$, and $u$ satisfies (1.1). If, in addition, $\Phi$ belongs to $L(Z, H^{\frac{1}{2}}(\Omega))$, then $U_2 \in C^2(]0,+\infty[; H^2(\Omega)) \times C^1([0,+\infty[; H^2(\Omega))$. Hence, $u$ belongs to

$$C^2(]0,+\infty[; L^2(\Omega)) \cap C([0,+\infty[; H^1(\Omega)),$$

and $u+\rho u'$ belongs to $C(]0,+\infty[; H^2(\Omega))$, and the statement concerning problem (1.1) is proved.

Let us consider now problem (1.2). If $\Phi$ belongs to $L(Z, L^2(\partial\Omega))$, then $N\Phi g_2$ belongs to $C^2([0,+\infty[; H^{\frac{3}{2}}(\Omega)) \subset C^2([0,+\infty[; D(-A)^{\frac{3}{4}-\varepsilon})$ for every $\varepsilon \in ]0,\frac{3}{4}[$, so that the function

$$t \to F(t) = \begin{bmatrix} 0 \\ N\Phi g_2(t) + \rho(t)(-A)^{\frac{1}{2}}N\Phi g_2'(t) - N\Phi g_2''(t) \end{bmatrix}$$

belongs to $C([0,+\infty[; \{0\} \times D(-A)^{\frac{1}{2}-\varepsilon}) \subset C([0,+\infty[; D_{\mathcal{B}(0)}(\frac{1}{2}-\varepsilon, 2))$. The rest of the proof is similar to the proof in the case of problem (1.1), the unique difference being that now $U_1 \in C([0,+\infty[; \mathcal{X}) \cap C^1(]0,+\infty[; \mathcal{X}) \cap C(]0,+\infty[; \mathcal{D})$ (since $D(\mathcal{B}(t)) = \mathcal{D}$ for every $t$), so that

$$u_1 \in C^2(]0,+\infty[; L^2(\Omega)) \cap C([0,+\infty[; H^1(\Omega)) \cap C(]0,+\infty[; H^1(\Omega)). \quad \square$$

Let us consider now problems (1.3), (1.4). If $g_3 \in C^2([0,+\infty[, Z)$, $u_0 \in D(A)$, $v_0 \in L^2(\Omega)$, and $u$ is a solution of (1.3), then, setting $z = u(t,x) - (I - A)^{-1}N\Phi g_3(t)(x)$, $z$ satisfies

$$\begin{aligned}
z_{tt}(t,x) &= -\Delta^2 z(t,x) - \rho(t)\Delta^2 z_t(t,x) - [(I-A)^{-1} - 2]N\Phi g_3(t) \\
&\quad + \rho(t)N\Phi g_3'(t)(x) - (I-A)^{-1}N\Phi g_3'(t)(x),\ t > 0, x \in \Omega, \\
z(0,x) &= u_0(x) - (I-A)^{-1}N\Phi g_3(0)(x),\ x \in \Omega, \\
z_t(0,x) &= v_0(x) - (I-A)^{-1}N\Phi g_3'(0)(x),\ x \in \Omega, \\
\frac{\partial z(t,x)}{\partial \nu} &= 0,\ \frac{\partial \Delta z(t,x)}{\partial \nu} = 0\ t > 0, x \in \partial\Omega,
\end{aligned} \qquad (2.19)$$

so that

$$\begin{aligned}
\begin{bmatrix} u \\ u_t \end{bmatrix} &= U_\mathcal{L}(t,0) \begin{bmatrix} u_0 - (I-A)^{-1}N\Phi g_3(0) \\ v_0 - (I-A)^{-1}N\Phi g_3'(0) \end{bmatrix} + + \begin{bmatrix} (I-A)^{-1}N\Phi g_3(t) \\ (I-A)^{-1}N\Phi g_3'(t) \end{bmatrix} \\
&\quad + \int_0^t U_\mathcal{L}\begin{bmatrix} 0 \\ -[(I-A)^{-1} - 2][N\Phi g_3(s) + \rho(s)N\Phi g_3'(s)] \end{bmatrix} ds \\
&\quad - \int_0^t U_\mathcal{L}\begin{bmatrix} 0 \\ (I-A)^{-1}N\Phi g_3''(s) \end{bmatrix} ds,\ t \geq 0.
\end{aligned} \qquad (2.20)$$

Similarly, if $u$ is a solution to (1.4), then $z$ satisfies

$$\begin{aligned}
z_{tt}(t,x) &= -\Delta^2 z(t,x) + \rho(t)\Delta z_t(t,x) \\
&\quad - [(I-A)^{-1} - 2]N\Phi g_4'(t) + \rho(t)A(I-A)^{-1}N\Phi g_4''(t)(x) \\
z(0,x) &= u_0(x) - (I-A)^{-1}N\Phi g_4(0)(x),\ x \in \Omega, \\
z_t(0,x) &= v_0(x) - (I-A)^{-1}N\Phi g_4'(0)(x),\ x \in \Omega, \\
\frac{\partial z(t,x)}{\partial \nu} &= 0,\ \frac{\partial \Delta z(t,x)}{\partial \nu} = 0\ t > 0, x \in \partial\Omega,
\end{aligned} \qquad (2.21)$$

so that

$$\begin{aligned}
\begin{bmatrix} u \\ u_t \end{bmatrix} &= G_\mathcal{M}(t,0) \begin{bmatrix} u_0 - (I-A)^{-1}Ng_4(0) \\ v_0 - (I-A)^{-1}Ng_4'(0) \end{bmatrix} + \begin{bmatrix} u_0 - (I-A)^{-1}Ng_4(0) \\ v_0 - (I-A)^{-1}Ng_4'(0) \end{bmatrix} \\
&\quad + \int_0^t \begin{bmatrix} 0 \\ -[(I-A)^{-1} - 2]N\Phi g_4(s) + \rho(s)A(I-A)^{-1}N\Phi g_4'(s) \end{bmatrix} ds \\
&\quad + \int_0^t \begin{bmatrix} 0 \\ -(I-A)^{-1}N\Phi g_4''(s) \end{bmatrix} ds,\ t \geq 0
\end{aligned} \qquad (2.22)$$

The functions defined in formulas (2.20) and (2.22) are in fact solutions of problems (1.3) and (1.4), as the following proposition states. Since the proof is very similar to the one of Proposition 2.3, we omit it.

**Proposition 2.4.** Let $g_i \in C^2([0, +\infty[; Z), i = 1, 2$. If $\Phi$ belongs to $L(Z, L^2(\Omega))$, then the function $u$ given by (2.20) (resp. (2.22)) belongs to $C^2(]0, +\infty[; L^2(\Omega)) \cap C([0, +\infty[; H^2(\Omega))$, $u + \rho u'$ belongs to $C(]0, +\infty[; H^4(\Omega'))$ for every $\Omega' \subset\subset \Omega$ (resp. $u \in C^2(]0, +\infty[; L^2(\Omega)) \cap C([0, +\infty[; H^2(\Omega)) \cap C(]0, +\infty[; H^4(\Omega'))$ for every $\Omega' \subset\subset \Omega$), and $u$ is a solution of problem (1.3) (resp. (1.4)). If $\Phi$ belongs to $L(Z, H^{\frac{1}{2}}(\partial\Omega))$, then $u + \rho u'$ belongs to $C(]0, +\infty[; H^4(\Omega))$ (resp. $u \in C(]0, +\infty[; H^4(\Omega))$).

Concerning asymptotic behavior, the following proposition holds true (see [8], [5]).

**Proposition 2.5.** Let $X$ be a Banach space, and let $\{A(t)\}$ be a $T$-periodic family of operators satisfying the assumptions of Proposition 2.1. Let $H(t, s)$ be the evolution operator generated by $\{A(t)\}$, and assume that the circle $\{z \in \mathbb{C} : |z| = e^{-\omega t}\}$ does not intersect the spectrum of $H(s + T, s)$ for every $s$. Fix any $V_0 \in X$, and any function $F \in C([0, \infty[; X)$ such that $\sup_{t>0} \|F(t)e^{\omega t}\|_X < \infty$. Then the function

$$V(t) = H(t, 0)V_0 + \int_0^t H(t, s)F(s)ds, \; t \geq 0 \qquad (2.23)$$

is such that $\sup_{t>0} \|V(t)e^{\omega t}\| < +\infty$ if and only if

$$P(0)V_0 = -\int_0^{+\infty} H(0, s)P(s)F(s)ds, \qquad (2.24)$$

where $P(s) = \frac{1}{2\pi i} \int_\gamma (z - H(s+T, s))^{-1} dz$, $\gamma$ being the circle centered at 0 with radius $e^{-\omega T}$. If in addition $F(t)$ belongs to $D_A(\theta, 2)$ for some $\theta \in ]0, 1[$, with $\sup_{t>0} \|F(t)e^{\omega T}\|_{D_A(\theta, 2)} < +\infty$, and (2.24) holds, then $V'(t) = A(t)V(t) + F(t)$ for $t > 0$, $\sup_{t>a} \|A(t)V(t)e^{\omega t}\|_X < +\infty$ for every $a > 0$.

## 3. Stabilization

First we consider systems (1.2) and (1.4). We use notation from Section 1.

**Proposition 3.1.** Let $\Phi \in L(Z, L^2(\partial\Omega))$, and let $0 < \omega < -\frac{1}{T} \ln \sup\{|z| : z \in \sigma(G_\mathcal{B}(T, 0))\}$ (resp. $G_\mathcal{M}(T, 0), z \neq 1\}$). The following statements are equivalent:

(a) For each $(u_0, v_0) \in \mathcal{X}$ ( resp. $\mathcal{Y}$) there is $g \in C^2([0, +\infty[; Z)$ such that

$$\sup_{t>0} \|g(t)e^{\omega t}\|_Z + \sup_{t>0} \|g'(t)e^{\omega t}\|_Z + \sup_{t>0} \|g''(t)e^{\omega t}\|_Z < +\infty$$

and

$$\sup_{t>0} \|u_t(t, \cdot)e^{\omega t}\|_{L^2(\Omega)} + \sup_{t>0} \|u(t, \cdot)e^{\omega t}\|_{H^1(\Omega)} < +\infty$$

( resp. $\sup_{t>0} \|u_t(t,\cdot)e^{\omega t}\|_{L^2(\Omega)} + \sup_{t>0} \|u(t,\cdot)e^{\omega t}\|_{H^2(\Omega)} < +\infty$),

where $u$ is the solution of (1.2) (resp. (1.4)).

(b) There exists $z \in Z$ such that $\int_{\partial\Omega} (\Phi z)(y) d\sigma_y \neq 0$.

If, in addition, $\Phi$ belongs to $L(Z, H^{\frac{1}{2}}(\partial\Omega))$, and either (a) or (b) holds, then for every $a > 0$ we have:

$$\sup_{t>a} \|u_t(t,\cdot)e^{\omega t}\|_{H^1(\Omega)} + \sup_{t>a} \|u(t,\cdot)e^{\omega t}\|_{H^2(\Omega)} < +\infty$$

( resp. $\sup_{t>a} \|u_t(t,\cdot)e^{\omega t}\|_{H^2(\Omega)} + \sup_{t>a} \|u(t,\cdot)e^{\omega t}\|_{H^4(\Omega)} < +\infty$).

**Proof.** First we consider system (1.2). Since $N\Phi g$ and $N\Phi g'$ decay exponentially in the norm of $H^{\frac{3}{2}}(\Omega)$, from Proposition 2.5 we get that

$$\sup_{t>0} \|u_t(t,\cdot)e^{\omega t}\|_{L^2(\Omega)} + \sup_{t>0} \|u(t,\cdot)e^{\omega t}\|_{H^1(\Omega)} < +\infty$$

if and only if

$$\begin{bmatrix} P & 0 \\ 0 & P \end{bmatrix} \begin{bmatrix} u_0 - N\Phi g(0) \\ v_0 - N\Phi g'(0) \end{bmatrix} = -\int_0^\infty G_B(0,s) \begin{bmatrix} P & 0 \\ 0 & P \end{bmatrix} \\ \times \begin{bmatrix} 0 \\ N\Phi g(s) + \rho(s)(-A)^{\frac{1}{2}} N\Phi g'(s) - N\Phi g''(s) \end{bmatrix} ds \quad (3.2)$$

Now, it is easy to see that $PN = P'$ and $P(-A)^{\frac{1}{2}}N = 0$ (recall that $P$ and $P'$ are defined in (1.12)). By using also (2.11), we find that (3.2) is equivalent to

$$\begin{bmatrix} Pu_0 - P'\Phi g(0) \\ Pv_0 - P'\Phi g'(0) \end{bmatrix} = -\int_0^\infty \begin{bmatrix} -sP'\Phi g(s) + sP'\Phi g''(s) \\ P'\Phi g(s) - P'\Phi g''(s) \end{bmatrix} ds. \quad (3.3)$$

Let us show now that also system (1.4) leads to equation (3.3). Since $N\Phi g$ and $N\Phi g'$ decay exponentially in the norm of $H^{\frac{3}{2}}(\Omega)$, then $(I-A)^{-1} N\Phi g$ and $(I-A)^{-1} N\Phi g'$ decay exponentially in the norm of $H^{\frac{5}{2}}(\Omega)$. From Proposition 1.5 we get that

$$\sup_{t>0} \|u_t(t,\cdot)e^{\omega t}\|_{L^2(\Omega)} + \sup_{t>0} \|u(t,\cdot)e^{\omega t}\|_{H^2(\Omega)} < +\infty$$

if and only if

$$\begin{bmatrix} P & 0 \\ 0 & P \end{bmatrix} \begin{bmatrix} u_0 - (I-A)^{-1} N\Phi g(0) \\ v_0 - (I-A)^{-1} N\Phi g'(0) \end{bmatrix} = -\int_0^\infty G_{\mathcal{M}}(0,s) \begin{bmatrix} P & 0 \\ 0 & P \end{bmatrix} \\ \times \begin{bmatrix} 0 \\ 2N\Phi g_4(s) - N(I-A)^{-1}\Phi(g_4(s) - \rho(s)g_4'(s) + g_4''(s) \end{bmatrix} ds \quad (3.4)$$

holds. Using now the equalities $P(I - A)^{-1}N = P'$ and, together with (2.11), we find that also (3.4) is equivalent to (3.3).

After some integrations by parts, one can see that (3.3) is equivalent to (1.11). Of course, if (1.11) is solvable for every couple $(u_0, v_0)$, then condition (b) holds. Conversely, if (b) holds, for each $(u_0, v_0)$ one can find easily a solution of (1.11) in the form $g(t) = \gamma(t)z$, where $\gamma$ is a scalar $C^\infty$ function with compact support and $\Phi z$ has nonzero mean value. The equivalence of (a) and (b) is so proved.

If $\Phi$ belongs to $L(Z, H^{\frac{1}{2}}(\partial\Omega))$, then for every $g \in C^1([0, \infty[; Z)$ such that $\sup_{t>0} \|g(t)e^{\omega t}\|_Z + \sup_{t>0} \|g'(t)e^{\omega t}\|_Z < +\infty$, the function

$$t \to \begin{bmatrix} N\Phi g(t) \\ N\Phi g'(t) \end{bmatrix} \quad \left( \text{resp. } t \to \begin{bmatrix} (I-A)^{-1}N\Phi g(t) \\ (I-A)^{-1}N\Phi g'(t) \end{bmatrix} \right)$$

belongs to

$$C^1([0, +\infty[; H^2(\Omega) \times H^2(\Omega)) \quad (\text{resp. to } C^1([0, +\infty[; H^4(\Omega) \times H^4(\Omega)))$$

and decays exponentially in the $H^2(\Omega) \times H^2(\Omega)$ (resp. $H^4(\Omega) \times H^4(\Omega)$) norm as $t \to \infty$. If (a) or (b) holds, by Proposition 2.2 we get that

$$t \to \begin{bmatrix} u - N\Phi g(t) \\ u' - N\Phi g'(t) \end{bmatrix} \quad \left( \text{resp. } t \to \begin{bmatrix} u - (I-A)^{-1}N\Phi g(t) \\ u' - (I-A)^{-1}N\Phi g'(t) \end{bmatrix} \right)$$

decays exponentially in the $\mathcal{D}$-norm (resp. in the $\mathcal{E}$-norm). This implies that

$$\sup_{t>a} \|u_t(t, \cdot)e^{\omega t}\|_{H^2(\Omega)} + \sup_{t>a} \|u(t, \cdot)e^{\omega t}\|_{H^4(\Omega)} < +\infty$$

for every $a > 0$. $\square$

Let us consider now systems (1.1) and (1.3).

**Proposition 3.2.** Let $\Phi \in L(Z, L^2(\partial\Omega))$, and let $0 < \omega < -\frac{1}{T}\ln\sup\{|z| : z \in \sigma(G_\mathcal{A}(T,0))\}$ (resp. $G_\mathcal{L}(T,0)$), $z \neq 1\}$. The following statements are equivalent:

(a) For each $(u_0, v_0) \in \mathcal{X}$ (resp. $\mathcal{Y}$) there is $g \in C^2([0, +\infty[; Z)$ such that

$$\sup_{t>0} \|g(t)e^{\omega t}\|_Z + \sup_{t>0} \|g'(t)e^{\omega t}\|_Z + \sup_{t>0} \|g''(t)e^{\omega t}\|_Z < +\infty$$

and

$$\sup_{t>0} \|u_t(t, \cdot)e^{\omega t}\|_{L^2(\Omega)} + \sup_{t>0} \|u(t, \cdot)e^{\omega t}\|_{H^1(\Omega)} < +\infty$$

(resp.

$$\sup_{t>0} \|u_t(t, \cdot)e^{\omega t}\|_{L^2(\Omega)} + \sup_{t>0} \|u(t, \cdot)e^{\omega t}\|_{H^2(\Omega)} < +\infty),$$

where $u$ is the solution of (1.1) (resp. (1.3) ).

(b) There exists $z \in Z$ such that $\int_{\partial\Omega} (\Phi z)(y) d\sigma_y \neq 0$.

*If, in addition, $\Phi$ belongs to $L(Z, H^{\frac{1}{2}}(\partial\Omega))$, and either (a) or (b) holds, then for every $a > 0$ we have:*

$$\sup_{t>a} \|u_t(t,\cdot)e^{\omega t}\|_{H^1(\Omega)} + \sup_{t>a} \|u(t,\cdot)e^{\omega t}\|_{H^2(\Omega)} < +\infty$$

*(resp.*

$$\sup_{t>a} \|u_t(t,\cdot)e^{\omega t}\|_{H^2(\Omega)} + \sup_{t>a} \|u(t,\cdot)e^{\omega t}\|_{H^4(\Omega)} < +\infty).$$

**Proof.** First we consider system (1.1). For every $g \in C^2([0,+\infty[;Z)$, $N\Phi g$ belongs to $C^2([0,+\infty[; H^{\frac{3}{2}}(\Omega))$. Since $\{0\} \times D(A)$ is contained in $\mathcal{D}$, then $\{0\} \times H^{\frac{3}{2}}(\Omega)$ is contained in $D_{A(t)}(\theta,\infty)$ for every $\theta < \frac{3}{4}$ (see [6]). Therefore, the function

$$f(s) = \begin{bmatrix} 0 \\ N\Phi g(s) + \rho(s)N\Phi g'(s - N\Phi g''(s)) \end{bmatrix}$$

belongs to $C([0,\infty[; \mathcal{X})$ and $\sup_{t>0} \|e^{\omega t} f(t)\|_{D_{A(t)}(\theta,\infty)} < \infty$. By Proposition 2.5 we get that

$$\sup_{t>0} \|u_t(t,\cdot)e^{\omega t}\|_{L^2(\Omega)} + \sup_{t>0} \|u(t,\cdot)e^{\omega t}\|_{H^2(\Omega)} < +\infty$$

if and only if

$$\begin{bmatrix} P & 0 \\ 0 & P \end{bmatrix} \begin{bmatrix} u_0 \\ v_0 \end{bmatrix} = -\int_0^\infty G_A(0,s) \begin{bmatrix} P & 0 \\ 0 & P \end{bmatrix}$$
$$\times \begin{bmatrix} 0 \\ N\Phi g(s) + \rho(s)N\Phi g'(s) - N\Phi''(s) \end{bmatrix} ds \quad (3.5)$$

holds. Due to (2.11) and to the fact that $PN = P'$, (3.5) is equivalent to

$$\begin{bmatrix} Pu_0 - P'\Phi g(0) \\ Pv_0 - P'\Phi g,(0) \end{bmatrix}$$
$$= -\int_0^\infty \begin{bmatrix} -sP'\Phi g(s) - s\rho(s)P'\Phi g'(s) + sP'\Phi g''(s) \\ P'\Phi g(s) + \rho(s)P'\Phi g'(s) - P'\Phi g''(s) \end{bmatrix} ds. \quad (3.6)$$

Let us show now that also system (1.3) leads to equation (3.6). For every exponentially decaying $g \in C^2([0,\infty[; Z)$, $N\Phi g$ belongs to $C^2([0,\infty[; H^{\frac{3}{2}}(\Omega))$, and $(I-A)^{-1}N\Phi g$ belongs to $C^2([0,\infty[; H^{\frac{5}{2}}(\Omega))$. Moreover, $(I-A)^{-1}N\Phi g$ and $(I-A)^{-1}N\Phi g$ decay exponentially in the norm of $H^{\frac{5}{2}}(\Omega)$.

Since $\{0\} \times D(A^2)$ is contained in $\mathcal{E}$, then $\{0\} \times H^{\frac{3}{2}}(\Omega)$ is contained in $D_{A(t)}(\theta, \infty)$ for every $\theta < \frac{3}{8}$ (see [6]). Therefore, the function

$$F(s) = \begin{bmatrix} 0 \\ -[(I-A)^{-1} - 2][N\Phi g_3(s) + \rho(s)N\Phi g_3'(s) - N\Phi g_3''(s)] \end{bmatrix}$$

belongs to $C([0, +\infty[; \mathcal{X})$ and $\sup_{t>0} \|e^{\omega t} F(t)\|_{D_{\mathcal{L}(t)}(\theta, \infty)} < \infty$. By Proposition 2.5 we get that $\begin{bmatrix} u \\ u_t \end{bmatrix}$ decays exponentially in the $H^2(\Omega) \times L^2(\Omega)$ norm if and only if

$$\begin{bmatrix} P & 0 \\ 0 & P \end{bmatrix} \begin{bmatrix} u_0 - (I-A)^{-1} N\Phi g_3(0) \\ v_0 - (I-A)^{-1} N\Phi g_3'(0) \end{bmatrix}$$
$$= -\int_0^\infty U_{\mathcal{L}}(0, s) \begin{bmatrix} P & 0 \\ 0 & P \end{bmatrix} F(s) ds \quad (3.7)$$

holds. Due to (2.11) and to the equalities $PN = P(I-A)^{-1}N = P'$, we find that also (3.7) is equivalent to (3.6).

After some integration by parts, one can see that (3.6) is equivalent to (1.10). In its turn, (1.10) is solvable if and only if condition (b) holds. Also in this case, for each $(u_0, v_0)$ one can find easily a solution of (1.10) in the form $g(t) = \gamma(t)z$, where $\Phi z$ has nonzero mean value, and $\gamma$ is a scalar $C^\infty$ function with compact support, with $\gamma(0) = 0$. The equivalence of (a) and (b) is so proved. The proof of the last part of the proposition is the same as the last part of the proof of Proposition 3.1. □

## References

[1] P. ACQUISTAPACE, *Evolution operators and strong solutions of abstract linear parabolic equations*, Diff. Int. Equations **1** (1988), 433–457.

[?] P. ACQUISTAPACE, B. TERRENI, *A Unified Approach to Abstract Linear Nonautonomous Parabolic Equations*, Rend. Sem. Mat. Univ. Padova **78** (1987), 47–107.

[3] S. CHEN, R. TRIGGIANI, *Proof of extension of two conjectures on structural damping for elastic systems: the case $\frac{1}{2} \leq \alpha \leq 1$*, Pacific J. Math. **136** (1989), 15–55.

[4] G. DA PRATO, A. LUNARDI *Floquet exponents and stabilizability in time periodic parabolic systems*, Appl. Math. Optimiz., (to appear).

[5] M. FURHMAN, *Bounded solutions for abstract time periodic parabolic equations with nonconstant domains*, Preprint.

[6] P. GRISVARD, *Equations différentielles abstraites*, Ann. Scient. Ec. Norm. Sup. **2** (1969), 311–395.

[7] T. KATO, H. TANABE, *On the abstract evolution equations*, Osaka Math. J. **14** (1962), 107–133.

[8] A. LUNARDI, *Bounded Solutions of Linear Periodic Abstract Parabolic Equations*, Proc. Royal Soc. Edinburgh **110A** (1988), 253–279.

[9] P.E. SOBOLEVSKII, *Equations of parabolic type in a Banach space*, Amer. Math. Soc. Transl. **49** (1966), 1–62.

[10] H. TANABE, *On the equations of evolution in a Banach space*, Osaka J. Math. **12** (1960), 363–376.

Alessandra Lunardi
Dipartimento di Matematica
Università di Cagliari
Via Ospedale 72
I-09124 Cagliari, Italy

# Convergence in Lotka-Volterra Systems with Diffusion and Delay

R.H. Martin
H.L. Smith

Department of Mathematics, North Carolina State University
Department of Mathematics, Arizona State University

In this note we establish a basic result on the behavior of solutions as $t \to \infty$ to the diffusion-reaction-delay system having the form

$$\partial_t u^i(x,t) = d_i \Delta u^i(x,t) + b_i u^i(x,t) G_i(u(x,\cdot)_t) \quad \text{on} \quad \Omega \times (0,\infty), \tag{1}$$
$$\partial_\nu^i u^i(x,t) = 0 \quad \text{on} \quad \partial\Omega \times (0,\infty), \quad u^i(x,s) = \chi_i(x,s) \quad \text{on} \quad \Omega \times [-\tau,0].$$

where $i = 1, \ldots, n$, $u = (u^1, \ldots, u^n)$, and $\Omega$ is a bounded region in $\mathbb{R}^N$ with $\partial\Omega$ smooth. Furthermore, $\Delta$ denotes the Laplacian on $\Omega$, $\partial_\nu$ denotes the outward normal derivative on $\partial\Omega$, and $d_i$ and $b_i$ are positive constants. The initial functions $\chi_i : \overline{\Omega} \times [-\tau, 0] \to \mathbb{R}$ are assumed continuous and nonnegative. Also $C[-\tau, 0]^n$ denotes the space of continuous functions from $[-\tau, 0]$ into $\mathbb{R}^n$ and for each $x \in \overline{\Omega}$, $u(x,\cdot)_t = (u^i(x,\cdot)_t)_1^n$ denotes the member of $C[-\tau, 0]^n$ defined by $u(x,\theta)_t \equiv u(x, t+\theta)$ for $\theta \in [-\tau, 0]$. The functionals $G_i$ in (1) map $C[-\tau, 0]^n$ into $\mathbb{R}$ and are assumed to have the form

$$G_i(\phi) = 1 - a_i c_{ii} \phi_i(0) - \sum_{j=1}^n c_{ij} \int_{-\tau}^0 \phi_j(\theta) dv_{ij}(\theta) \tag{2}$$

for all $\phi = (\phi_i)_1^n \in C[-\tau, 0]^n$ and $i = 1, \ldots, n$, where

(a) $c_{ij}$, $r_i$ and $a_i$ are real constants with $\frac{1}{2} < a_i \leq 1$,

(b) for $i \neq j$, $v_{ij}$ is a real-valued Borel measure with $v_{ij}([-\tau, 0]) = 1$, $\quad$ (3)

(c) $v_{ii}$ is a positive Borel measure with $v_{ii}([-\tau, 0]) = (1 - a_i)$.

Finally, for notational convenience define the $n \times n$ matrix $\hat{C}$ by

$$\hat{C} = (\hat{c}_{ij}) \quad \text{where} \quad \begin{cases} \hat{c}_{ii} = (2a_i - 1)c_{ii} \\ \hat{c}_{ij} = -|c_{ij}| \end{cases} \quad \text{for } i \neq j. \tag{4}$$

where $\gamma_{ij} = |v_{ij}|([-\tau, 0])$ and $|v_{ij}|$ is the total variation of $v_{ij}$. Under these circumstances we have the following result:

**Theorem 1.** *Suppose that (2) and (3) hold and that each eigenvalue of the matrix $\hat{C}$ in (4) has positive real part (or, equivalently, $\hat{C}$ is an M-matrix). Then there is a unique $\eta^* = (\eta_i^*)_1^n \in \mathbb{R}^n$ such that*

$$\eta_i^* \geq 0, \quad \sum_{j=1}^n c_{ij}\eta_j^* \geq r_i \quad \text{and} \quad \eta_i^*(r_i - \sum_{j=1}^n c_{ij}\eta_j^*) = 0 \quad \text{for} \quad i = 1, \ldots, n. \tag{5}$$

*Furthermore, the solution $u = (u^i)_1^n$ to (1) exists on $\overline{\Omega} \times [0, \infty)$ and if*

$$\sup\{\chi_i(x,0) : x \in \overline{\Omega}\} > 0 \quad \text{whenever} \quad \eta_i^* > 0,$$

*then*

$$\lim_{t \to \infty} \max\{|u^i(x,t) - \eta_i^*| : x \in \overline{\Omega}, \ i = 1, \ldots, n\} = 0. \tag{6}$$

**Remark 1.** Since $\hat{C}$ is a nonsingular $M$-matrix, there is a matrix $D = \operatorname{diag}(\delta_1, \ldots, \delta_n)$ such that $\delta_i > 0$ for all $i$ and $\hat{C}D$ is strictly diagonally dominant (see Berman and Plemmons [1, $M_{35}$, p. 137]). In particular from (4) we have

$$(2a_i - 1)c_{ii} - \sum_{j \neq i} \delta_i |c_{ij}| \gamma_{ij} \delta_j^{-1} > 0 \quad \text{for} \quad i = 1, \ldots, n. \tag{7}$$

Since $1/2 < a_i \leq 1$ we see that $c_{ii} > 0$ and $(2a_i - 1)c_{ii} \leq c_{ii}$. Also, $\gamma_{ij} \geq 1$ by (3b) so

$$c_{ii} - \sum_{j \neq i} \delta_i |c_{ij}| \delta_j^{-1} > 0 \quad \text{for} \quad i = 1, \ldots, n$$

and it follows from [1, Theorem 2.3, p. 134, $M_{35} \Rightarrow A_1$] that all of the principal minors of $C = (c_{ij})$ are positive. This implies in particular that the linear complementarity problem: given $\zeta \in \mathbb{R}^n$ there exists a unique $\eta \in \mathbb{R}^n$ such that

$$\eta \geq 0, \quad C\eta - \zeta \geq 0 \quad \text{and} \quad (C\eta - \zeta) \cdot \eta = 0 \tag{LCP}$$

has a solution (see [1, Theorem 2.15, p. 274]). If $\zeta_i = r_i$ for all $i$ and we let $\eta^*$ be the solution $\eta$ to (LCP), then it is clear that (5) holds. Therefore, the hypothesis of the theorem guarantees a unique solution $\eta^*$ to (5) and also the existence of $\delta_i > 0$, $i = 1, \ldots, n$, such that (7) holds. Furthermore, since (7)

continues to hold for small changes in the $\delta_i$'s, we may assume without loss of generality that
$$\delta_i \eta_i^* \neq \delta_j \eta_j^* \quad \text{if} \quad i \neq j \quad \text{and} \quad \eta_i^* > 0. \tag{8}$$

**Remark 2.** The results in Dunbar, Rybakowski and Schmitt [3] have considerable overlap with our Theorem, but the reaction terms in [3] have no delays. The crucial assumptions in [3] is that there is a diagonal matrix $D$ with strictly positive diagonal entries such that $DC + C^*D$ is negative semidefinite and that the equilibrium $\eta^*$ to (1) has strictly positive components. Since $a_i = 1$ is the non-delay case, the assumption that $\hat{C}$ be an $M$-matrix implies the existence of such a diagonal $D$ so that $DC + C^*D$ is negative indefinite [1, Theorem 2.3, p. 134]. However, the equilibrium $\eta^*$ need not have strictly positive components in this case. In particular one can conclude from our theorem that if $\hat{C}$ is an $M$-matrix then the solution to (1) is *persistent*, that is

$$\inf_{x \in \Omega} \lim_{t \to \infty} u^i(x,t) > 0 \quad \text{for all} \quad i = 1, \ldots, n$$

if and only if

$$\sup_{x \in \Omega} \chi_i(x,0) > 0 \quad \text{for all} \quad i = 1, \ldots, n \quad \text{and}$$

$$\eta = C^{-1} \rho \quad \text{has strictly positive components}$$

where $\rho$ denotes the member of $\mathbb{R}^n$ with the i-th component equal $r_i$.

The result in Brown [2, Theorem 3.3], which considers system (1) without time delays, is included in these techniques. A criteria analogous to that used in [2] is

$$\sum_{j=1}^n c_{ij}(c_{jj}a_j)^{-1} < 2 \quad \text{for} \quad i = 1, \ldots, n$$

(it is also assumed $c_{ij} \geq 0$ in [2]). Taking $\delta_i = a_i c_{ii}$ for $i = 1, \ldots n$, shows that this inequality implies that (7) holds, and Lemma 3.2 in [2] implies that $\eta^* > 0$ for all $i = 1, \ldots, n$. These ideas are expanded to the delay case in Martin and Smith [6]. The important connection between the global attractiveness of an equilibrium and the linear complementarity problem (LCP) was introduced in Hofbauer [4], where a system of ordinary differential equations is analyzed.

Local existence and nonnegativeness of (mild) solutions to (1) follows with standard techniques (see Martin and Smith [5]). If $u = (u^i)^n$, is a solution on $\Omega \times [-\tau, T]$ and

$$M_i \equiv \sup\{|G_i(u(x, \cdot)_t| : (x,t) \in \overline{\Omega} \times [0, T]\}$$

then

$$d_i \Delta u^i - b_i M_i u_i \leq \partial_t u^i \leq d_i \Delta u^i + b_i M_i u_i$$

and it follows by comparison and the maximum principle that

(a) $\chi_i(x,0) \equiv 0$ implies $u^i(x,t) \equiv 0$ and

(b) $\chi_i(x_o,0) > 0$ for some $x_o \in \Omega$ implies $u^i(x,t) > 0$ for all $x \in \overline{\Omega}$, $t > 0$.
$$\tag{9}$$

In particular by our assumptions we have

$$u^i(x,t) > 0 \quad \text{for all} \quad x \in \overline{\Omega},\ t > 0 \quad \text{and all } i \text{ such that} \quad \eta_i^* > 0. \tag{10}$$

Let $\delta_i$, $i = 1,\ldots,n$, be as in (7) and for each $\xi \in \mathbb{R}^n$ define

$$V[\xi] = \max\{\delta_i|\xi_i - \eta_i^*| : i = 1,\ldots,n\} \quad \text{and} \quad \mathcal{N}(\xi) = \{i : V[\xi] = \delta_i|\xi_i - \eta_i^*|\}.$$

Also let $C(\overline{\Omega})^n$ be the space of continuous functions $y : \overline{\Omega} \to \mathbb{R}^n$ and define $W$ and $\mathcal{M}$ on $C(\overline{\Omega})^n$ by

$$W[y] = \max\{V[y(x)] : x \in \overline{\Omega}\} \quad \text{and} \quad \mathcal{M}[y] = \{x : W[y] = V[y(x)]\}.$$

One may easily check that if

$$D_-W[y](z) \equiv \lim_{h \to 0+} \frac{W[y - hz] - W[y]}{-h}$$

for all $y, z \in C(\overline{\Omega})^n$, then for $y(x) \not\equiv \eta^*$ on $\overline{\Omega}$,

$$D_-W[y](z) = \min\{\delta_i \operatorname{sgn}(y_i(x) - \eta_i^*) z_i(x) : x \in \mathcal{M}(y),\ i \in \eta(y(x))\} \tag{11}$$

where $\operatorname{sgn}(r)$ is 1 if $r > 0$ and -1 if $r < 0$.

Now let $\mathcal{C}[-\tau,0]^n$ be the space of continuous functions $\phi : [-\tau,0] \to C(\overline{\Omega})^n$ and identify it with the space of continuous functions $\phi : \overline{\Omega} \times [-\tau,0] \to \mathbb{R}^n$. Therefore, if $\phi \in \mathcal{C}[-\tau,0]^n$ then $\phi(\cdot,\theta) \in C(\overline{\Omega})^n$ for each $\theta \in [-\tau,0]$ and $\phi(x,\cdot) \in C[-\tau,0]^n$ for each $x \in \overline{\Omega}$. Let the linear operator $A$ be the closure in $C(\overline{\Omega})^n$ of the linear operator $B$ defined by

$$By = (d_i \Delta y_i)_1^n \quad \text{for all} \quad y = (y_i)_i^n \in D(B),$$
$$D(B) = \{y : y \in C^2(\overline{\Omega}) \quad \text{and} \quad \partial_\nu y = 0 \quad \text{on } \partial\Omega\}.$$

Also, define the map $F = (F_i)_i^m : \mathcal{C}[-\tau,0]^n \to C(\overline{\Omega})^n$ by

$$[F_i(\phi)](x) = b_i \phi_i(x,0) G_i(\phi(x,\cdot)), \qquad x \in \overline{\Omega},\ i = 1,\ldots,n, \tag{12}$$

and observe that equation (1) can be written in the abstract form

$$\partial_t u = Au + F(u_t), \quad u_o = \chi, \tag{1}'$$

where $u_t \in \mathcal{C}[-\tau, 0]^n$ is defined by $u_t(x, \theta) \equiv u(x_1 t + \theta)$ for $(x, \theta) \in \overline{\Omega} \times [-\tau, 0]$.

The results in Rothe [7, p.15] show that $A$ is the generator of an analytic semigroup $T = \{T(t) : t \geq 0\}$ on $C(\overline{\Omega})^n$, and so by variation of constants (1)' can be integrated and written in the form

$$u(t) = T(t-s)u(s) + \int_s^t T(t-r)F(u_r)dr \quad \text{for} \quad t \geq s \geq 0. \qquad (1)''$$

Solutions to (1)'' are called mild solutions to (1) and the results in [5] establish the existence of mild solutions as well as the comparison results necessary to show that (9) is valid (we may assume the solution to (1) is classical only when $t > \tau$ for general initial values $\chi \in \mathcal{C}[-\tau, 0]^n$). Noting that $T(t)\eta^* \equiv \eta^*$ for $t \geq 0$ and applying the maximum principle shows that

$$W[T(t)y] \leq W[y] \quad \text{for all} \quad t \geq 0 \quad \text{and} \quad y \in C(\overline{\Omega})^n.$$

If $t > 0$ and $0 < h < t$, we have from (1)'' and the continuity of $T$ and $F$ that

$$u(t) - hF(u_t) = T(h)u(t-h) + o(h)$$

where $h^{-1}|o(h)| \to 0$ as $h \to 0+$. Hence

$$W[u(t) - hF(u_t)] \leq W[T(h)u(t-h)] + o(h)$$
$$\leq W[u(t-h)] + o(h)$$

and it follows that if $d^-/dt$ denotes the lower left Dini derivate, then

$$\frac{d^-}{dt}W[u(t)] \leq D_-W[u(t)](F(u_t)) \qquad (13)$$

for all $t > 0$ and all mild solutions $u$ to (1).

Combining (13) and (11) gives the following:

**Lemma 1.** *Suppose that $t > 0$ and $u$ is a mild solution to (1) such that $W[u(t)] \geq W[u(t+\theta)]$ for all $-\tau \leq \theta \leq 0$. Then*

$$\frac{d^-}{dt}W[u(t)] \leq 0. \qquad (14)$$

*Moreover, if $m \in \{1, \ldots, n\}$ and $x_o \in \overline{\Omega}$ are such that $W[u(t)] = \delta_m|u^m(x_o, t) - \eta_m^*| > 0$ and $u^m(x_o, t) > 0$, then*

$$\frac{d^-}{dt}W[u(t)] < 0. \qquad (15)$$

**Proof.** Assume $W[u(t)] = \delta_m |u^m(x_o, t) - \eta_m^*|$ and define $\phi \in C[-\tau, 0]^m$ by $\phi(\theta) = u(x_o, t+\theta)$ for $-\tau \leq \theta \leq 0$. Then

$$F_m(u(x_o, \cdot)_t) = b_m \phi_m(0) G_m(\phi)$$

and formulas (11) and (13) show that

$$\frac{d^-}{dt} W[u(t)] \leq \delta_m \operatorname{sgn}(\phi_m(0) - \eta_m^*) b_m \phi_m(0) G_m(\phi).$$

If $\phi_m(0) = 0$ or if $u(t) \equiv \eta^*$, then clearly (14) must hold and so it suffices to show that if $\phi_m(0) > 0$ and $\phi_m(0) \neq \eta_m^*$, then

$$\operatorname{sgn}(\phi_m(0) - \eta_m^*) G_m(\phi) < 0.$$

If $\eta_m^* = 0$ we have $\operatorname{sgn}(\phi_m(0) - \eta_m^*) = 1$ and $\sum_{j=1}^n c_{mj} \eta_j^* \geq 1$, and if $\eta_m^* > 0$ we have $\sum_{j=1}^n c_{mj} \eta_j^* = 1$ [see (5)]. Thus (2) and (3b) imply

$$\operatorname{sgn}(\phi_m(0) - \eta_m^*) G_m(\phi)$$

$$\leq \operatorname{sgn}(\phi_m(0) - \eta_m^*) \left\{ \sum_{j=1}^n c_{mj} \eta_j^* - a_m c_{mm} \phi_m(0) \right.$$

$$\left. - \sum_{j=1}^n c_{mj} \int_{-\tau}^o \phi_j(\theta) dv_{mj}(\theta) \right\}$$

$$= \operatorname{sgn}(\phi_m(0) - \eta_m^*) \left\{ -a_m c_{mm}(\phi_m(0) - \eta_m^*) \right.$$

$$\left. - \sum_{j=1}^m c_{mj} \int_{-\tau}^o (\phi_j(\theta) - \eta_j^*) dv_{mj}(\theta) \right\}$$

$$= |\phi_m(0) - \eta_m^*| \left\{ -a_n c_{mm} - \frac{\sum_{j=1}^m c_{mj} \int_{-\tau}^o (\phi_j(\theta) - \eta_j^*) dv_{mj}(\theta)}{\phi_m(0) - \eta_m^*} \right\}.$$

However, the assumption $W[u(t+\theta)] \leq W[u(t)]$ for $-\tau \leq \theta \leq 0$ implies

$$\frac{-c_{mj} \int_{-\tau}^o (\phi_j(\theta) - \eta_j^*) dv_{mj}}{(\phi_m(0) - \eta_m^*)} \leq \frac{|c_{mj}| \int_{-\tau}^o \delta_m \delta_j^{-1} (\delta_j |\phi_j(\theta) - \eta_j^*|) d|v_{mj}|}{\delta_m |\phi_m(0) - \eta_m^*|}$$

$$\leq \delta_m |c_{mj}| \delta_j^{-1} \frac{\int_{-\tau}^o W[u(t+\theta)] d|v_{mj}|}{W[u(t)]}$$

$$\leq \delta_m |c_{mj}| \delta_j^{-1} \int_{-\tau}^o d|v_{mj}|$$

$$= \begin{cases} \delta_m |c_{mj}| \delta_j^{-1} \gamma_{mj} & \text{if } j \neq m, \\ (1 - a_m) |c_{mm}| & \text{if } j = m. \end{cases}$$

Substituting into the preceding inequality and using (7) establishes the lemma.

**Lemma 2.** *Suppose $u$ is a (mild) solution to (1) and*

$$\mathcal{W}[u_t] \equiv \max\{W[u(t+\theta)] : -\tau \leq \theta \leq 0\}$$

*for all $t \geq 0$. Then $u$ exists on $[0, \infty)$ and the map $t \to \mathcal{W}[u_t]$ is nonincreasing on $[0, \infty)$.*

**Proof.** Since $\mathcal{W}[u_t] = W[u(t)]$ implies $d^-/dt\, W[u(t)] \leq 0$ by Lemma 1, it follows from routine arguments that $t \to \mathcal{W}[u_t]$ is nonincreasing so long as $u$ exists. Since $\mathcal{W}[u_t] \leq \mathcal{W}[\chi]$ implies $u$ remains uniformly bounded, the global existence assertion also follows using standard techniques.

**Proof of the Theorem.** By Lemma 2 set

$$c = \lim_{t \to \infty} \mathcal{W}[u_t].$$

It suffices to show that $c = 0$; so assume for contradiction that $c > 0$. Since the semigroup $T$ in $(1)''$ is compact, the $\omega$-limit set $\omega(\chi)$ of $u$ is nonempty, compact and invariant for the dynamical system generated by mild solutions to (1). In particular, if $\psi \in \omega(\chi)$ and $v$ is the mild solution to $(1)'$ with $\xi$ replaced by $\psi$, then

$$0 < c = \mathcal{W}[v_t] \quad \text{for all} \quad t \geq 0.$$

Now for each $t > 0$ select $x_t \in \overline{\Omega}$ and $m(t) \in \{1, \ldots, m\}$ so that

$$W[v(t)] = \delta_{m(t)} |v^{m(t)}(x_t, t) - \eta^*_{m(t)}|.$$

If $t > 0$ is such that $W[v(t)] = c$ and $v^{m(t)}(x_t, t) > 0$, then Lemma 1 implies

$$\liminf_{h \to 0+} \frac{W[v(t-h)] - W[v(t)]}{-h} < 0.$$

But this implies that for some sufficiently small $h$,

$$\mathcal{W}[v_{t-h}] \geq W[v(t-h)] > W[v(t)] = c.$$

Since this contradicts the fact that $\mathcal{W}[v_s] \equiv c$, we have

$$\begin{aligned}&\text{if } t > 0 \text{ and } W[v(t)] = \mathcal{W}[v_t], \text{ then} \\ &v^{m(t)}(x, t) = 0, \quad \text{and hence} \quad c = \delta_{m(t)} \eta^*_{m(t)}.\end{aligned} \qquad (16)$$

Because of (8) in Remark 1, there must exist exactly one $m \in \{1,\ldots,n\}$ such that $m(t) \equiv m$ in (16). Noting that

$$\mathcal{W}[v_{2\tau}] = \max\{W[v(t)] : \tau \le t \le 2\tau\} = c$$

shows there must exist some $s \in (\tau, 2\tau]$ such that $v^m(x,s) \equiv 0$, and hence $v^m(x,t) \equiv 0$ for all $t \ge 0$ [see (9)]. Therefore,

$$c = \delta_m \eta_m^* \quad \text{and} \quad v^m(x,t) \equiv 0 \text{ on } \overline{\Omega} \times [0,\infty). \tag{17}$$

But (16) implies that

$$\delta_i |v^i(x,t) - \eta_i^*| < \delta_m \eta_m^* \quad \text{for all} \quad (x,t) \in \overline{\Omega} \times [s-\tau, s]$$
and $i = 1,\ldots,n$ with $i \ne m$

and so by continuity and compactness we have

$$\max\{\delta_i |v^i(x,t) - \eta_i^*| : (x,t) \in \overline{\Omega} \times [s-\tau, s] \quad \text{and} \quad i \ne m\} < \delta_m \eta_m^*. \tag{18}$$

Also, $v_s \in \omega(\chi)$ and so there is a sequence $t_k \to \infty$ such that $u_{t_k} \to v_s$ in $\mathcal{C}[-\tau, 0]^n$. Therefore, (18) implies that there is a $K$ such that

$$\max_{i \ne m}\{\delta_i |u^i(x, t_k - \theta) - \eta_i^*| : x \in \overline{\Omega},\ -\tau \le \theta \le 0\} < \delta_m \eta_m^*$$

for all $k \ge K$. Also, $\mathcal{W}[u_{t_k}] \downarrow \delta_m \eta_m^*$ by Lemma 2, so

$$\delta_m \eta_m^* \le \mathcal{W}[u_{t_k}] = \max\{\delta_m |u^m(x, t_k - \theta) - \eta_m^*| : x \in \overline{\Omega},\ -\tau \le \theta \le 0\} \tag{19}$$

for all $k \ge K$. But $u_{t_k}^m \to v_s^m \equiv 0$ as $k \to \infty$, and so we may also assume $K$ is sufficiently large so that $0 \le u^m(x, t_k - \theta) \le \eta_m^*$ for all $(x, \theta) \in \overline{\Omega} \times [-\tau, 0]$ and $k \ge K$. Furthermore, $\eta_m^* > 0$ so $u^m(t,x) > 0$ on $\overline{\Omega} \times (0, \infty)$ by (10) and we have an obvious contradiction to (19). This contradiction shows that $c$ must equal 0 and completes the proof.

## References

[1] A. Berman and R. Plemmons, *Nonnegative Matrices in the Mathematical Sciences*, Academic Press, New York, 1979.

[2] P. N. Brown, *Decay to uniform states in ecological interactions*, SIAM J. App. Math. **38** (1980 pages 22–37).

[3] S. Dunbar, K. Rybakowski and K. Schmitt, *Persistence in models of predator-prey populations with diffusion*, J. Diff. Eq. **65** (1986 pages 117–138).

[4] J. Hofbauer, *Saturated Equilibria, Permanence and Stability for Ecological System*, Math. Ecology Proceedings, Trieste 1986, Eds. L. J. Gross, T. G. Hallam, S. A. Levin, World Scientific, 1988.

[5] R. Martin and H. Smith, *Abstract functional differential equations and reaction-diffusion systems*, Trans. A.M.S. (to appear).

[6] R. Martin and H. Smith, *Reaction-diffusion systems with time delays: monotonicity, invariance, comparison and convergence* (to appear).

[7] F. Rothe, *Global Solutions of Reaction-Diffusion Systems*, Lecture Notes in Mathematics **1072**, Springer-Verlag, Berlin, 1984.


R.H. Martin
Department of Mathematics
North Carolina State University
Raleigh, NC 27695, USA

H.L. Smith
Department of Mathematics
Arizona State University
Tempe, AZ 85287, USA

# Exact Finite Dimensional Representations of Models for Physiologically Structured Populations.I: The Abstract Foundations of Linear Chain Trickery

J.A.J. METZ
O. DIEKMANN

Institute of Theoretical Biology, University of Leiden
Center for Mathematics and Computer Science, Amsterdam

## 1. Prelude: A low dimensional representation of the population dynamics of generalized ectotherms

Suppose we want to model a population of ectothermic animals, *e.g.* the water flea *Daphnia magna*. Experimentally it appears that reproduction depends on the size of the individual animals and this observation motivated KOOIJMAN & METZ (1984) to introduce a size structured model. As the biological assumptions underlying the model are described already in some detail, in METZ & DIEKMANN (1986; I.3), METZ et al. (1988), DE ROOS et al. (preprint) and DE ROOS & METZ (preprint), we restrict ourselves here to its mathematical formulation:

$$\frac{\partial}{\partial t}n(t,\ell) + \frac{\partial}{\partial \ell}(\nu(s,\ell)n(t,\ell)) = -\mu(s,\ell)n(t,\ell),$$

$$\nu(s,\ell_b)n(t,\ell_b) = \int_{\ell_b}^{\ell_{\max}} \beta(s,\ell)n(t,\ell)d\ell. \qquad (1.1)$$

Here $\ell$ denotes length and $s$ substrate (more precisely: concentration of algae). The individual growth, death and reproduction rates are denoted by, respectively, $\nu$, $\mu$ and $\beta$. The density $n$ describes the concentration of *Daphnia* as well as their distribution with respect to length. All individuals are born with length $\ell_b$ and $\ell_{\max}$ is the maximal attainable length under abundant food conditions.

To describe some experiments one should consider $s$ as a given function of time but to describe others one has to specify the dynamics of $s$ as well. In the latter case we take

$$\frac{ds}{dt} = h(s) - \int_{\ell_b}^{\ell_{\max}} \gamma(s,\ell(n(t,\ell)d\ell, \qquad (1.2)$$

where $h$ corresponds to the rate of change of the algae concentration in the absence of daphnids and $\gamma$ is the per capita consumption rate. Under appropriate assumptions on the ingredients $\nu$, $\mu$, $\beta$, $\gamma$ and $h$, (1.1) and (1.2) together generate an infinite dimensional nonlinear dynamical system.

Since daphnids are filters feeders it is reasonable to assume that the consumption rate $\gamma$ is proportional to the surface area which in turn is proportional to $\ell^2$. So we put

$$\gamma(s,\ell) = f(s)\ell^2. \qquad (1.3)$$

If a constant fraction of the ingested energy is allotted to reproduction we may put

$$\beta(s,\ell) = \alpha f(s)\ell^2 \qquad (1.4)$$

(at this point we deliberately ignore the experimental fact that daphnids don't reproduce if they are still too small; see METZ & DIEKMANN, METZ et al. DE ROOS et al. and DE ROOS & METZ (op. cit.) for a formulation which does take into account a juvenile period characterized by $\ell < \ell_j$). If the remainder of the ingested energy is allotted to individual growth and maintenance and if maintenance is proportional to weight, which in turn is propertional to $\ell^3$, we may take

$$\frac{d}{dt}\ell^3 = 3\delta f(s)\ell^2 - 3\varepsilon\ell^3,$$

and therefore

$$\nu(s,\ell) = \frac{d}{dt}\ell = \delta f(s) - \varepsilon\ell. \qquad (1.5)$$

Finally we take

$$\mu(s,\ell) = \mu, \quad \text{a constant.} \qquad (1.6)$$

To anyalze (1.1) together with (1.2) for the special constitutive relations (1.3) to (1.6) we introduce

$$N_i(t) = \int_{\ell_b}^{\ell_{\max}} \ell^i n(t,\ell)dt, \quad i = 0,1,2, \qquad (1.7)$$

and find, using (1.1) – (1.7) and some straightforward integrations (by parts), that $(N, s)$ satisfies the *closed* system of ODE's

$$\begin{aligned}
\frac{dN_0}{dt} &= \alpha f(s)N_2 - \mu N_0, \\
\frac{dN_1}{dt} &= \ell_b \alpha f(s)N_2 - \delta f(s)N_0 - (\mu + \varepsilon)N_1, \\
\frac{dN_2}{dt} &= \ell_b^2 \alpha f(s)N_2 + 2\delta f(s)N_1 - (\mu + 2\varepsilon)N_2, \\
\frac{ds}{dt} &= h(s) - f(s)N_2.
\end{aligned} \qquad (1.8)$$

The powerful qualitative theory of finite dimensional dynamical systems now can be brought to bear on (1.8). Moreover one can choose from a multitude of well established schemes to study (1.8) numerically. As one example of the exploitation of these facts we point to DE ROOS (1988), who uses the relationship between (1.8) and (1.1) to investigate the accuracy of the 'escalator boxcar train', a new, efficient method developed by him for the numerical solution of the usual combinations of first order PDE's and non–local side conditions appearing in the theory of physiologically structured populations.

Of course neglecting the juvenile period has consequences, the main difference between the present model and the full one being that the latter not only allows the occurrence of predator prey oscillations due to the lag in recovery of the food population, but in addition oscillations related to the development lag (see METZ *et al.* 1988; DE ROOS *et al.* 1988; DE ROOS *et al.* preprint, and DE ROOS & METZ, preprint).

## 2. Introduction

The *Daphnia* example shows that it is sometimes possible to faithfully represent a full physiologically structured population model in a low dimensional manner, provided an appropriate choice of the constitutive relations, *viz.* the velocity and mortality functions and birth kernel, is made. The idea to search specifically for modelling approximations allowing such low dimensional representations is affectionately called 'linear chain tickery' by its practitioners. The name arose in the context of delay differential equations, where particular distributed delays can be represented as linear, i.e. unbranched, chains of coupled single ODE's (see *e.g.* MAC DONALD, 1978).

The earliest references to a systematic use of linear chain tricks that we are aware of are by VOGEL and by REPIN (1965) who applied them in the context of respectively Volterra integral and delay differential equations. The first analysis of necessary and sufficient conditions for linear chain trickability in the context of systems with hereditary action seems to have been given by FARGUE (1973, 1974). Good general references in this context with a slant towards biological applications are MAC DONALD (1978,1979). GURTIN & MAC CAMY (1974, 1979) were the first to use linear chain trickery for well specified age structured population model. GURNEY et al. (1986) provided the extension to physiological age, and MURPHY (1983) and EDELSTEIN & HADAR (1983), to size.

Another, practically very useful, extension of the idea of linear chain trickery, which, however, is less amenable to an abstract characterization, is provided by the stage structured models pioneered by the University of Strathclyde group of ecological modellers. Basically these are physiologically structured population models which can be represented in a fairly straightforward manner as systems of delay differential equations with a few, though possibly variable, discrete delays, and hence allow a rapid exploration of their dynamics using only slight extensions of the standard numerical techniques for ODE's. The advantage of aiming at using delay instead of ordinary differential equations in one's modelling approximations is the greater flexibility allowed, in particular if one wishes to keep the number of differential equations involved fairly small. A good introduction to the biological assumptions underlying the stage structure concept can be found in NISBET & GURNEY (1986). The first papers on the subject are GURNEY et al. (1983), which treats the case of fixed delay only, and NISBET & GURNEY (1983) which deals with the variable delay case (the symposium paper GURNEY & NISBET (1983) provides a nice summary). Various useful further extensions can be found in BLYTHE et al. (1984), GURNEY et al. (1986), and NISBET et al. (1985).

In the present contribution we report our attempts at elucidating for general population models the structural properties underlying the machinery of deriving faithful finite dimensional representations. This work forms part of an ongoing program, started in METZ & DIEKMANN (1986), aimed at clarifying the abstract mathematical structure inherent in our ways of thinking about the mechanistic basis of population dynamics. Some of the results reported in the present paper, in particular the characterization results form subsection 5.1.2, already appeared in DIEKMANN & METZ (1988,89).

## 3. An abstract formulation of physiologically structured population models

Let the individuals of a population be characterized by finitely many variables, which together we call the $i$-state. So the set of feasible $i$-states $\Omega$ is a nice subset of $\mathbb{R}^n$, for some $n$. At the individual level a model amounts to a specification of (i) the rate of $i$-state change, $\nu$, (ii) the death rate, $\mu$, (iii) the birth rate, $\beta$, and in particular how (i), (ii) and (iii) depend on the $i$-state $x$ and the prevailing environmental conditions. The latter are described by a (possibly even infinite dimensional) variable $E$. In the case of the birth rate we have to specify the (distribution of the) state at birth as well.

Once we have a model at the individual level we can immediately derive balance laws doing the necessary bookkeeping. These balance laws generate the time evolution at the population level. There are two types of balance laws, related to each other by duality. We can use duality since for $E$ a given function of time the equations are linear as a result of our previous assumption that for a given course of $E$ individuals are fully state–determined. The Kolmogorov backward equation is concerned with the clan mean of a continuous function on $\Omega$ (see below). The Kolmogorov forward equation describes infinitesimal changes in the measure which assigns to every measurable subset of $\Omega$ the concentration of individuals which have at that instant an $i$–state which belongs to that particular subset. This measure is called the $p$–state ($p$ for population) and the space $M(\Omega)$ of regular Borel measures on $\Omega$ is called the $p$–state space. Frequently (but not always) we can restrict our attention to densities, as we did in the case of the *Daphnia* example, and formulate the Kolmogorov forward equation for $L_1(\Omega)$.

Let for a particular course of $E$ the population state at $t$ deriving from an initial condition at $t_0$ corresponding to a unit mass at $x_0$ be denoted as $n(t, t_0, \mathbf{1}_{x_0})$. Then the clan mean of $\psi : \Omega \to \mathbb{R}$ is defined as

$$\nu(t_0, t, \psi)(x_0) := \int_\Omega \psi(x) n(t, t_0, \mathbf{1}_{x_0})(\{dx\}).$$

The Kolmogorov backward equation of a general physiologically structured population model is

$$-\frac{d}{dt_0}\nu(t_0, t, \psi) = A(E(t_0))\nu(t_0, t, \psi) \tag{3.1}$$

with 'final' condition

$$\nu(t, t, \psi) = \psi, \tag{3.2}$$

where

$$A(E) = A_0(E) + B(E) \tag{3.3}$$

with

$$(A_0(E)\psi)(x) = \frac{d\psi}{dx}\nu(x, E) - \mu(x, E)\psi(x) \tag{3.4}$$

the $i$–state movement *cum* death operators, and

$$(B(E)\psi)(x) = \int_\Omega \psi(y)\beta(a, E; \{dy\}) \tag{3.5}$$

with birth operator. To derive this equation from first principles one only has to consider what will and/or may befall an individual who at time $t_0 - dt$ has $i$–state $x_0$, during the next short time interval to $t_0$, and then perform the usual averaging at $t$ of $\psi$, first within and then over the clans generated by (i) what by $t_0$ has become of her and (ii) her offspring present at $t_0$.

The Kolmogorov forward equation can best be introduced as the formal adjoint of the backward equation:

$$\frac{dn}{dt}(t, t_0, n_0) = A(E(t))^* n(t, t_0, n_0). \tag{3.6}$$

The main use of the general decomposition (3.3) derives from the fact that for $B = 0$ we can write down explicit solutions to either (3.1) or (3.6) by the simple expedient of integration along characteristics. Biologically this is equivalent to the following of cohorts.

The description of our population is completed by specifying any outputs, such as total population size, total biomass, or total resources consumption, to be derived from it:

$$y(t, t_0, n_0) = C(E(t))n(t, t_0, n_0). \tag{3.7}$$

When the range of $y$ is finite dimensional, as is usually, but not always, the case, we can write

$$C(E)m = \langle \Gamma(E), m \rangle = \int_\Omega \Gamma(E)(x) m(\{dx\}) \tag{3.8}$$

with $\Gamma(E) : \Omega \to \mathbb{R}^h$. Given any specific initial condition, $t_0$, $n_0$, the previous description should be such as to enable us in principle to calculate $y$ as a function of $t > t_0$ for any sufficiently well behaved environmental input $E$.

From an applied point of view the main usefulness as well as interest of the previous considerations derives from the fact that many environmental variables, like food, are in turn influenced by the population, *e.g.* through consumption. Thus nonlinear evolution problems arise in a natural manner through the specification of the feedbacks through the environment.

The mathematical theory of provide a rigorous justification and interpretation of the general framework embodied in equation (3.1) to (3.8) is still in its infancy. Some first steps towards a functional analytic underpinning have been made in CLEMENT *et al.* (1987, 1988, 1989a, 1989b; see DIEKMANN, 1989, for a survey), but much work remains to be done. In the present contribution we restrict ourselves to formal manipulations, ignoring all problems related to the existence and uniqueness of solutions and to the precise interpretation of the differential equations (3.1) and (3.6).

## 4. An abstract formulation of linear chain trickery

From now on we shall always assume that the required output from the population model is finite (possible zero) dimensional, and that $E$ itself is the output from a dynamical system allowing a finite dimensional state representation.

### 4.1. The most general case.

Since our population equations (3.6) and (3.7) are linear in the state we do not loose any generality by assuming that any potential finite dimensional representation of them is linear in the state as well, and that the full model and its finite dimensional representation are related by a linear map $P : M(\Omega) \to \mathbb{R}^k$. In order that

$$N(t) = Pn(t) \tag{4.1}$$

provides us with a fully selfcontained description of the dynamical relationship between population input $E$ and output $y$

$$\frac{dN}{dt} = K(E)N, \tag{4.2}$$

$$y = Q(E)N, \tag{4.3}$$

we should have
$$C(E) = Q(E)P \tag{4.4}$$

and
$$PA(E)^* = K(E)P \tag{4.5}$$

for some family of $h \times k$ matrices $Q$ and some family of $k \times k$ matrices $K$.

**Remark.** It is not possible to attain greater generality by letting $P$ depend on $E$ as this will lead to a additional term $[\frac{d}{dE}P(E)\frac{dE}{dt}]n$ in (4.2). □

If and only if (4.4) and (4.5) are fulfilled the dynamics of $E$ and $N$ can be described by a coupled finite dimensional system of ODE's. Once $E$ is determined by solving this reduced system we can treat

$$\frac{dn}{dt} = A(E)^*n \tag{4.6}$$

as a non-autonomous (i.e. time dependent) but linear equation. If for example one can conclude from the $(N, E)$–system that $E$ approaches a limit (or a periodic solution) for $t \to \infty$, the linear equation for $n$ is asymptotically autonomous (periodic) and one can base further conclusions on the known asymptotic behaviour for these special cases.

If we are willing to assume that

$$Pm = \langle \Phi, m \rangle \tag{4.7}$$

for some vector $\Phi$ with components which are continuous functions of $\Omega$ we can reformulate (4.5) as
$$A(E)\Phi = K(E)\Phi, \tag{4.8}$$

provided $\Phi \in \mathcal{D}(A(E))$ for all $E$.

**Remark.** Actually $\cap_E \mathcal{D}(A(E))$ may be empty. However, within the context of dual semigroups one can extend $A(E)$ to an operator $A(E)^{\odot *}$ which has its range in a larger space $X^{\odot *}$ and therefore has larger domain as well (see CLEMENT *et al.* 1987, 1988, 1989a, 1989b, or DIEKMANN, 1989). One can then replace (4.8) by
$$A(E)^{\odot *}\Phi = K(E)\Phi.$$

In the following we shall not go into the distinction between this formulation and (4.8) (in fact we shall omit the precise definition of domains of unbounded operators). □

Furthermore we can use (3.8), to replace (4.4) by

$$\Gamma(E) = Q(E)\Phi. \tag{4.9}$$

(4.8) and (4.9) together provide us with an easy practical recipe for checking whether a particular combination of $\nu$, $\mu$, $\beta$ and $\Gamma$ allows a finite dimensional representation. First of all it should be possible to write $\Gamma(E)(x)$ as $Q_1(E)\Phi_1(x)$ for some vector $\Phi_1 = (\varphi_1, \ldots, \varphi_{k_1})^T$ of linearly independent functions $\phi_i$ and some $h \times k$ matrix family $Q_1$. If this is the case our problem is linear chain trickable if and only if the space spanned by all possible combinations $A(E_p) \ldots A(E_1)\phi_i$ for $i = 1, \ldots, k_1$, $p = 0, 1, \ldots$, is finite dimensional.

### 4.2. Two examples.

**Example 1.** Consider a cell population with size structure and assume that a mother cell divides into two parts without any mass loss, (see HEIJMANS, 1984 and METZ & DIEKMANN, 1986 (sub)section I.4, III.3.3.1, and VI.5, and the references given there). Then

$$(B(E)\psi)(x) = d(x,E)[-\psi(x) + 2\int_0^1 \psi(\theta x)p(x, \{d\theta\})],$$

where $d$ is the division rate and $p(x, \cdot)$ is the probability distribution of the sizes of the daughters relative to the size of their mother. The assumption of no mass loss implies that $p(x, \cdot)$ is symmetrical about $\theta = 1/2$. Now assume that the uptake of nutrient $E$ by a cell is proportional to its biomass. In that case

$$C(E) = g(E)\langle \phi, \cdot \rangle,$$

with $\phi(x) = x$, i.e. $\langle \phi, \cdot \rangle$ is the total biomass functional. Next we observe that necessarily
$$B(E)\phi = 0$$

in accordance with the initial assumption that biomass is conserved in the division process. Finally we observe that we get

$$A(E)\phi = A_0(E)\phi = ((f-(E) - \mu(E))\phi$$

if we make in additional assumptions that

$$\nu(x,E) = f(E)x \quad \text{and} \quad \mu(x,E) = \mu(E).$$

The first condition is *i.a.* fulfulled when basal metabolism is proportional to biomass, and cell growth is proportional to nutrient uptake minus loss through basal metabolism:
$$f(E) = \alpha(g(E) - m).$$
The second condition is *i.a.* fulfulled when the only cause of cell loss is washout. If finally we assume chemostat dynamics, so that $\mu(E) = D$, the dilution rate, we arrive at
$$\frac{dN}{dt} = \alpha(g(E) - m)N - DN,$$
$$\frac{dE}{dt} = D(E^i - E) - g(E)N,$$
where $E^i$ is the concentration of the limiting substrate in the inflowing nutrient both. Under appropriate conditions on $g$ the resulting ODE system has a globally stable steady state. □

**Example 2.** This example is more contrived. Assume again that individuals acquire food at a rate $g(E)x$ where $E$ is the surrounding food concentration and $x$ is their size. Assume moreover that the acquired food is partitioned into a fraction $\kappa(x)$ which is spent on reproduction and a fraction $1 - \kappa(x)$ spent on basal metabolism and growth, and that the cost of producing offspring biomass equals that of producing parent biomass. Finally assume agin that basal metabolism is proportional to size and that the death rate is size independent. In that case
$$(A_0(E)\psi)(x) = (g(E)(1 - \kappa(x)) - m)x\psi'(x) - \mu(E)\psi(x)$$
and
$$(B(E)\psi)(x) = x_b^{-1}g(E)\kappa(x)x\psi(x_b),$$
where $x_b$ is the size of the young. If we choose again $\phi(x)$ to be equal to $x$ we find
$$A(E)\phi = (g(E) - m - \mu(E))\phi. \quad □$$

### 4.3. 'Ordinary' LCT.

Usually the term linear chain trickery if reserved for a special subclass of the general class of tricks discussed in the previous subsections, the restriction being that it should also be possible to calculate the birth rate into the population from the resulting finite-dimensional representation. The reason for the special importance of this smaller class of problems is that once we know the birth rate as a function of time we can easily construct the full population trajectory by using a variation of constants formula involving the explicit solution $\tilde{n}$ of
$$\frac{d\tilde{n}(t, t_0, n_0)}{dt} = A_0(E(t))^* \tilde{n}(t, t_0, n_0) \quad \text{with} \quad \tilde{n}(t_0, t_0, n_0) = n_0.$$

The 'ordinary' LCT problem is characterized by the conditions that there exist a map $P : M(\Omega) \to \mathbb{R}^k$, a family of maps $R(E) : \mathbb{R}^k \to M(\Omega)$, and families of $k \times k$ matrices $H$ and $h \times k$ matrices $Q$ such that

$$B(E)^* = R(E)P, \qquad (4.10)$$

$$PA_0(E)^* = H(E)P, \qquad (4.11)$$

$$C(E) = Q(E)P. \qquad (4.12)$$

The resulting system of ODE's is

$$\frac{dN}{dt} = H(E)N + PR(E)N. \qquad (4.13)$$

If we may in addition make the special assumption (4.7), i.e $P = \langle \Phi, \cdot \rangle$, (4.10) to (4.12) may be replaced by

$$\beta(x, E, \cdot) = \sum b_i(E; \cdot)\phi_i(x), \qquad (4.14)$$

$$A_0(E)\Phi(x) = H(E)\Phi(x), \qquad (4.15)$$

$$\Gamma(E)(x) = Q(E)\Phi(x) \qquad (4.16)$$

for all $x$.

**Remark.** In the case of generalized LCT nothing can be said about the component of the $p$–state in the kernel of the map $P$. This is unfortunate as a slight perturbation of the model usually brings it out of the LCT class. If unpleasant things happen in the kernel of $P$ this would result is an extreme non–robustness of the conclusions derived from the LCT variants. It is clear from the discussion at the start of this subsection that the situation is much better for ordinary LCT as usually it is quite easy to prove that $\tilde{n}(t, t_0, n_0) \to 0$ for all $n_0$ in a very fast manner. As a consequence for example the local linearization about an equilibrium of a model in the ordinary LCT class always leads to a polynomial characteristic equation, corresponding to a decomposition of the $p$–state space into a finite number of (generalized) eigenvectors and a remaining component consisting entirely of 'fast descenders'. □

## 5. Necessary and sufficient conditions for linear chain trickery

We shall in this section proceed from (4.14) – (4.16) on the assumption that $\nu$, $\mu$, $\beta$ and $\gamma$ are sufficiently smooth in $x$. Moreover, we shall only consider minimal representations, in the sense that $k$ is as small as possible.

### 5.1. One dimensional $i$–state spaces.

Assume that the $i$–state space is one dimensional. Then (3.4) reduces to

$$((A_0(E))\psi(x) = \nu(x, E)\psi'(x) - \mu(x, E)\psi(x). \qquad (5.1)$$

### 5.1.1. The case of but one single resulting ODE.
We first restrict ourselves to the special case where $P$ has one–dimensional range, i.e. our population model can be represented by just a single ODE. The question then is 'Under which conditions on $\nu$ and $\mu$ can we find a (continuous) function $\phi(x)$ and a function $\lambda(E)$ such that

$$\nu(x, E)\phi'(x) - \mu(x, E)\phi(x) = \lambda(E)\phi(x)?' \qquad (5.2)$$

If we rewrite (5.2) in the form $\frac{\mu(x,E)+\lambda(E)}{\nu(x,E)} = \frac{\phi'(x)}{\phi(x)}$ we see that a necessary as well as sufficient condition for the family $A_0(E)$ to allow linear chain trickery population models is that there exists a function $\lambda(E)$ such that

$$\frac{\mu(x, E) + \lambda(E)}{\nu(x, E)} = f(x) \qquad (5.3)$$

independent of $E$. For the full population model to be linear chain trickable moreover (4.14) and (4.16) should apply with

$$\phi(x) = \exp[\int^x f(\xi)d\xi]. \qquad (5.4)$$

**Example 1.** Let $\nu(x, E) = \nu(E)$, i.e. $x$ is physiological age. In the case $A_0$ allows linear chain trickable population models iff

$$\mu(x, E) = \nu(E)\mu_1(x) + \mu_2(E). \qquad (5.5)$$

Moreover $\phi$ should be of the form

$$\phi(x) = \exp[\int^x \mu_1(\xi)dx] \cdot \exp[-\alpha x] \qquad (5.6)$$

where $\alpha$ still is a free parameter which can be chosen to comply with the conditions on the birth and output operators. □

**Example 2.** Let $\mu(x, E) = \mu(E)$, i.e. the $i$–state of an individual does not influence its chances of dying. In that case $A_0(E)$ allows linear chain trickable population models iff

$$\nu(x, E) = \nu_1(x)\nu_2(E), \qquad (5.7)$$

which after a rescaling of $x$ brings us back to the previous example, or

$$\phi(x) = 1 \quad \text{and} \quad \lambda(E) = -\mu(E). \qquad (5.8)$$

Note that in the latter case the conditions (4.14) and (4.16) imply that both the *per capita* birth rate and the *'per capita* resource consumption rate' are independent of the $i$–state, i.e. the classification of individuals by $x$ is population dynamically irrelevant. □

### 5.1.2. Physiological age models.

Let us now make the special assumption that $\nu(x, E) = \nu_1(x)\nu_2(x)E$. Without loss of generality we may set $\nu_2(E_0) = 1$ for some (arbitrarily chosen) $E_0$, and $\nu_1(x) = 1$: Just rescale to physiological age

$$\tilde{x} := \int^x \frac{d\xi}{\nu_1(\xi)}. \tag{5.9}$$

In this new variable condition (4.15) becomes (from now on we drop the index 2 and the tilda)

$$\nu(E)\Phi'(x) - \mu(x, E)\Phi(x) = H(E)\Phi(x), \tag{5.10}$$

from which we deduce that $\Phi$ should take the form

$$\Phi(x) = \exp[\int_0^x \mu(\xi, E_0)d\xi] \cdot \exp[H(E_0)x] \cdot \Phi(0). \tag{5.11}$$

Substitution of (5.11) and (5.10) gives

$$[\nu(E)\mu(x, E_0) - \mu(x, E)]\Phi(x) = [H(E) - \nu(E)H(E_0)]\Phi(x), \tag{5.12}$$

i.e. $\Phi(x)$ is an eigenvector of $H(E) - \nu(E)H(E_0)$. For fixed $E$ the eigenvalues of $H(E) - \nu(E)H(E_0)$ form a discrete set. On the other hand it is reasonable to assume that the map $x \mapsto \nu(E)\mu(x, E_0) - \mu(x, E)$ is continuous. A continuous function taking values in a discrete set is constant. Therefore we can conclude that we should have

$$\mu(x, E) = \nu(E)\mu(x, E_0) - \lambda(E), \tag{5.13}$$

where $\lambda(E)$ is only subject to the consistency condition $\lambda(E_0) = 0$, and

$$H(E) = \nu(E)H(E_0) + \lambda(E)I, \tag{5.14}$$

where $H(E_0)$ may still be chosen freely to comply with (4.14) and (4.16).

As a final consideration we note that a function $\phi(x)$ can be written as $q^T \exp[H(E_0)x]\Phi(0)$ if and only if it can be written as a weighted sum of polynomials times (complex) exponentials. This tells us what freedom we have in choosing birth and output operators.

### 5.1.3. Death rate independent of the $i$-state.

If we try to generalize the approach from the previous subsection to $i$-states moving in a less restricted manner we end up with

$$\left[\frac{\nu(x, E)}{\nu(x, E_0)}\mu(x, E_0) - \mu(x, E)\right]\Phi(x) = \left[H(E) - \frac{\nu(x, E)}{\nu(x, E_0)}H(E_0)\right]\Phi(x) \tag{5.15}$$

as the analogue of (5.10), and our argument breaks down since the matrix on the right hand side is no longer independent on $x$. The case of one resulting ODE discussed in the previous subsection and the *Daphnia* example from section 1 make clear that this indeed makes an essential difference.

The results from subsection 5.1.1 indicate that there will always exist a possibility for a trade off between the rate of $i$-state change $\nu$ and the death rate $\mu$, mucking up any attempt at getting nice clean result. Except in certain special cases, like the one of physiological age, it is difficult to see which biological mechanisms could ever cause in general precisely the required relationships. Therefore we shall make our lives easy and stick here to the case where $\mu$ does not depend on $x$.

**Result.** If $\mu(x, E) = \mu(E)$ the combinations

$$\nu(x, E) = \nu(E) \quad \text{with}$$
$$\Phi(x) = (e^{\lambda_1 x}, xe^{\lambda_1 x}, \ldots, x^{k_1-1}e^{\lambda_1 x}, \ldots, e^{\lambda_r x}, \ldots, x^{k_r-1}e^{\lambda_r x})^T \qquad (5.16)$$

and

$$\nu(x, E) = f(E) + g(E)x \quad \text{with} \quad \Phi(x) = (1, x, \ldots, x^{k-1})^T \qquad (5.17)$$

are, up to a scale change for $x$ and a change of basis for the range of $P$ (or rather a linear equivalence of the triples $(P, R(E), Q(E))$), the only one satisfying condition (4.15), with respectively

$$H(E) = \nu(E)\Lambda - \mu(E)I \qquad (5.18)$$

with

$$\Lambda = \begin{pmatrix} \lambda_1 & & & & & & \\ 1 & \lambda_1 & & & & \emptyset & \\ & \ddots & \ddots & & & & \\ & & k_1 - 1 & \lambda_1 & & & \\ & & & 0 & \lambda_2 & & \\ & \emptyset & & & \ddots & \ddots & \\ & & & & & k_r - 1 & \lambda_r \end{pmatrix} \qquad (5.19)$$

and

$$H(E) = \begin{pmatrix} 0 & & & & \emptyset & \\ f(E) & g(E) & & & & \\ & \ddots & \ddots & & & \\ \emptyset & & & (k-1)f(E) & (k-1)g(E) \end{pmatrix} - \mu(E)I. \quad \square \qquad (5.20)$$

Note that (5.16) corresponds to the physiological age case with which we dealt in the previous subsection, and that (5.17) is but a slight extension of the *Daphnia* example from section 1. Note also that (5.20) definitely does not belong to the family (5.18), in accordance with the remark made at the start of this subsection.

To prove our result we first choose a environment value $E_0$ and rescale $x$ so that $\nu(x, E_0) = 1$ (we assume that a value of $E_0$ exists such that $\nu(x, E_0) > 0$ on the whole interior of $\Omega$). Next we rearrange (5.15) into

$$\tilde{H}(E)\Phi(x) = \nu(x, E)\tilde{H}(E_0)\Phi(x) \tag{5.21}$$

with

$$\tilde{H}(E) = H(E) + \mu(E)I. \tag{5.22}$$

Moreover

$$\Phi(x) = \exp[\tilde{H}(E_0)x]\Phi(0). \tag{5.23}$$

As a next step we observe that our choice of $\phi_i$ is to a large extent arbitrary as long as the set of $\phi_i$'s spans one and the same subspace of the continuous functions on $\Omega$. Therefore we may without loss of generality write

$$\Phi(x) = (e^{\lambda_1 x}, xe^{\lambda_1 x}, \ldots, x^{k_1-1}e^{\lambda_1 x}, e^{\lambda_2 x}, \ldots, x^{k_r-1}e^{\lambda_r x})^T, \tag{5.24}$$

where the $\lambda_i$ are the eigenvalues of $\tilde{H}(E_0)$. Note that (5.24) corresponds to the particular choice $\tilde{H}(E_0) = \Lambda$. Note also that all possible $\tilde{H}(E_0)$ can be obtained from this particular choice by a change of basis for $N = Pn$. Restriction of our attention to minimal representations moreover guarantees that all the $\lambda_i$ are different.

Substitution of (5.24) into (5.21) yields

$$\nu(x, E)(qx^{q-1} + x^q\lambda_p)e^{\lambda_p x} = \sum_{i=1}^{r}\sum_{j=0}^{k_i-1}\tilde{h}_{(p,q)(i,j)}(E)x^j e^{\lambda_i x}, \tag{5.25}$$

where the symbols $(p, q)$ and $(i, j)$ relate in an obvious manner to the indices characterizing the components of $\Phi$. To proceed further we need several lemmas.

**Lemma 1a.** *Let $\lambda_i \in \mathbb{C}$ for $i = 1, \ldots, r$ be all different and let $U_p := \{\lambda_p - \lambda_i | i = 1, \ldots, r\}$ then $\cap_{p=1}^{1} U_p = \{0\}$*

**Proof.** $\cap_{p=1}^{r} U_p \neq \{0\}$ iff there exists a complex number $\alpha \neq 0$ common to all $U_p$. Assume that such an $\alpha$ exists. This allows us to define a relation $\to$ on $E_r := \{1, \ldots, r\}$ by $i \to p :\Leftrightarrow \lambda_i - \lambda_p = \alpha$. Under $\to$ every element of $E_r$ connects in the forward and backward direction to at most one other element of $E_r$ since (i) $\lambda_i - \lambda_{p'} = \alpha\lambda_i - \lambda_{p''} \Rightarrow \lambda_{p'} = \lambda_{p''}$ and (ii) $\lambda_{i'} - \lambda_p = \alpha = \lambda_{i''} - \lambda_p \Rightarrow \lambda_{i'} = \lambda_{i''}$. Since we have $r$ sets $U_p$ we should have at least $r$ connections under $\to$. As $E_r$ has but $r$ elements this would mean that there has to exist as least one cycle. But this is inconsistent with the geometrical interpretation (in $\mathbb{C}$) of the relation $\to$. (Note that the existence of a nonzero common element to only $r-1$ of the $U_p$ implies that the $\lambda_i$ lie at fixed distance on a straight line in $\mathbb{C}$.) $\square$

Exactly the same argument yields

**Lemma 1b.** *Let $\lambda_i \in \mathbb{C}$ for $i = 1, \ldots, r$ be all different and let $U_p := \{\lambda_p - \lambda_i | i = 1, \ldots, r\}$. Assume $\lambda_1 = 0$. Then either*

$$\bigcap_{p=2}^{r} U_p = \{0\}$$

*or, possibly after renumbering the $\lambda_i$'s,*

$$\lambda_i = (i-1)\alpha \quad \text{and} \quad \bigcap_{p=2}^{r} U_p \{0, +\alpha\}$$

*for some $\alpha \in \mathbb{C}$.*

**Lemma 2a.** *Let $k \geq 1$ be a given integer. Suppose there exist complex numbers $\lambda \neq 0$ and $\alpha_{jq}$, $j, q \in \{0, \ldots, k-1\}$ such that*

$$R(q, x) = \sum_{j=0}^{k-1} \frac{\alpha_{jq} x^j}{q x^{q-1} + \lambda x^q}, \quad q = 0, \ldots, k-1,$$

*is independent of $q$. Then $R$ is independent of $x$ as well.*

**Proof.** By taking $q = 0$ we find that $R$ is a polynomial in $x$ of degree $\leq k - 1$. By taking $q = k - 1$ we obtain that $((k-1)x^{k-2} + \lambda x^{k-1}) R(q, x)$ is a polynomial degree $\leq k - 1$. Therefore the degree of $R$ is necessarily zero. □

**Lemma 2b.** *Let $k \geq 2$ be a given integer. Suppose there exist complex numbers $\alpha_{jq}$, $j, q \in \{1, \ldots, k-1\}$ such that*

$$R(q, x) = \sum_{j=0}^{k-1} \frac{\alpha_{jq} x^j}{q x^{q-1}}, \quad q = 1, \ldots, k-1,$$

*is independent of $q$. Then $R$ is necessarily of the form $a + bx$.*

**Proof.** By taking $q = 1$ we find that $R$ is a polynomial in $x$ of degree $\leq k - 1$. By taking $q = k - 1$ we obtain that $(k-1)x^{k-2} R(q, x)$ is a polynomial of degree $\leq k - 1$. Therefore the degree of $R$ is necessarily $\leq 1$. □

**Lemma 3.** Let for $j, q \in \mathbb{N}$, $\theta \in \mathbb{C}$

$$U(j, q, \beta, \theta) := \frac{x^j}{qx^{q-1} + \beta x^q} e^{\theta x}$$

then a necessary condition for $U(j_0, q_0, \beta_0, \theta_0)$ to be in the linear span of $\{U(j_i, q_i, \beta_i, \theta_i) | i = 1, \ldots, k-1\}$ is that $\theta_0 \in \{\theta_i | i = 1, \ldots, k-1\}$.

**Proof.** Suppose that $U(j_0, q_0, \beta_0, \theta_0) = \sum_{i=1}^{k-1} \xi_i U(j_i, q_i, \beta_i, \theta_i)$. Multiply both sides with $\prod_{i=0}^{k-1}(q_i x^{q_i-1} + \beta_i x^{q_i})$. At the left and right hand side we now only have polynomials times exponentials in $x$. Any collection of functions $x^{m_i} e^{\theta_i}$ for which the pairs $(m_i, \theta_i)$ are all different are linearly independent. Therefore the factor $e^{\theta_0 x}$ has to appear on both sides of the equal sign. $\square$

If either $\lambda_p \neq 0$ or $q \neq 0$ we can rewrite (5.25) in the form

$$\nu(x, E) = \sum_{i=1}^{r} \sum_{j=0}^{k_i-1} \tilde{h}_{(p,q)(i,j)}(E) \frac{x^j}{qx^{q-1} + x^q \lambda_p} e^{(\lambda_i - \lambda_p)x}. \quad (5.26)$$

If for all $p$ either $\lambda_p \neq 0$ or $k_p > 1$ we thus find at least $r$ (in fact $k = \sum_{i=1}^{r} k_i$) expressions for $\nu$.

First assume that for all $p$ either $\lambda_p \neq 0$ or $k_p > 1$. In that case (5.26), Lemma 3 and Lemma 1a together imply that

$$\tilde{h}_{(p,q)(i,j)}(E) = 0 \quad \text{for} \quad i \neq p$$

and therefore that

$$\nu(x, E) = \sum_{j=0}^{k_r-1} \tilde{h}_{(p,q)(p,j)}(E) \frac{x^j}{qx^{q-1} + x^q \lambda_p}. \quad (5.27)$$

We can now apply Lemma 2a to conclude that $\nu$ is independent of $x$ provided 0 is not the only $\lambda$. We are then in the situation described by (5.16) and (5.19). When $\lambda = 0$ is the only eigenvalue we apply Lemma 2b to conclude that $\nu$ is linear in $x$. This brings us to the situation described by (5.17) and (5.20).

Next we assume that $r \geq 2$ and, say, $\lambda_1 = 0$, $k_1 = 1$. We still obtain (5.26) for $p = 2, \ldots, r$. When not $\lambda_i = (i-1)\alpha$ for some $\alpha \neq 0$ Lemma 1b tells us that we are in the first of the two situations encountered before. When, on the other hand, $\lambda_i = (i-1)\alpha$ we deduce from Lemma 3 together with Lemma 1b that

$$\nu(x, E) = \sum_{j=0}^{k_p-1} \tilde{h}_{(p,q)(p,j)}(E) \frac{x^j}{qx^{q-1} + x^q \lambda_p}$$

$$+ \sum_{j=0}^{k_{p-1}-1} \tilde{h}_{(p,q)(p-1,j)}(E) \frac{x^j e^{-\alpha x}}{qx^{q-1} + \lambda_p x^q} \quad (5.28)$$

for $p \geq 2$. Applying Lemma 2a to each of the sums we infer that

$$\nu(x,E) = \frac{1}{\alpha}(g(E) + f(E)e^{-\alpha x}) \tag{5.29}$$

(the reason for this particular 'parameterization' with $g$, $f$ and $1/\alpha$ will become clear below). We claim that in this situation necessarily $k_p = 1$ for all $p$. We proceed by induction. Suppose $k_2 > 1$ then we can take $p = 2$, $q = 1$ in (5.28) to obtain

$$\nu(x,E) = \sum_{j=0}^{k_2-1} \tilde{h}_{(2,1)(2,j)}(E)\frac{x^j}{1+\lambda_2 x} + \tilde{h}_{(2,1)(1,0)}(E)\frac{e^{-\alpha x}}{1+\lambda_2 x}.$$

Since $\lambda_2 \neq 0$ this is incompatible with (5.29). We conclude that $k_2 = 1$. We then use the same argument for $p = 3$ etc.

Finally we transform to $\tilde{x} = e^{\alpha x}$. This yields $\tilde{\nu}(\tilde{x}, E) = f(E) + g(E)\tilde{x}$ and $\tilde{\Phi}(\tilde{x}) = (1, \tilde{x}, \ldots, \tilde{x}^{r-1})$ which, modulo tilda's and $r \to k$, is precisely (5.17). □

**Remark 1.** When judging the generality of the linear growth low (5.17) one should keep in mind that one can still employ an $E$–independent change of $i$–state variable to bring a particular biological growth law in that form. For example, the growth laws most commonly encountered in the literature
(i)     von Bertalanffy:    $\frac{dy}{dt} = \alpha y^{2/3} - \beta y$
(ii)    logistic:              $\frac{dy}{dt} = \alpha y - \beta y^2$
(iii)   Gompertz:         $\frac{dy}{dt} = \alpha y - \beta y \log y$

can all be linearized:
(i)    $x = y^{1/3} \Rightarrow$           $\frac{dx}{dt} = \frac{1}{3}(\alpha - \beta x)$
(ii)   $x = \frac{1}{y} \Rightarrow$             $\frac{dx}{dt} = \beta - \alpha x$
(iii)  $x = \log y \Rightarrow$         $\frac{dx}{dt} = \alpha - \beta x$

(we thank Y. Iwasa for reminding us to (ii) and (iii)). □

**Remark 2.** If we set $\mu(x, E) = \nu(x, E)\mu_1(x) + \mu_2(E)$ the combinations (5.16) and (5.17) with the old $\Phi(x)$ replaced by $\Psi(x) = \exp(\int^x \mu_1(\xi)d\xi)\Phi(x)$ still satisfy (4.15) with the same $H(E)$ as when $\mu_1 = 0$. □

## 5.2. Higher dimensional $i$–state spaces.

We do not have any general results for the case where $\Omega$ is higher dimensional. What we do have is a whole zoo of weird and wonderful examples. We just give three of them.

**Example 1.** Let $\Omega$ be two–dimensional and let $\nu$ be given by

$$\nu(x,E) = \begin{bmatrix} a(E) + b(E)x_1 \\ c(E) \end{bmatrix}.$$

Define
$$\Phi(x) = \begin{bmatrix} 1 \\ x_1 \\ x_1^2 \\ e^{-kx_2} \\ x_1 e^{-kx_2} \\ x_1^2 e^{-kx_2} \end{bmatrix}$$

and

$$L(E) = \begin{bmatrix} 0 & 0 & 0 & 0 & 0 & 0 \\ a & b & 0 & 0 & 0 & 0 \\ 0 & 2a & 2b & 0 & 0 & 0 \\ 0 & 0 & 0 & -kc & 0 & 0 \\ 0 & 0 & 0 & a & (b-kc) & 0 \\ 0 & 0 & 0 & 0 & 2a & 2b-kc \end{bmatrix}.$$

A straightforward calculation then shows that

$$\frac{d\Phi}{dx}(x) \cdot \nu(x, E) = L(E)\Phi(x)$$

which is the required relation $A_0(E)\Phi = H(E)\Phi$ for $\mu = 0$. When $\mu$ is nonzero but still independent of $x$, $L(E)$ has to be replaced by $H(E) = L(E) - \mu(E)I$.

The biological interest of this example is that we may interpret $x_1$ as size and $x_2$ as physiological age. Moreover $\Phi$ is chosen in such a way that we can choose

$$\beta(x, E) = f(E)(1 - e^{-kx_2})x_1^2$$

as an age and size dependent birth rate of individuals. □

The next two examples do not allow immediate biological applications. They do show, however, that in the case of higher dimensional $i$-state spaces there exist also cases with nonlinear $i$-state dynamics which are yet linear chain trickable.

**Example 2.** Let again $\Omega$ be two dimensional, and let

$$\nu(x, E) = \begin{pmatrix} a(E) + b(E)x_1 \\ c(E)x_1^2 \end{pmatrix}, \quad \Phi(x) = \begin{pmatrix} 1 \\ x_1 \\ x_1^2 \\ x_2 \end{pmatrix},$$

$$L(E) = \begin{pmatrix} 0 & 0 & 0 & 0 \\ a & b & 0 & 0 \\ 0 & 2a & 2b & 0 \\ 0 & 0 & c & 0 \end{pmatrix}. \quad \square$$

**Example 3.** Let $\Omega$ be three dimensional and let

$$\nu(x,E) = \begin{pmatrix} a_1(E) \\ a_2(E) \\ c_1(E)e^{\lambda_1 x_1} + c_2(E)e^{\lambda_2 x_2} + c_3(E)e^{\lambda_1 x_1 + \lambda_2 x_2} \end{pmatrix},$$

$$\Phi(x) = \begin{pmatrix} e^{\lambda_1 x} \\ e^{\lambda_2 x} \\ e^{\lambda_1 x_1 + \lambda_2 x_2} \\ x_3 \end{pmatrix},$$

$$L(E) = \begin{pmatrix} \lambda_1 a_1 & 0 & 0 & 0 \\ 0 & \lambda_2 a_2 & 0 & 0 \\ 0 & 0 & \lambda_1 a_1 + \lambda_2 a_2 & 0 \\ c_1 & c_2 & c_3 & 0 \end{pmatrix}. \quad \square$$

## 6. Discussion

Understanding the precise nature of the necessary and sufficient conditions for linear chain trickery to be possible is of interest of three reasons. First of all there is the intrinsic esthetic appeal of the problem. Secondly its solution amounts to a *complete* catalogue of cases for which a reduction of finite dimension is possible. No doubt this catalogue will contain useful cases which thus far escaped our attention (like the first example from section 5.2). Thirdly solving the general linear chain trickery problem will tell us which (classic) ODE models can be reinterpreted reduced structured models. (In our, admittedly somewhat biased, opinion the justification of any ODE population model should derive from the fact that such an interpretation is possible).

In this paper we to a large extent have solved the ordinary, or special, linear chain trickery problem for the case of a one–dimensional $i$–state space. A full characterization of linear chain trickable models with higher dimensional $i$–state spaces is still lacking. And we have only scratched the surface of the generalized linear chain trickery problem. However, we plan to keep working on these problems.

**Acknowledgments.** Hans Metz wishes to thank the Department of Physics and Applied Physics of the University of Strathclyde, Glasgow, for its hospitality during part of the research reported here.

**Note added in print.** In the meantime we have also solved the 'ordinary' LCT characterization problem for one dimensional $i$–state spaces in a general manner, i.e., without assuming any restrictions on either the rate of $i$–state change $\nu$ or the death rate $\mu$. The result is bizarre.

## References

[1] S.P. Blythe, R.M. Nisbet, W.S.C. Gurney (1984), *The dynamics of population models with distributed maturation period*, Theor. Pop. Bio. **25**, 289–311.

[2] Ph. Clement, O. Diekmann, M. Gyllenberg, H.J.A.M. Heijmans, C.H.R. Thieme (1987, 1988, 1989a, 1989b), *Perturbation theory for dual semigroups*,
I. *The sun-reflexive case*, Math. Ann. **277**, 709–725.
II. *Time-dependent perturbations in the sun-reflexive case*, Proc. Roy. Soc. Edinburgh A **109**, 145–172.
III. *Nonlinear Lipschitz continuous perturbations in the sun-reflexive case*, in "Volterra Integro-Differential Equations in Banach Spaces and Applications", G. Da Prato, M. Iannelli, Eds., Longman, 67–89.
IV. *The intertwining formula and the canonical pairing*, in "Trends in Semigroup Theory and Applications", Ph. Clement, S. Invernizzi, E. Mitidieri, I.I. Vrabie, Eds., Marcel Dekker, 95–116.

[3] O. Diekmann (1989), *On semigroups and populations*, Advanced Topics in the Theory of Dynamical Systems, G. Fusco, M. Iannelli, L. Salvadori, Eds., Academic Press, 125–135.

[4] O. Diekmann, H. Metz (1988,89), *Exploring linear chain trickery for physiologically structured populations*, TW in Beeld, CWI, Amsterdam, 73–84. Also: CWI Quarterly **2**, 3–14.

[5] L. Edelstein, Y. Hadar (1983), *A model for pellet size distribution in submerged mycelial cultures*, J. Theor. Biol. **105**, 427–457.

[6] D. Fargue (1983), *Reducibilité des systèmes héréditaires à des systémes dynamiques (régis des équations différentielles ou aux dérivés partielles)*, C.R. Acad. Sc. Paris B **277**, 471–473.

[7] D. Fargue (1984), *Reducibilité des systèmes héréditaires*, Int. J. Non–linear Mechanics **9**, 331–338.

[8] W.S.C. Gurney, R.M. Nisbet (1983), *The systematic formulation of delay–differential models of age or size structured populations*, Population Biology, M.I. Freeman & C. Strobeck, Eds., Springer Lect. Notes in Biomath. **52**, 163–172.

[9] W.S.C. Gurney, R.M. Nisbet, S.P. Blythe (1986), *The systematic formulation of models of stage-structured populations*, The Dynamics of Physiologically Structured Populations, J.A.J. Metz & O. Diekmann, Eds., Springer Lect. Notes in Biomath. **68**, 474–494.

[10] W.S.C. Gurney, R.M. Nisbet, J.H. Lawton (1983), *The systematic formulation of tractable single species models incorporating age-structure*, J. Anim. Ecol. **52**, 479–496.

[11] M.E. Gurtin, R.C. MacCamy (1974), *Non-linear age-dependent population dynamics*, Archive for Rational Mechanics and Analysis **54**, 281–300.

[12] M.E. Gurtin, R.C. MacCamy (1979), *Some simple models for non-linear age-dependent population dynamics*, Math. Biosc. **43**, 199–211.

[13] H.J.A.M. Heijmans (1984), *On the stable size distribution of populations reproducing by fission into two unequal parts*, Math. Biosc. **72**, 19–50.

[14] S.A.L.M. Kooijman, J.A.J. Metz, *On the dynamics of chemically stressed populations: the deduction of population consequences from effects on individuals*, Ecotox. Env. Saf. **8**, 254–274.

[15] M. MacDonald (1978), *Time lags in biological models*, Springer Lect. Notes in Biomath. **27**.

[16] M. MacDonald (1989), *Biological delay systems: linear stability theory*, Cambridge U Press, Cambridge.

[17] J.A.J. Metz, O. Diekmann (Eds.) (1986), *The Dynamics of Physiologically Structured Populations*, Springer Lect. Notes in Biomath. **68**.

[18] J.A.J. Metz, A.M. De Roos, F. van den Bosch (1988), *Population models incorporating physiological structure: a quick survey of the basic concepts and an application to size-structured population dynamics in waterfleas*, Size-structured Populations: Ecology and Evolution, E. Ebenman & L. Persson, Eds, Springer, Berlin, Heidelberg, 106–124.

[19] L.F. Murphy (1983), *A non-linear growth mechanism in size structured population dynamics*, J. Theor. Biol. **104**, 493–506.

[20] R.M. Nisbet, W.S.C. Gurney (1983), *The systematic formulation of population models for insects with dynamically varying instar duration*, Theor. Pop. Biol. **23**, 114–135.

[21] R.M. Nisbet, W.S.C. Gurney (1986), *The formulation of age-structured models*, Mathematical Ecology, T.G. Hallam & S.A. Levin, Eds., Springer, Berlin, Heidelberg, 95–115.

[22] R.M. Nisbet, S.P. Blythe, W.S.C. Gurney, J.A.J. Metz (1985), *Stage-structure models of populations with distinct growth and development processes*, IMA J. Math. Appl. in Med. & Biol. **2**, 57–68.

[23] Y.M. Repin (1965), *On the approximate replacement of systems with lag by ordinary differential equations*, J. Appl. Math. and Mech **29**, 254–264.

[24] A.M. De Roos, O. Diekmann, J.A.J. Metz (1988), *The escalator boxcar train: basic theory and an application to Daphnia population dynamics*, Report AM-R8814, Center for Mathematics and Computer Science, Amsterdam.

[25] A.M. De Roos, J.A.J. Metz (preprint), *Stabilizing and destabilizing mechanisms in a planktonic system: modelling Daphnia population dynamics*.

[26] A.M. De Roos, J.A.J. Metz, E. Evers, A. Leipoldt (preprint), *A size dependent predator-prey interaction; who pursues whom?*.

[27] T. Vogel (1965), *Théorie des Systèmes Evolutifs*, Gauthier Villars, Paris.


J.A.J. Metz
Institute of Theoretical Biology
University of Leiden
Kaiserstraat 63
NL-2311 GP Leiden, The Netherlands


O. Diekmann
Center for Mathematics and Computer Science
Kruislaan 413
NL-1098 SJ Amsterdam, The Netherlands

# The Nonrelativistic Limit of Klein–Gordon and Dirac Equations

BRANKO NAJMAN

Department of Mathematics, University of Zagreb

The two linear equations of relativistic quantum mechanics to be considered are

1. the Klein–Gordon equation

$$\frac{1}{2c^2}\psi_{tt} - \frac{1}{2}\Delta\psi + \frac{c^2}{2}\psi = \tilde{f}_c(t), \qquad (0.1)$$

and

2. the Dirac equation

$$i\psi_t = -ic\alpha\nabla\psi + c^2\beta\psi + \tilde{F}_c(t). \qquad (0.2)$$

In both equations $c$ is the speed of light and the physical measurement scale has been adjusted so that both the Planck constant $\hbar$ and the mass $m$ are equal to one. The functons $\tilde{f}_c$ and $\tilde{F}_c$ have no physical meaning in the linear case; however, it is useful to consider them as a preparation for the nonlinear theory.

Throughout this note, all the functions are defined on $\mathbb{R}^n$ and $t$ varies over $\mathbb{R}$. In case of the Klein–Gordon equation $n$ is arbitrary and in the case of the Dirac equation $n = 3$.

The nonrelativistic limit of the equations (0.1) and (0.2) is the limit $c \to \infty$. It is known ([2],[8],[13]) that under appropriate assumptions on the initial data the solutions of these converge to the solutions of the appropriate Schrödinger equation; cf. also [3],[4],[7],[10],[11],[12] and the references therein for related results.

The purpose of this note is twofold. First, by using the explicit solution operators from [2] we extend to convergence results to the spaces $H^2$ (for any function space $X$ stands for $X(\mathbb{R}^n)$); in all the previous references only $L^2$ (i.e. $s=0$) convergence was proved. Moreover, the obtained results are useful in the treatment of the nonrelativistic limit of the nonlinear Dirac equation, which will be given elsewhere ([6]). We briefly mention the results in Section 3. It turns out that the nonrelativistic limit of the nonlinear Dirac equation is a coupled system of nonlinear Schrödinger equations.

For the nonrelativistic limit of the nonlinear Klein–Gordon equation, we refer to [5] (see also [9]).

## 1. The linear Klein–Gordon equation

Consider the equation (0.1); substituting $\varepsilon = \frac{1}{2c^2}$, $A = -\frac{1}{2}\Delta$, $f_\varepsilon(t) = e^{-\frac{it}{2\varepsilon}}\tilde{f}_\varepsilon(t)$, $\varphi(t) = e^{\frac{it}{2\varepsilon}}\psi(t)$, we find the equation for $\varphi$:

$$\varepsilon\varphi_{tt} - i\varphi_t + A\varphi = f_\varepsilon(t) \tag{1.1}$$

which we consider together with the initial conditions

$$\varphi(0) = \varphi_{0\varepsilon}, \quad \varphi_t(0) = \varphi_{1\varepsilon}. \tag{1.2}$$

The formal limit as $\varepsilon \to 0$ of the initial value problem (1.1) and (1.2) is the initial value problem

$$\begin{aligned} i\varphi_t &= A\varphi - f_0(t) \\ \varphi(0) &= \varphi_{00}. \end{aligned} \tag{1.3}$$

In the next theorem a mild solution of an inhomogeneous initial value problem is the function given by the variation of parameters formula; it is a classical solution if the initial data and the inhomogeneous term are smooth enough. As mentioned before, $H^s$ stands for $H^s(\mathbb{R}^n)$.

**Theorem 1.1.** *Let $I \subset \mathbb{R}$ be a bounded interval containing zero, $\lambda$ and $s$ nonnegative numbers such that $\lambda \leq \min\{1, s\}$. Assume that*

$$\varphi_{0\varepsilon} \in H^2 \ (\varepsilon \geq 0), \ \varphi_{1\varepsilon} \in H^{s-\lambda} \ (\varepsilon > 0), \tag{1.4}$$

$$f_\varepsilon \in L^1(I, H^s) \ (\varepsilon \geq 0), \tag{1.5}$$

$$\lim_{\varepsilon \to 0} \varphi_{0\varepsilon} = \varphi_{00} \text{ in } H^s, \tag{1.6}$$

$$\lim_{\varepsilon \to 0} \varepsilon^{1-\frac{\lambda}{2}}\varphi_{1\varepsilon} = 0 \text{ in } H^{s-\lambda}, \tag{1.7}$$

$$\lim_{\varepsilon \to 0} f_\varepsilon = f_0 \text{ in } L^1(I, H^s). \tag{1.8}$$

Then the initial value problems (1.1), (1.2) and (1.3) have mild solutions $\varphi_\varepsilon$ and $\varphi_0$, respectively, which belong to $C(\bar{I}, H^s)$. Moreover

$$\lim_{\varepsilon \to 0} \varphi_\varepsilon = \varphi_0 \text{ in } C(\bar{I}, H^s). \tag{1.9}$$

**Proof.** The existence of $\varphi_\varepsilon$ and $\varphi_0$ is well known.

Let $A_\varepsilon = \frac{1}{\varepsilon}(\varepsilon A + \frac{1}{4})^{1/2}$ with the domain $H^1$ in the space $L^2$. We define following bounded operators on $L^2$:

$$I_\varepsilon(t) = e^{\frac{it}{2\varepsilon}}(\cos t A_\varepsilon - \frac{i}{2\varepsilon} A_\varepsilon^{-1} \sin t A_\varepsilon,$$

$$J_\varepsilon(t) = e^{\frac{it}{2\varepsilon}} A_\varepsilon^{-1} \sin t A_\varepsilon,$$

$$I_0(t) = e^{-iAt}.$$

Then (see [2] or [5])

$$\varphi_\varepsilon(t) - \varphi_0(t) = \sum_{i=1}^{5} \ell_\varepsilon^{(i)}(t),$$

with

$$\ell_\varepsilon^{(1)}(t) = [I_\varepsilon(t) - I_0(t)]\varphi_{00},$$

$$\ell_\varepsilon^{(2)}(t) = I_\varepsilon(t)(\varphi_{0\varepsilon} - \varphi_{00}),$$

$$\ell_\varepsilon^{(3)}(t) = J_\varepsilon(t)\varphi_{1\varepsilon},$$

$$\ell_\varepsilon^{(4)}(t) = \int_0^t [\frac{1}{\varepsilon} J_\varepsilon(t-s) - i I_0(t-s)] f_0(s) ds,$$

$$\ell_\varepsilon^{(5)}(t) = \frac{1}{\varepsilon} \int_0^t J_\varepsilon(t-s)[f_\varepsilon(s) - f_0(s)] ds.$$

Instead of (1.9) it is sufficient to prove

$$\lim_{\varepsilon \to 0} \ell_\varepsilon^{(i)} = 0 \text{ in } C(\bar{I}, H^s) \tag{1.10$_i$}$$

for $i = 1, \ldots, 5$. From

$$\|\ell_\varepsilon^{(1)}(t)\|_{H^s} \leq \|[I_\varepsilon(t) - I_0(t)](I + 2A)^{s/2}\varphi_{00}\|_{L^2},$$

$$\|\ell_\varepsilon^{(2)}(t)\|_{H^s} \leq \|[I_\varepsilon(t)(I + 2A)^{s/2}(\varphi_{0\varepsilon} - \varphi_{00})\|_{L^2},$$

from the results of [2] and from the assumptions (1.4) and (1.6) it follows that $(1.10_i)$ holds for $i = 1$ and 2. Further

$$\|\ell_\varepsilon^{(3)}(t)\|_{H^s} \leq \|(I + 2A)^{\lambda/2} J_\varepsilon(t)\|_{\mathcal{L}(L^2)} \|\varphi_{1\varepsilon}\|_{H^{s-\lambda}}.$$

Since

$$\|(I + 2A)^{\lambda/2} J_\varepsilon(t)\|_{\mathcal{L}(L^2)} = \sup_\mu \left| \frac{(1+\mu)^{\lambda/2} \sin \frac{t}{2\mu} \sqrt{4\varepsilon\mu + 1}}{\frac{1}{2\varepsilon}\sqrt{4\varepsilon\mu + 1}} \right|,$$

$$\sup_\mu \left| 2\varepsilon \left( \frac{1+4\mu}{1+4\varepsilon\mu} \right)^{\lambda/2} \frac{1}{(1+4\varepsilon\mu)^{1-\lambda/2}} \right| \leq 2\varepsilon^{1-\lambda/2} \sup_\mu \left| \frac{1+4\mu}{\frac{1}{\varepsilon}+4\mu} \right|^{\lambda/2} \leq 2\varepsilon^{1-\lambda/2},$$

the identity $(1.10_3)$ is a consequence of (1.7). Further $(1.10_4)$ follows directly from [2] and (1.5).

Finally $\|\frac{1}{\varepsilon} J_\varepsilon(t)\|_{\mathcal{L}(L^2)} \leq C$ independently of $\varepsilon$ and $t$, hence $(1.10_5)$ follows from (1.8).

**Theorem 1.2.** *Let $I \subset \mathbb{R}$ be a bounded interval containing zero, $\lambda$ and $s$ nonnegative numbers such that $\lambda \leq \min\{1, s\}$. Assume that in addition to (1.6) the following conditions are satisfied:*

$$\varphi_{1\varepsilon} \in H^s \ (\varepsilon \geq 0), \tag{1.11}$$

$$f_\varepsilon \in C(\bar{I}, H^s), \ \frac{df_\varepsilon}{dt} \in L^1(I, H^s) \ (\varepsilon \geq 0), \tag{1.12}$$

$$A\varphi_{0\varepsilon} - f_\varepsilon(0) \in H^{s-\lambda} \ (\varepsilon > 0), \tag{1.13}$$

$$\lim_{\varepsilon \to 0} \varphi_{1\varepsilon} = -iA\varphi_{00} + if_0(0) \text{ in } H^s, \tag{1.14}$$

$$\lim_{\varepsilon \to 0} \varepsilon^{\lambda/2}(\varphi_{1\varepsilon} + iA\varphi_{0\varepsilon} - if_\varepsilon(0)) = 0 \text{ in } H^{s-\lambda}, \tag{1.15}$$

$$\lim_{\varepsilon \to 0} \frac{df_\varepsilon}{dt} = \frac{df_0}{dt} \text{ in } L^1(I, H^s). \tag{1.16}$$

*Then $\psi_\varepsilon \in C^1(I, H^s)$. Moreover*

$$\lim_{\varepsilon \to 0} \varphi_\varepsilon = \varphi_0 \text{ in } C^1(\bar{I}, H^s). \tag{1.17}$$

**Proof.** Note that (1.12) and (1.13) imply $\varphi_{0\varepsilon} \in H^{s-\lambda}$, hence $\varphi_{0\varepsilon} \in H^{s+2-\lambda}$. It follows that all the assumptions of Theorem 1.1 are fulfilled. The differentiability of $\varphi_\varepsilon$ is well known, so we only have to prove (1.17). Let $\theta_\varepsilon = \frac{d\varphi_\varepsilon}{dt}$; from (1.1), (1.2) and (1.3) it follows that $\theta_\varepsilon$ ($\varepsilon \geq 0$) is the (mild) solution of the following initial value problem:

(i) if $\varepsilon > 0$ then $\theta_\varepsilon$ is the solution of

$$\begin{aligned} \varepsilon \theta_{tt} - i\theta_t + A\theta &= g_\varepsilon(t), \\ \theta(0) &= \theta_{0\varepsilon}; \ \theta_t(0) = \theta_{1\varepsilon} \end{aligned} \tag{1.18}$$

where $g_\varepsilon(t) := \frac{df_\varepsilon}{dt}$, $\theta_{0\varepsilon} := \varphi_{1\varepsilon}$, $\theta_{1\varepsilon} := \frac{1}{\varepsilon}(f_\varepsilon(0) - A\varphi_{0\varepsilon} + i\varphi_{1\varepsilon})$;

(ii) if $\varepsilon = 0$ then $\theta_0$ is the solution of

$$i\theta_t = A\theta - g_0(t), \qquad (1.19)$$
$$\theta(0) = \theta_{00},$$

where $g_0(t) := \frac{df_0}{dt}$, $\theta_{00} := -iA\varphi_{00} + if_0(0)$.

Now we can apply Theorem 1.1 to the initial value problems (1.18) and (1.19). Note that (1.11) and (1.13) reduce to (1.4), (1.12) to (1.5), (1.14) to (1.6), (1.15) to (1.7) and (1.16) to (1.8). From Theorem 1.1 we conclude that $\frac{d\varphi_\varepsilon}{dt}$ converge to $\frac{d\varphi_0}{dt}$ in $C(\bar{I}, H^s)$. Together with the convergence of $\varphi_\varepsilon$ to $\varphi_0$ (already proved in Theorem 1.1) this proves (1.17).

**Remark.** Without any additional effort we can treat the case of the Klein-Gordon equation with a potential $V$, replacing $A = -\frac{1}{2}\Delta$ by $\tilde{A} = -\frac{1}{2}\Delta + V$. All the statements, as well as the proofs, hold true if $H^s$ is replaced by $\mathcal{D}(\tilde{A}^{s/2})$. In particular, if $V$ is bounded then $\mathcal{D}(\tilde{A}^{s/2}) = H^s$ for $s \geq 2$; if $V$ is smooth with bounded derivatives then this identity holds for large $s$ too. Hence the convergence results hold as stated if a sufficiently smooth $V$ is added into the equation.

## 2. The linear Dirac equation

Consider the Dirac equation (0.2); as mentioned in the introduction, we consider it in $\mathbb{R}^3$ only. In (0.2) $\alpha\nabla = \sum_{j=1}^{3} \alpha_j \partial_j$, $\psi$ is a function from $\mathbb{R}^3$ into $\mathbb{C}^4$ and $\alpha_j$, $\beta$ are $4 \times 4$ matrices satisfying the anitcommutation rules

$$\alpha_j \alpha_k + \alpha_k \alpha_j = 2\delta_{jk} I, \quad \alpha_j \beta + \beta \alpha_j = 0, \quad \beta^2 = I.$$

By substitution $\varepsilon = \frac{1}{2c^2}$ and $\Phi = 2e^{\frac{i\beta t}{2\varepsilon}} \beta\psi$ we are led to consider the initial value problem

$$\Phi_t = \sqrt{\frac{1}{2\varepsilon}} e^{\frac{i\beta t}{\varepsilon}} \alpha\nabla\Phi - iF_\varepsilon(t), \qquad (2.1)$$
$$\Phi(0) = \Phi_{0\varepsilon},$$

where $F_\varepsilon(t) = \frac{1}{2} e^{-\frac{i\beta t}{2\varepsilon}} \beta \tilde{F}_{1/\sqrt{2\varepsilon}}(t)$. Differentiating (2.1) we obtain the initial value problem

$$\Phi_{tt} - \frac{1}{2\varepsilon}[\Delta\Phi + 2i\beta\Phi_t + B_\varepsilon(t)] = 0, \qquad (2.2)$$
$$\Phi(0) = \Phi_{0\varepsilon}, \quad \Phi_t(t) = \Phi_{1\varepsilon}$$

with $B_\varepsilon(t) := -2\beta F_\varepsilon(t) - i\sqrt{2\varepsilon} e^{\frac{i\beta t}{\varepsilon}} \alpha\nabla F_\varepsilon(t) - 2i\varepsilon \partial_t F_\varepsilon(t)$, $\Phi_{1\varepsilon} := \sqrt{\frac{1}{2\varepsilon}} \alpha\nabla\Phi_{0\varepsilon} - iF_\varepsilon(0)$.

The limit ($\varepsilon = 0$) problem is

$$\Phi_t = \frac{1}{2} i\beta \Delta \Phi - iF_0(t),$$
$$\Phi(0) = \Phi_{00}. \qquad (2.3)$$

The space $H^s(\mathbb{R}^3)^4$ will be denoted $H^s$.

**Theorem 2.1.** *Let $I \subset \mathbb{R}$ be a bounded interval containing zero, $\lambda$ and $s$ nonnegative numbers such that $\lambda \leq \min\{1, s\}$.*
  *Assume that*

$$\Phi_{0\varepsilon} \in H^{s-\lambda+1} \ (\varepsilon > 0), \qquad (2.4)$$

$$\Phi_{00} \in H^s, \qquad (2.5)$$

$$F_\varepsilon \in L^1(I, H^{s+1}), \ \frac{\partial F_\varepsilon}{\partial t} \in L^1(I, H^s) \ (\varepsilon > 0), \qquad (2.6)$$

$$F_0 \in L^1(I, H^s), \qquad (2.7)$$

$$\lim_{\varepsilon \to 0} \Phi_{0\varepsilon} = \Phi_{00} \text{ in } H^s, \qquad (2.8)$$

$$\lim_{\varepsilon \to 0} \varepsilon^{1-\frac{\lambda}{2}} \left[ \sqrt{\frac{1}{2\varepsilon}} \alpha \nabla \Phi_{0\varepsilon} - iF_\varepsilon(0) \right] = 0 \text{ in } H^{s-\lambda}, \qquad (2.9)$$

$$\lim_{\varepsilon \to 0} F_\varepsilon = F_0 \text{ in } L^1(I, H^s), \qquad (2.10)$$

$$\lim_{\varepsilon \to 0} \varepsilon^{1/2} F_\varepsilon = 0 \text{ in } L^1(I, H^{s+1}), \qquad (2.11)$$

$$\lim_{\varepsilon \to 0} \varepsilon \frac{\partial F_\varepsilon}{\partial t} = 0 \text{ in } L^1(I, H^s). \qquad (2.12)$$

*Then the initial value problems (2.1) and (2.3) have mild solutions $\Phi_\varepsilon \in C(\bar{I}, H^s)$. Moreover*

$$\lim_{\varepsilon \to 0} \Phi_\varepsilon = \Phi_0 \text{ in } C(\bar{I}, H^s). \qquad (2.13)$$

**Proof.** The existence part is easy; for $\varepsilon > 0$ one shows the existence of the mild solution of the initial value problem (2.2) and then shows that this solution actually solves (2.1). We shall not go into details here; we proceed to prove (2.13). Since $\beta$ and $\Delta$ commute, the proof of Theorem 1.1 can be repeated. However, in estimation of $\ell_\varepsilon^{(i)}$, $i = 1, \ldots, 4$, we can not use the results from [2]. Instead we use a variation of Fattorini's results: Let $A$ be the "diagonal" operator $-\frac{1}{2}\Delta$ on $L^2(\mathbb{R}^3)^4$; i.e. $\mathcal{D}(A) = H^2(\mathbb{R}^3)^4$, $(A\psi)_i = -\frac{1}{2}\Delta \psi_i$, $i = 1, \ldots, 4$ for $\psi = (\psi_1, \psi_2, \psi_3, \psi_4) \in \mathcal{D}(A)$. Let $I_\varepsilon(t)$ and $J_\varepsilon(t)$ be defined as in Section 1 and let

$$I_0(t) = e^{-i\beta A t} = e^{\frac{1}{2}i\beta \Delta t}.$$

**Lemma 2.2.** Let $T > 0$, $\bar{I} = [-T, T]$.

(a) If $\Phi \in L^2$ then
$$\lim_{\varepsilon \to 0} \|[I_\varepsilon(\cdot) - I_0(\cdot)]\Phi\|_{C(\bar{I}, L^2)} = 0.$$

(b) Let $f \in L^1(I, L^2)$ and denote
$$g_\varepsilon(t) = \int_0^t [\frac{1}{\varepsilon} J_\varepsilon(t-s) - i\beta I_0(t-s)] f(s) ds.$$

Then
$$\lim_{\varepsilon \to 0} g_\varepsilon = 0 \text{ in } C(\bar{I}, L^2).$$

We omit the standard proof.

Once we have this result, we can repeat the proof of Theorem 1.1 in the present situation. The assumptions (2.4) – (2.12) imply that the assumptions of Theorem 1.1 are satisfied: (2.4) and (2.5) imply (1.4) (noting that (2.6) implies $F_\varepsilon(0) \in H^s$), (2.6) and (2.7) imply (1.5) and finally the convergence assumptions (2.9) – (2.12) imply (1.6) – (1.8).

**Remark.** In a similar way, Theorem 1.2 can be applied to find sufficient conditions in order that
$$\lim_{\varepsilon \to 0} \Phi_\varepsilon = \Phi_0 \text{ in } C^1(\bar{I}, H^s).$$

The details are left to the reader.

## 3. The nonlinear Dirac equation

The nonlinear Dirac equation
$$i\psi_t = -ic\alpha \nabla \psi + c^2 \beta \psi + 2\lambda(\beta\psi|\psi)\beta\psi \tag{3.1}$$

was investigated in the physical literature; it was considered in mathematical literature only recently by L. Vasquez, T. Cazenave and their co-workers (see e.g. [1] and the references therein).

In (3.1) the parameter $\lambda$ is a positive constant.

Introducing $\varepsilon = \frac{1}{2c^2}$, $\Phi = 2e^{\frac{i\beta t}{2\varepsilon}} \beta\psi$ as in Section 2, we obtain the initial value problem
$$\Phi_t = \sqrt{\frac{1}{2\varepsilon}} e^{\frac{i\beta t}{\varepsilon}} \alpha \nabla \Phi - \frac{i\lambda}{2}(\beta\Phi|\Phi)\beta\Phi, \tag{3.2}$$
$$\Phi(0) = \Phi_{0\varepsilon}$$

Differentiating we find the second order (Klein–Gordon type) initial value problem
$$\Phi_{tt} - \frac{1}{2\varepsilon}[\Delta\Phi + 2i\beta\Phi_t - \lambda(\beta\Phi|\Phi)\Phi] = F_\varepsilon(t,\Phi), \qquad (3.3)$$
$$\Phi(0) = \Phi_{0,\varepsilon}, \quad \Phi_t(0) = \Phi_{1\varepsilon},$$
where $\Phi_{1\varepsilon} := \sqrt{\frac{1}{2\varepsilon}}\alpha\nabla\Phi_{0\varepsilon} - \frac{i\lambda}{2}(\beta\Phi_{0\varepsilon}|\Phi_{0\varepsilon})\beta\Phi_{0\varepsilon}$ and
$$F_\varepsilon(t,\Phi) := -\frac{\lambda^2}{2}(\beta\Phi|\Phi)^2\Phi$$
$$- \frac{i\lambda}{\sqrt{2}}\varepsilon^{-1/2}\sum_j \left\{e^{\frac{i\beta t}{\varepsilon}}[\partial_j(\beta\Phi|\Phi)]\alpha_j\Phi + 2Re(\beta\Phi|e^{\frac{i\beta t}{\varepsilon}}\alpha_j\partial_j\Phi)\beta\Phi\right\}.$$

The formal limit as $\varepsilon \to 0$ of the initial value problems (3.3) is the coupled system of nonlinear Schrödinger equations
$$\Phi_t = \frac{1}{2}i\beta\Delta\Phi - \frac{1}{2}i\lambda(\beta\Phi|\Phi)\beta\Phi, \qquad (3.4)$$
$$\Phi(0) = \Phi_{00}.$$

We state without proofs the local existence result for the problems (3.3) and (3.4) and the convergence result for the solutions $\Phi_\varepsilon$:

**Theorem 3.1.** *Let $\varepsilon_0 > 0$. Assume $\Phi_{0\varepsilon} \in H^2$ ($0 < \varepsilon < \varepsilon_0$) and*
$$\sup_{\varepsilon \leq \varepsilon_0} \|\Phi_{0\varepsilon}\|_{H^2} < \infty.$$
*There exists $T > 0$ such that for every $\varepsilon$ with $0 < \varepsilon \leq \varepsilon_0$ there exists a unique solution*
$$\Phi_\varepsilon \in C^2([-T,T],L^2) \cap C^1([-T,T],H^1) \cap C([-T,T],H^2)$$
*of the initial value problem (3.2).*

**Theorem 3.2.** *Assume $\Phi_{00} \in H^2$. There exists $T > 0$ such that the initial value problem (3.3) has a unique solution $\Phi_0 \in C^1([-T,T],L^2) \cap C([-T,T],H^2)$.*

**Theorem 3.3.** *Let $\varepsilon_0 > 0$. Assume $\Phi_{0\varepsilon} \in H^2$ ($0 \leq \varepsilon \leq \varepsilon_0$) and*
$$\sup_{\varepsilon \leq \varepsilon_0} \|\Phi_{0\varepsilon}\|_{H^2} < \infty;$$
*moreover for some $\alpha \in [0,1]$ it holds*
$$\lim_{\varepsilon \to 0} \Phi_{0\varepsilon} = \Phi_{00} \text{ in } H^\alpha.$$

*Let $T > 0$ be such that there exist unique solutions $\Phi_\varepsilon$ of the initial value problem (3.3) for $\varepsilon > 0$ and $\Phi_0$ of the initial value problem (3.4). Then*
$$\lim_{\varepsilon \to 0} \Phi_{0\varepsilon} = \Phi_0 \text{ in } C([-T,T],H^\alpha).$$

The proof of these theorems can be found in [6].

## References

[1] Balabane, M., Cazenave, T., Douady, A., Merle, F., *Existence of excited states for a nonlinear Dirac field*, Comm. Math. Phys. **119** (1988), 153–176.

[2] Fattorini, H.O., *Second Order Linear Differential Equations in Banach Spaces*, North Holland, 1985.

[3] Hunziker, W., *On the nonrelativistic limit of the Dirac theory*, Comm. Math. Phys. **40** (1975), 215–222.

[4] Ingolfsson, K., *Notes on the classical and the nonrelativistic limits in quantum mechanics*, Lett. Math. Phys. **1** (1976), 315–359.

[5] Najman, B., *The nonrelativistic limit of the Klein–Gordon Equation*, Nonl. Anal. T.M.A. (to appear).

[6] Najman, B., *The nonrelativistic limit of the nonlinear Dirac equation*, in preparation.

[7] Schoene, A., *Semigroups and a class of singular perturbation problems*, Indiana Univ. Math. J. **20** (1970), 247–263.

[8] Schoene, A., *On the nonrelativistic limit of the Klein–Gordon and Dirac equations*, J. Math. Anal. Appl. **71** (1979), 36–47.

[9] Tsutsumi, M., *Nonrelativistic approximation of the nonlinear Klein–Gordon equations in two space dimensions*, Nonl. Anal. T.M.A. **8** (1984), 637–643.

[10] Veselić, K., *The nonrelativistic limit of the Dirac equation and the spectral concentration*, Glasnik Math. 4(24) (1969), 231–241.

[11] Veselić, K., *Perturbation of pseudoresolvents and analyticity in 1/c in relativistic quantum mechanics*, Comm. Math. Phys **22** (1971), 27–43.

[12] Veselić, K., *On the nonrelativistic limit of the bound states of the Klein–Gordon equation*, J. Math. Anal. Appl. **96** (1983), 63–84.

[13] Veselić, K., Weidmann, J., *Existenz der Wellenoperatoren für eine allgemeine Klasse von Operatoren*, Math. Z. **134** (1973), 255–274.


Branko Najman
Department of Mathematics
University of Zagreb
P.O.Box 187
YU-41000 Zagreb, Yugoslavia

# Spatially Degenerate Diffusion with Periodic-Like Boundary Conditions

Gisèle Ruiz Rieder

Department of Mathematics, Louisiana State University

## 1. Introduction

We wish to study the following initial boundary value problem

$$u_t = \varphi(t,x,u,u_x)u_{xx} + \psi(t,x,u,u_x), \qquad 0 \leq x \leq 1,\ 0 \leq t \leq T,$$
$$u(0) - u(1) = g_1(t); \quad u'(0) - u'(1) = g_2(t), \qquad (1.1)$$
$$u(0,x) = u_0(x).$$

Here $T > 0$, $\varphi, \psi : [0,T] \times [0,1] \times \mathbb{R}^2 \to \mathbb{R}$ are continuous, and $g_i : [0,T] \to \mathbb{R}$ is continuously differentiable for $i = 1, 2$. In addition we make the following assumptions on $\varphi$ and $\psi$:

(1) $\varphi(t,x,p,q) \geq \varphi_0(x)$ for $0 \leq x \leq 1$ where $\varphi_0 \in C[0,1]$, $\varphi_0(x) > 0$ for $0 < x < 1$ and $\varphi_0^{-1} \in L^1[0,1]$.

(2) There exist constants $K, L > 0$ and a continuous nondecreasing function $H : [0,T] \to \mathbb{R}$ such that

$$|\varphi(t,x,p,q) - \varphi(s,x,\widetilde{p},q)| \leq K\varphi_0(x)\{|H(t) - H(s)| + |p - \widetilde{p}|\} \qquad (1.2)$$

and

$$|\psi(t,x,p,q) - \psi(s,x,\widetilde{p},q)| \leq L\{|H(t) - H(s)| + |p - \widetilde{p}|\}. \qquad (1.3)$$

---

Partially supported by a LaSER grant.

(3) There exist nondecreasing nonnegative functions $\mathcal{M}$ and $\mathcal{L}$ on $[0, \infty)$ with $\frac{\mathcal{L}(r)}{r} \to 0$ as $r \to \infty$ such that

$$|\psi(t, x, p, q)| \leq \mathcal{M}(|p|)(1 + \varphi(t, x, p, q))\mathcal{L}(|q|). \tag{1.4}$$

If $g_i(t) = 0$ for $i = 1, 2$, the solution, as a function of $x$, can be viewed as a periodic function of period one defined on all of $\mathbb{R}$; for this reason we call such boundary conditions "periodic-like boundary conditions." In this case the problem may be viewed as a nonlinear heat conduction problem (with a source as convection term in a circular rod).

The following theorem is the main result.

**Theorem.** *Let $\varphi$ and $\psi$ be as above. Then the initial boundary value problem (1.1) has a unique solution (in the sense of a limit solution).*

We actually study the Banach space version of (1.1) using the Banach space $C[0, 1]$ of real continuous functions on $[0, 1]$. Define the operator $A(t)$ on $C[0, 1]$ by

$$A(t)u = \varphi(t, \cdot, u, u')u'' + \psi(t, \cdot, u, u')$$

with domain

$$\mathcal{D}(A(t)) = \{u \in C^2(0, 1) \cap C^1[0, 1]: \\ A(t)u \in C[0, 1], u(0) - u(1) = g_1(t), u'(0) - u'(1) = g_2(t)\}.$$

Then for any $u_0 \in \mathcal{D}(A(0))$, (1.1) is equivalent to the problem

$$\begin{aligned} u'(t) &= A(t)u(t), \\ u(0) &= u_0. \end{aligned} \tag{1.5}$$

We shall use techniques from the theory of nonlinear evolution equations in Banach spaces to solve (1.5); the work of Evans [5] and Dorroh and Rieder [4] will be crucial to many of our arguments.

Theorem 1 was proven by Goldstein and Lin [8] in the special case where $\varphi$ is independent of $t$ and the diffusion $u$, $\psi \equiv 1$, and time-independent boundary conditions, that is $g_1(t) = g_2(t) = 0$. In [4] Dorroh and Rieder solved the problem with the linear time-dependent boundary conditions

$$u(0) - \alpha u'(0) = g_1(t); \quad u(1) + \beta u'(1) = g_2(t).$$

With the dependence of the diffusion coefficient on the function $u$, the techniques of [8] no longer apply; $A(t)$ is not quasi-dissipative on all of $\mathcal{D}(A(t))$.

We denote the $L^p[0,1]$ norm by $\|\cdot\|_p$ for $1 \leq p < \infty$; the supremum norm is denoted by $\|\cdot\|$. For $f \in L^1[0,1]$ we define

$$\omega_L(f, \delta) = \sup\{\|f\|_{L^1(\Omega)} : \Omega \text{ is a subinterval of } [0,1] \text{ with } |\Omega| < \delta\}.$$

For $0 \leq \delta \leq 1$ we define

$$\omega_C(f, \delta) = \sup\{|f(x) - f(y)| : x, y \in [0,1], |x - y| \leq \delta\}.$$

Clearly, if $f \in C^1[0,1] \cap C^2(0,1)$ and $\|f''\|_1 < \infty$, then $\omega_C(f', \delta) \leq \omega_L(f'', \delta)$. We shall often use the following lemma proven by Dorroh and Rieder [4].

**Lemma 1.** *Let $f \in C^1[0,1] \cap C^2(0,1)$ and $f'' \in L^1[0,1]$. Then for $0 < \delta \leq 1$*

$$\|f'\| \leq 2\|f\| + \|f''\|_1;$$
$$\|f'\| \leq \frac{2}{\delta}\|f\| + \omega_L(f'', \delta).$$

We denote the duality map on the Banach space $X$ by $\mathcal{J}$, i.e. $\mathcal{J} : X \to 2^{X^*}$ is defined by

$$\mathcal{J}(x) = \{x^* \in X^* : \|x^*\| = 1, \langle x, x^* \rangle = \|x\|\}.$$

For $f \in X = C[0,1]$, $\mathcal{J}(f) \ni \pm \delta_\xi$ where $\|f\| = \pm f(\xi)$. If, in addition, $f \in C^1[0,1] \cap C^2(0,1)$ $f(0) = f(1)$ and $f'(0) = f'(1)$ and we regard $f$ as a periodic function on all of $\mathbb{R}$ with period one, every point is an interior point, and by the maximum principle.

$$f'(\xi) = 0 \quad \text{and} \quad f''(\xi) \leq 0$$

if $\|f\| = f(\xi)$ and $0 < \xi < 1$. If $\xi = 0$, we may conclude $f'(0) = 0$ and $f'(\varepsilon) \leq 0$ for $\varepsilon$ small, whence $f''_+(0) \leq 0$. Similarly if $\xi = 1$, we have $f'(1) = 0$ and $f''_-(1) \leq 0$. If $\|f\| = -f(\eta)$, we have

$$f'(\eta) = 0 \quad \text{and} \quad f''(\eta) \geq 0,$$

again with the second derivative replaced by the appropriate one-sided derivative if $\eta = 0$ or $\eta = 1$.

Finally, we define the sets

$$J^+(f) = \{\xi \in [0,1] : \|f\| = f(\xi)\},$$
$$J^-(f) = \{\xi \in [0,1] : \|f\| = -f(\xi)\}.$$

Since $f$ is a continuous function on $[0,1]$, either $J^+(f) \neq \emptyset$ or $J^-(f) \neq \emptyset$.

## 2. Reduction to a problem with time independent boundary conditions

**Definition 2.1.** Let $\{B(t) : 0 \leq t \leq T\}$ be a family of multivalued operators in a Banach space $X$, and let $f \in L^1_{loc}([0,T];X)$. A function $u \in C([0,T];X)$ is a *limit solution* of

$$u'(t) \in B(t)u(t) + f(t), \qquad (2.2)$$
$$u(0) = u_0$$

on $[0, T_0]$, $0 < T_0 \leq T$ if for every $n \in \mathbb{N}$ and $k = 0, 1, \ldots, N(n)$, (i) $t_0^n = 0 < t_1^n < t_2^n < \cdots < t_{N(n)-1}^n \leq T_0 \leq t_{N(n)}^n \leq T$ and $\lim_{n\to\infty} \max(t_k^n - t_{k-1}^n) = 0$, (ii) there exist $x_k^n, f_k^n \in X$ such that $\sup_{n,k} \|x_k^n\| < \infty$, $\lim_{n\to\infty} \sum_{k=1}^{N(n)} (t_k^n - t_{k-1}^n)\|f_k^n\| = 0$ and

$$\frac{x_k^n - x_{k-1}^n}{t_k^n - t_{k-1}^n} \in B(t_k^n)x_k^n + f_k^n, \qquad (2.3)$$

and (iii) the sequence of functions $\{u_n(t)\}$ defined by

$$u_n(t) = \begin{cases} x_k^n & \text{if } t \in (t_{k-1}^n, t_k^n] \\ x_0^n & \text{if } t = 0 \end{cases}$$

converges uniformly to $u(t)$ on $[0, T_0]$.

One can show that (1.1) has a limit solution if and only if

$$v'(t) \in B(t)v(t) - \omega v(t), \qquad (2.4)$$
$$v(0) = u_0 - z(0)$$

has a limit solution where

$$v(t) = e^{-\omega t}u(t) - z(t),$$
$$B(t)v(t) = \widetilde{\varphi}(t, \cdot, v, v')v'' + \widetilde{\psi}(t, \cdot, v, v'),$$
$$\widetilde{\varphi}(t, x, p, q) = \varphi\Big(t, x, e^{\omega t}(p + z(t,x)), e^{\omega t}(q + z_x(t,x))\Big),$$
$$\widetilde{\psi}(t, x, p, q) = \varphi\Big(t, x, e^{\omega t}(p + z(t,x)), e^{\omega t}(q + z_x(t,x))\Big)z_0''(x)$$
$$+ e^{-\omega t}\psi\Big(t, x, e^{\omega t}(p + z(t,x)), e^{\omega t}(q + z_x(t,x))\Big)$$
$$- z_t(t,x) - \omega z(t,x),$$

$z(t) = z_0 + Z(t, \cdot) - Z(0, \cdot)$, $Z$ is the solution of

$$Z_{xx}(t,x) = 0,$$
$$Z(t,0) - Z(t,1) = e^{-\omega t}g_1(t),$$
$$Z_x(t,0) - Z_x(t,1) = e^{-\omega t}g_2(t),$$

and the function $z_0 \in \mathcal{D}(A(0))$ is arbitrary. The functions $\widetilde{\varphi}$ and $\widetilde{\psi}$ satisfy the same sort of conditions as $\varphi$ and $\psi$ do, but with different $K$, $L$, $\mathcal{M}$, $\mathcal{L}$ and $H$. One can easily verify that the function $v$ satisfies the homogeneous boundary conditions

$$v(t,0) = v(t,1), \quad v_x(t,0) = v_x(t,1)$$

for each $t \geq 0$. Henceforth we replace $\varphi$ by $\widetilde{\varphi}$ and $\psi$ by $\widetilde{\psi}$, i.e. we replace $u$ and (1.1) by $v$ and (2.4), but for typographical convenience we do not show it in the notation.

## 3. Quasidissipativity and properties of $A(t)$.

The preceding section we showed that the boundary conditions in (1.1) may be replaced by the homogeneous boundary conditions

$$u(0) = u(1); \quad u'(0) = u'(1). \tag{3.1}$$

In this section we find sets on which the operators $A(t)$ are quasidissipative and on which we have some control on the resolvents $J_\lambda(t)$. (This will be made more precise later.)

For each $v \in C[0,1]$ define the operator

$$A^v(t)u = \varphi(t,\cdot,v,u')u'' + \psi(t,\cdot,v,u') \tag{3.2}$$

with domain

$$\mathcal{D}(A^v(t)) = \{u \in C^1[0,1] \cap C^2(0,1) :$$
$$A^v(t)u \in C[0,1], \, u(0) = u(1), \, u'(0) = u'(1)\}.$$

We consider the initial value problem

$$u'(t) = A^v(t)u$$
$$u(0) = u_0$$

for $u_0 \in \mathcal{D}(A^v(0))$.

**Lemma 2.** *The operator $A^v(t)$ is dissipative.*

**Proof.** Let $u_1, u_2 \in \mathcal{D}(A^v(t))$, and assume $J^+(u_1 - u_2) \neq \emptyset$. If $\xi \in J^+(u_1 - u_2)$ and $0 < \xi < 1$, then

$$\langle A^v(t)u_1 - A^v(t)u_2, \delta_\xi \rangle = \varphi(t,\xi,v(\xi),u_1'(\xi))u_1''(\xi)$$
$$+ \psi(t,\xi,v(\xi),u_1'(\xi)) - \varphi(t,\xi,v(\xi),u_2'(\xi))u_2''(\xi)$$
$$- \psi(t,\xi,v(\xi),u_2'(\xi)) \leq 0$$

since $u_1'(\xi) = u_2'(\xi)$ and $u_1''(\xi) \leq u_2''(\xi)$. If $\xi \in \{0,1\}$, we replace the second derivative by the appropriate one-sided second derivative, and the same proof works. A similar argument holds if $J^-(u_1 - u_2) \neq \emptyset$. □

**Proposition 1.** *The operator $A^v(t)$ is m-dissipative.*

We must show $\mathcal{R}(I - \lambda A^v(t)) = C[0,1]$ for some $\lambda > 0$. Let $\alpha_n = e^{i2n\pi x}$. Clearly, $\{\alpha_n\}_{n=-\infty}^{\infty}$ is an orthonormal basis for $L^2[0,1]$. If we define $Bu = u''$, then $B\alpha_n = -(2\pi n)^2 \alpha_n$, $(I - \lambda B)^{-1}\alpha_n = (1 + 4\lambda\pi^2 n^2)^{-1}\alpha_n$ and

$$(I - \lambda B)^{-1} f = \sum_{n=-\infty}^{\infty} (1 + 4\lambda\pi^2 n^2)^{-1} <f, \alpha_n> \alpha_n$$

for $f \in C[0,1]$. Define

$$g_\lambda(x,y) = \sum_{n=-\infty}^{\infty} (1 + 4\pi^2 \lambda n^2)^{-1} e^{i2n\pi(x-y)}.$$

Then $g_\lambda$ is the Green's function for

$$Cu = u - \lambda u'' \quad u(0) = u(1), \quad u'(0) = u'(1); \tag{3.3}$$

that is, if $Cu = f$ and $u(0) = u(1)$, $u'(0) = u'(1)$, then

$$u(x) = \int_0^1 g_\lambda(x,y) f(y) dy. \tag{3.4}$$

From (3.4) it is clear that $(I - \lambda B)^{-1}$ is continuous and compact on $C[0,1]$ for $\lambda > 0$.

For each $h \in C[0,1]$ and $t \in [0,T]$ we define the operator $S_\lambda^{h,v} : C^1[0,1] \to C^1[0,1]$ by

$$(S_\lambda^{h,v} u)(x) = \int_0^1 g_\lambda(x, \cdot) \left[ u - \frac{u - h - \lambda\psi(t, \cdot, v, u')}{\varphi(t, \cdot, v, u')} \right] \tag{3.5}$$

Then $\mathcal{R}(I - \lambda A^v(t)) = C[0,1]$ is equivalent to $S_\lambda^{h,v}$ having a fixed point. Hence, Proposition 1 follows immediately from the next lemma.

**Lemma 3.** *Choose $\lambda > 8(\frac{\pi^2}{6} + 1)(1 + \|\varphi_0^{-1}\|_1)$. Then for each $h, v \in C[0,1]$ and $t \in [0,T]$, $S_\lambda^{h,v}$ has a fixed point.*

**Proof.** Let $u \in C^1[0,1]$, and let $w = S_\lambda^{h,v} u$. Recall that $\sum_{n=1}^{\infty} n^{-2} = \frac{\pi^2}{6}$. An elementary calculation shows $\|g_\lambda\| \leq \frac{\pi^2}{6\lambda}$. Furthermore,

$$\|w\| \leq \|g_\lambda\| \{\|u\| + (\|u\| + \|h\|)\|\varphi_0^{-1}\|_1 + \lambda M(\|v\|)(1 + \|\varphi_0^{-1}\|_1)\mathcal{L}(\|u'\|)\} \tag{3.6}$$

and
$$w'' = \lambda^{-1}(w-u) + \frac{u - h - \lambda\psi(t,\cdot,v,u')}{\lambda\varphi(t,\cdot,v,u')}.$$

It follows that
$$\|w''\|_1 \leq \lambda^{-1}(\|g_\lambda\| + 1)\{\|u\| + (\|u\| + \|h\|)\|\varphi_0^{-1}\|_1 \\ + \lambda \mathcal{M}(\|v\|)(1 + \|\varphi_0^{-1}\|_1)\mathcal{L}(\|u'\|)\}. \tag{3.7}$$

Since $\omega_C(w',\delta) \leq \omega_L(w'',\delta) \leq \|w\|_1$, we have that $\{(S_\lambda^{h,v}u)' : u \in \Lambda\}$ is an equicontinuous collection for any bounded set $\Lambda$ in $C^1[0,1]$. We seek a closed, convex bounded set $\Lambda$ such that $S_\lambda^{h,v}(\Lambda) \subseteq \Lambda$. By the Arzela-Ascoli theorem it will follow that $S_\lambda^{h,v}(\Lambda)$ is compact in $C^1[0,1]$; hence by the Schauder fixed point theorem we will conclude that $S_\lambda^{h,v}$ fixes a point of $\Lambda$.

Choose $N_0$ so large that $N_0 > 3\|h\|$, $(\frac{\pi^2}{6}+1)\mathcal{M}(\|v\|)(1+\|\varphi_0^{-1}\|_1)\mathcal{L}(N_0) \leq \frac{1}{8}N_0$, and $M_0 = \|h\| + N_0$. We define the set $\Lambda$ by
$$\Lambda = \{u \in C^1[0,1] : \|u\| \leq M_0, \|u'\| \leq N_0\}.$$

Then $\Lambda$ is closed, bounded and convex. Moreover, using (3.5) and (3.6) for $u \in \Lambda$ we have
$$\|w\| \leq \lambda^{-1}(\frac{\pi^2}{6}+1)(1+\|\varphi_0^{-1}\|_1)(M_0 + \|h\|) + \frac{1}{8}N_0$$
$$\leq \frac{1}{8}(M_0 + \|h\|) + \frac{1}{8}N_0 = \frac{1}{4}M_0,$$
$$\|w''\|_1 \leq \frac{1}{8}(M_0 + \|h\|) + \frac{1}{8}N_0 = \frac{1}{4}M_0.$$

Hence, by Lemma 1 it follows that
$$\|w'\| \leq 2\|w\| + \|w''\|_1 \leq \frac{3}{4}M_0$$
$$= \frac{3}{4}(\|h\| + N_0) \leq N_0.$$

Consequently $w \in \Lambda$. □

By the dissipativity of $A^v(t)$, we see that $J_\lambda^v(t) = (I - \lambda A^v(t))^{-1}$ is nonexpansive and onto for every $\lambda > 0$.

Next we prove that for some $h \in C[0,1]$, we can solve the nonlinear elliptic problem
$$u - \lambda\varphi(t,\cdot,u,u')u'' - \lambda\psi(t,\cdot,u,u') = h. \tag{3.8}$$

This is a key step in the proof of local existence for the problem (1.1). We solve (3.8) via another fixed point argument. For each $h \in C[0,1]$ and $t \in [0,T]$, define the map $S_\lambda^{h,v} : C[0,1] \to C[0,1]$ by
$$S_\lambda^h v = J_\lambda^v(t)h.$$

Then finding a fixed point of $S_\lambda^h$ is equivalent to solving (3.8).

**Proposition 2.** Let $r_0 > 0$, and let

$$\lambda_0 \leq \frac{r_0}{2\mathcal{M}(r_0)(1 + Q(r_0))\mathcal{L}(0)}$$

where $Q(s) = \sup\{\varphi(t, x, p, 0) : t \in [0, T], x \in [0, 1] \text{ and } |p| \leq s\}$. Then for each $t \in [0, T]$ and $0 < \lambda \leq \lambda_0$, $S_\lambda^h$ has a fixed point in the ball of radius $r_0$ for each $h$ satisfying $\|h\| \leq \frac{r_0}{2}$.

**Proof.** Assume $\|h\| \leq \frac{r_0}{2}$, $\lambda \in (0, \lambda_0]$, $t \in [0, T]$, $\|v\| \leq r_0$ and $u = S_\lambda^h v$, i.e. $u = J_\lambda^v(t)h$. Suppose $J^+(u) \neq \emptyset$, say $\xi \in J^+(u)$ and $0 < \xi < 1$. Then

$$\|u\| = h(\xi) + \lambda\varphi(t, \xi, v(\xi), u'(\xi))u''(\xi) + \xi\psi(t, \xi, v(\xi), u'(\xi))$$
$$\leq \|h\| + \lambda\mathcal{M}(r_0)(1 + Q(r_0))\mathcal{L}(0) \leq r_0.$$

(If $\xi \in \{0, 1\}$, we must replace $u''(\xi)$ by the appropriate one-sided derivative in the preceding equality.) Let $E_{r_0}$ be the closed ball of radius $r_0$ in $C[0, 1]$; we have shown $S_\lambda^h(E_{r_0}) \subseteq E_{r_0}$. Moreover, $u'' = \frac{u - h - \lambda\psi(t, \cdot, v, u')}{\lambda\varphi(t, \cdot, v, u')}$, so that

$$\omega_L(u'', \delta) \leq \frac{3r_0}{2\lambda}\omega_L(\varphi_0^{-1}, \delta) + \mathcal{M}(r_0)(\omega_L(\varphi_0^{-1}, \delta) + \delta)\mathcal{L}(\|u'\|). \tag{3.9}$$

Choose $N$ so large that

$$\mathcal{L}(r) \leq r \quad \text{if} \quad r \geq N, \tag{3.10}$$

and choose $\delta > 0$ so small that $N\mathcal{M}(r_0)(\omega_L(\varphi_0^{-1}, \delta) + \delta) < \frac{1}{2}$. Then

$$\omega_L(u'', \delta) \leq \frac{3r_0}{2\lambda}\omega_L(\varphi_0^{-1}, \delta) + \frac{1}{2}$$

and by Lemma 1

$$\|u'\| \leq \frac{2}{\delta}\|u\| + \omega_L(u'', \delta) \leq \frac{2r_0}{\delta} + \frac{3r_0}{2\lambda}\omega_L(\varphi_0^{-1}, \delta) + \frac{1}{2}.$$

Hence $\{S_\lambda^h v : v \in E_{r_0}\}$ is a pointwise bounded equicontinuous set in $C[0, 1]$. By Arzela-Ascoli it follows that this set is compact in $C[0, 1]$. Hence, by the Schauder fixed point theorem, $S_\lambda^h$ fixes a point of $E_{r_0}$. $\square$

Let $\widehat{A}(t) = A(t)\big|_{\mathcal{D}(A(t)) \cap E_{r_0}}$, and let $J_\lambda(t) = (I - \lambda\widehat{A}(t))$. Then for each $\lambda \in (0, \lambda_0]$ and $t \in [0, T]$, we have shown that $J_\lambda(t)$ exists, $B(\frac{r_0}{2}) \subseteq \mathcal{D}(J_\lambda(t))$ and $\mathcal{R}(J_\lambda(t)) \subseteq B(r_0)$. Here $B(r)$ is the ball of radius $r$. We now seek a set on which $J_\lambda(t)$ is single-valued. Since $A(t)$ and $\widehat{A}(t)$ agree on $\mathcal{D}(A(t)) \cap E_{r_0}$, we write $A(t)$ for $\widehat{A}(t)$ in the remaining portion of the paper. The remainder of the arguments in the paper follow closely those in Dorroh and Rieder [4]; hence we do not include the details here. For the full proofs the reader may see the aforementioned paper.

Choose $a > 0$ so that
$$|\psi(t,x,p,0)| < a \qquad (3.11)$$
for $t \in [0,T]$, $x \in [0,1]$ and $p \in [-r_0, r_0]$. Next we choose $b \in \mathbb{R}$ and a $\delta_0 > 0$ so that
$$0 < b \leq \frac{r_0}{2}, \qquad (3.12)$$
$$N\mathcal{M}(b)\big(\omega_L,(\varphi_0^{-1},\delta_0) + \delta_0\big) < \frac{1}{2} \qquad (3.13)$$
where $N$ is as in (3.10).

**Lemma 4.** Let $h \in \mathcal{D}(A(t)) \cap B(b)$ with $\|\varphi_\circ h''\| \leq M$ and $u_\lambda \in J_\lambda(t)h$. Choose $\omega \geq MK + L$. Then for all $\lambda$ satisfying $0 < \lambda < \min\{\lambda_0, \omega^{-1}, a^{-1}(b - \|h\|)\}$
  (i) $\|A(t)u_\lambda\| \leq (1-\lambda\omega)^{-1}\|A(t)h\|$
  (ii) $\|u_\lambda\| \leq \|h\| + \lambda a < b$
  (iii) $\|u_\lambda'\| \leq 1 + \frac{4b}{\delta_0} + 2(1-\lambda\omega)^{-1}\|A(t)h\|\omega_L(\varphi_0^{-1},\delta_0)$.

Parts (i) and (ii) follow primarily from the maximum principle; part (iii) follows from (1.2), (1.4) and Lemma 1.

Let $d > 0$, and choose $c$ so that
$$c > 1 + \frac{4b}{\delta_0} + 4d\omega_L(\varphi_0^{-1},\delta_0). \qquad (3.14)$$
Increase $a$ (if necessary) so that
$$|\psi(t,x,p,q)| < a \qquad (3.15)$$
for $t \in [0,T]$, $x \in [0,1]$, $p \in [-b,b]$ and $q \in [-c,c]$. Finally, choose $M$ and $\omega$ so that
$$M \geq 2d + a \quad \text{and} \quad \omega \geq MK + L. \qquad (3.16)$$

**Lemma 5.** Let $h \in \mathcal{D}(A(t)) \cap B(b)$ with $\|A(t)h\| < d$, $\|h'\| \leq c$ and $u_\lambda \in J_\lambda(t)h$ for $0 < \lambda < \min\{(2\omega)^{-1}, \lambda_0, a^{-1}(b-\|h\|)\}$. Then $\|\varphi_\circ h''\| \leq M$, $\|u_\lambda\| < a$, $\|u_\lambda'\| \leq c$, $\|\varphi_\circ u_\lambda''\| \leq M$, and
$$\|A(t)u_\lambda\| \leq (1-\lambda\omega)^{-1}\|A(t)h\|. \qquad (3.17)$$

The proof of this lemma is a simple calculation using Lemma 4 and the choice of $a$, $b$, $c$ and $d$. For $t \in [0,T]$ and $0 < \lambda < \min\{\lambda_0, (2\omega)^{-1}, \frac{b}{a}\}$, define
$$D_\lambda(t) = \{u \in \mathcal{D}(A(t)) : \|u\| < b - a\lambda, \|u'\| < c, \|A(t)u\| < (1-\lambda\omega)d\}.$$
By Lemma 4,
$$J_\lambda(t)D_\lambda(t) \subseteq D_0(t). \qquad (3.18)$$
Let $\widetilde{A}(t) = A(t)|_{D_0(t)}$ and $\widetilde{J}_\lambda(t) = (I - \lambda\widetilde{A}(t))^{-1}$. In particular, (3.18) holds with $J_\lambda(t)$ replaced by $\widetilde{J}_\lambda(t)$ and $D_\lambda(t) \subseteq \mathcal{D}(\widetilde{J}_\lambda(t))$.

**Lemma 6.** $\widetilde{A}(t)$ is $\omega$-dissipative for $\omega \geq MK + L$.

This lemma is a simple consequence of the maximum principle and assumptions (1.2) and (1.3). It follows that $\widetilde{J}_\lambda(t)$ is single-valued, and for $u, v \in D_\lambda(t)$

$$\|\widetilde{J}_\lambda(t)u - \widetilde{J}_\lambda(t)v\| \leq (1 - \lambda\omega)^{-1}\|u - v\|.$$

Define

$$A^\omega(t) = \widetilde{A}(t) - \omega I,$$
$$J_\lambda^\omega(t) = \bigl(I - \lambda A^\omega(t)\bigr)^{-1}.$$

Clearly, $A^\omega(t)$ is dissipative, and $J_\lambda^\omega(t)$ is single valued, nonexpansive and a simple calculation shows that

$$J_\lambda^\omega(t) = \widetilde{J}_\mu(t)(1 + \lambda\omega)^{-1}$$

for $\mu = \lambda(1 + \lambda\omega)^{-1}$. Choose

$$d = \frac{a(b^{-1} - K) - L}{2K},$$
$$M = 2d + a,$$
$$\omega = MK + L.$$

Then $\omega = \frac{a}{b}$ and $d > b\omega = a$. Choose $T$ small enough that

$$\omega\bigl(H(T) - H(0)\bigr) < d - b\omega$$

and $\widetilde{d} > 0$ so that

$$\widetilde{d} + \omega\bigl(H(T) - H(0)\bigr) < d - \omega b.$$

Combining the previous estimates and observations we obtain the next lemma.

**Lemma 7.** Let $0 < \lambda < \lambda_0$ and $\|u\| < b$. Then $\|J_\lambda^\omega(t)u\| < b$. If $0 < \lambda < \min\{\lambda_0, \omega^{-1}\}$, $\|u\| < b$, $\|u'\| < c$ and $\|A^\omega(t)u\| < \widetilde{d}$, then

$$\|\bigl(J_\lambda^\omega(t)u\bigr)'\| \leq c,$$
$$\|\bigl(J_\lambda^\omega(t)u\bigr)''\varphi_0\| \leq M,$$
$$\|A^\omega(t)J_\lambda^\omega(t)u\| \leq \|A^\omega(t)u\| < \widetilde{d}.$$

We define the sets

$$D^\omega(s) = \{u \in D_0(t) : \|u\| < b, \|u'\| < c, \|A^\omega(t)u\| < \widetilde{d} + \omega(H(t) - H(0))$$
$$\text{for } s \leq t \leq T\}.$$

If $0 < \lambda < \min\{\lambda_0, \omega^{-1}\}$, it is clear that

$$J_\lambda^\omega(s) : D^\omega(r) \longrightarrow D^\omega(s) \quad \text{for} \quad 0 \leq r \leq s \leq T. \tag{3.19}$$

Moreover, one can readily show that if $0 < \lambda < \min\{\lambda_0, \omega^{-1}\}$, $u \in D^\omega(s)$ and $v \in D^\omega(r)$

$$\|J_\lambda^\omega(s)u - J_\lambda^\omega(r)v\| \leq \|u - v\| + \lambda\omega|H(s) - H(r)|; \tag{3.20}$$

if $0 \leq r \leq s \leq t \leq T$, then

$$\|A^\omega(t)J_\lambda^\omega(s)v\| \leq \tilde{d} + \omega(H(t) - H(0)). \tag{3.21}$$

## 4. Existence of a limit solution

By the discussion in section 2 it suffices to prove the existence of a limit solution for the transformed initial value problem

$$u'(t) = A^\omega(t)u(t),$$
$$u(0) = u_0 - z(0).$$

The complication is that the operator $A^\omega(t)$ is dissipative but not $m$-dissipative. However, as Dorroh and Rieder [4] proved the results of Evans [5] may be extended to obtain a limit solution in this more general case.

We briefly describe the construction of the limit solution. For each $n$ we choose
(i) a partition $\{t_0^n, \ldots, t_{N(n)}^n\}$ of $[0, T(n)]$ for $T_0 \leq T(n) \leq T$ with $0 = t_0^n \leq t_1^n \leq \ldots \leq t_{N(n)}^n = T(n)$ such that $\delta_k^n = t_k^n - t_{k-1}^n < \lambda_0 \wedge \omega^{-1}$ and $\lim_{n \to \infty} \max_k \delta_k^n = 0$;
(ii) $\{x_0^n\} \subseteq D^\omega(0)$ such that $\lim_{n \to \infty} x_0^n = u_0 - z(0) = \tilde{u}_0$.

For $n = 1, 2, \ldots$ and $k = 0, 1, \ldots, N(n)$ define

$$x_k^n = J_{\delta_k^n}^\omega(t_k^n)x_{k-1}^n. \tag{4.1}$$

Since $x_0^n \in D^\omega(0)$ and by (3.18), the mapping in (4.1) is well-defined. Moreover (4.1) is equivalent to

$$\frac{x_k^n - x_{k-1}^n}{t_k^n - t_{k-1}^n} = A^\omega(t_k^n)x_k^n.$$

By Lemma 7, $\|x_k^n\| < b$ for all $n, k$. If we define the step functions

$$u^n(t) = \begin{cases} x_k^n & \text{on } (t_{k-1}^n, t_k^n], \\ x_0^n & \text{if } t = 0 \end{cases}$$

we must show that $\lim_{n\to\infty} u^n(t) = u(t)$ and that $u(t)$ is continuous.

This part of the proof requires that we check several technical conditions which follow essentially from (3.19) and (3.20). The interested reader should refer to [4], [5]. Various properties of limit solutions, such as uniqueness, Lipschitz continuity, differentiability, continuous dependence, etc., are well known and can be found primarily in Benilan [1].

The author is very grateful to Bob Dorroh for many helpful conversations. The author gratefully acknowledges the partial support of Louisiana Education Quality Support Fund, contract number 86-LBR-01604.

## References

[1] P. Benilan, *Equations d'evolution dans un espace de Banach quelconque et applications*, Thesis, Universitié de Paris XI, Orsay, 1972.

[2] M. G. Crandall and T. Liggett, *Generation of semigroups of nonlinear transformations on general Banach spaces*, Amer. J. Math. **93** (1971), 265–298.

[3] M. G. Crandall and A. Pazy, *Nonlinear evolution equations in Banach space*, Israel J. Math **11** (1972), 57–94.

[4] J. R. Dorroh and G. R. Rieder, *A singular quasilinear parabolic problem in one space dimension*, submitted.

[5] L. C. Evans, *Nonlinear evolution equations in an arbitrary Banach space*, Math. Res. Center Tech. Summary Report **1568**, August 1975, Madison, Wisconsin.

[6] L. C. Evans, *Nonlinear evolution equations in Banach Spaces*, Israel J. Math. **26** (1977), 1–42.

[7] D. Gilbarg and N. S. Trudinger, *Elliptic Partial Differential Equations of Second Order*, second ed., Grand. des Mat. Wiss., vol. 224, Springer Verlag, Berlin, 1983.

[8] J. A. Goldstein and C. Y. Lin, *Singular nonlinear parabolic boundary value problems in one space dimension*, J. Diff. Equations **68** (1987), 429–443.

Gisèle Ruiz Rieder
Department of Mathematics
Louisiana State University
Baton Rouge, Louisiana 70803, USA

# Scattering Theory of a Supersymmetric Dirac Operator

BERND THALLER

Institut für Mathematik, Universität Graz

## 1. Introduction

We are interested in relativistic scattering theory for a spin-1/2 particle moving in an external magnetic field. One of the basic problems is proving the existence of the wave operators

$$\Omega_\pm(H, H_0) \equiv \underset{t \to \pm\infty}{\text{s-lim}}\, e^{iHt} e^{-iH_0 t} P_{cont}(H_0), \tag{1}$$

where the self-adjoint operator $H$, which is to be defined in a suitable Hilbert space, generates the quantum mechanical time evolution under the influence of the external field, and $H_0$ generates the "free motion". $P_{cont}(H_0)$ denotes the projection operator onto the subspace belonging to the continuous spectrum of $H_0$, which is usually assumed to be absolutely continuous. Existence of (1) says, that any free motion is for $t \to \pm\infty$ asymptotic to the motion of an interacting particle.

In three space dimensions the magnetic field strength $B$ is usually described by a vector field $B(x)$ satisfying $\text{div}\, B = 0$. Hence we can write $B = \text{rot}\, A$ with a "magnetic vector potential" $A(x)$. We have to use this vector potential in order to set up a quantum mechanical description. The time evolution of a relativistic charged particle with mass $m$ in a magnetic field is generated by the following Dirac operator

$$H(A) \equiv \vec{\alpha} \cdot (p - A) + \beta m, \quad m > 0, \tag{2}$$

where $p = -i\nabla$, and where $\vec{\alpha} = (\alpha_1, \alpha_2, \alpha_3)$ and $\beta$ are the famous $4 \times 4$ "Dirac matrices",

$$\beta = \begin{pmatrix} 1 & 0 \\ 0 & -1 \end{pmatrix}, \quad \alpha_i = \begin{pmatrix} 0 & \sigma_i \\ \sigma_i & 0 \end{pmatrix}, \quad i = 1, 2, 3. \tag{3}$$

The $2 \times 2$ "Pauli matrices" are defined by

$$\sigma_1 = \begin{pmatrix} 0 & 1 \\ 1 & 0 \end{pmatrix}, \quad \sigma_2 = \begin{pmatrix} 0 & -i \\ i & 0 \end{pmatrix}, \quad \sigma_3 = \begin{pmatrix} 1 & 0 \\ 0 & -1 \end{pmatrix}. \tag{4}$$

If we assume that the fields $B$ and $A$ are infinitely differentiable functions on $\mathbb{R}^3$, then the Dirac operator (2) is essentially self-adjoint on $\mathcal{C}_0^\infty(\mathbb{R}^3)^4$ which is a dense subspace of the Hilbert space $\mathcal{H} = L^2(\mathbb{R}^3)^4$ [6]. This result is true even without restriction on the growth of $B$ or $A$ at infinity. We want to stress that the vector potential is not directly observable. If we replace $A$ by $A + \nabla g$ with $g \in \mathcal{C}^\infty(\mathbb{R}^3)$, then the magnetic field strength remains unchanged and the new Dirac operator is unitarily equivalent to the original one. Eventually, we shall use this gauge freedom to make the formalism as simple as possible. Therefore we choose $H(0)$ as the Dirac operator for a free particle, although the operator $H(\nabla g)$ would also describe a free motion, but obviously in a more complicated way.

Calculating the square of the Dirac operator we find

$$H(A)^2 = (p - A)^2 - \vec{\Sigma} \cdot B + m^2$$
$$= p^2 + m^2 - 2A \cdot p + i \operatorname{div} A + A^2 - \vec{\Sigma} \cdot B, \tag{5}$$
$$H(0)^2 = p^2 + m^2, \tag{6}$$

with

$$\Sigma_i = \begin{pmatrix} \sigma_i & 0 \\ 0 & \sigma_i \end{pmatrix}, \quad i = 1, 2, 3. \tag{7}$$

We see that $H(A)^2$ and $H(0)^2$ (apart from the trivial summand $m^2$ describing the rest energy) are just the operators for the corresponding nonrelativistic scattering problem, which usually is easier to handle. $H(A)^2$ is called "Pauli operator" in contrast to the Schrödinger operator $(p - A)^2$ for a spinless particle. We want to use this close relation between Dirac and Pauli operators to obtain some information on the relativistic scattering problem.

## 2. Supersymmetry

The general structure of our problem is given by supersymmetric quantum mechanics. In the Hilbert space $\mathcal{H}$ of our system a unitary involution $\tau$ is defined, i.e., a bounded operator satisfying

$$\tau^* \tau = \tau \tau^* = \tau^2 = 1. \tag{8}$$

In our case $\tau = \beta$. Moreover, there is a self-adjoint operator $Q$ (here given by $\vec{\alpha} \cdot (p - A)$) which anticommutes with $\tau$,

$$\tau \mathcal{D}(Q) = \mathcal{D}(Q), \quad \{\tau, Q\} \equiv \tau Q + Q\tau = 0 \text{ on } \mathcal{D}(Q). \tag{9}$$

A supersymmetric Dirac operator is a self-adjoint operator $H$ of the form

$$H = Q + m\tau, \quad \mathcal{D}(H) = \mathcal{D}(Q), \quad m > 0. \tag{10}$$

The square of a supersymmetric Dirac operator is simply given by $H^2 = Q^2 + m^2$ and commutes with $\tau$.

The unitary involution $\tau$ can only have the eigenvalues $\pm 1$ and the Hilbert space decomposes into a direct sum of the corresponding eigenspaces

$$\mathcal{H} = \mathcal{H}_+ \oplus \mathcal{H}_- . \tag{11}$$

With respect to this decomposition all operators in $\mathcal{H}$ with domain left invariant by $\tau$ are most naturally represented by $2 \times 2$ matrices. We have

$$\tau = \begin{pmatrix} 1 & 0 \\ 0 & -1 \end{pmatrix}, \quad Q = \begin{pmatrix} 0 & D^* \\ D & 0 \end{pmatrix}, \tag{12}$$

$$H = \begin{pmatrix} m & D^* \\ D & -m \end{pmatrix}, \quad H^2 = \begin{pmatrix} D^*D + m^2 & 0 \\ 0 & DD^* + m^2 \end{pmatrix}. \tag{13}$$

Here, $D$ is a suitable closed operator, densely defined in $\mathcal{H}_+$ with range in $\mathcal{H}_-$. In fact, any densely defined closed operator $D$ defines via (12) a self-adjoint operator $Q$ in a larger Hilbert space.

The supersymmetric Dirac operator can be diagonalized by a suitable unitary "Foldy-Wouthuysen" transformation.

**Theorem 1.** *Let $H = Q + m\tau$ be a supersymmetric Dirac operator. The unitary transformation $U$ given by*

$$U = a_+ + \tau (\operatorname{sgn} Q) a_-, \quad a_\pm = \frac{1}{\sqrt{2}} \sqrt{1 \pm m|H|^{-1}}, \tag{14}$$

*brings $H$ to the diagonal form*

$$UHU^{-1} = \begin{pmatrix} \sqrt{D^*D + m^2} & 0 \\ 0 & -\sqrt{DD^* + m^2} \end{pmatrix} = \tau |H|. \tag{15}$$

**Proof.** It is easy to verify the following formulas for the bounded operators $a_\pm$

$$a_+^2 + a_-^2 = 1, \quad a_+^2 - a_-^2 = m|H|^{-1}, \quad 2a_+ a_- = |Q||H|^{-1}. \tag{16}$$

Furthermore we note that $|H| = (Q^2 + m^2)^{1/2}$ commutes with $\tau$ and $Q$, and the following commutation relations hold on $\mathcal{D}(H) = \mathcal{D}(Q)$

$$[H, a_\pm] = [Q, a_\pm], \quad H\tau (\mathrm{sgn}\, Q) = -\tau (\mathrm{sgn}\, Q) H. \tag{17}$$

Now we can verify Eq. (14) in the following way

$$\begin{aligned} UHU^{-1} &= (a_+ + \tau (\mathrm{sgn}\, Q)\, a_-) H (a_+ - \tau (\mathrm{sgn}\, Q)\, a_-) = \\ &= (a_+^2 + 2\tau (\mathrm{sgn}\, Q)\, a_+ a_- - a_-^2) H \\ &= (m|H|^{-1} + \tau (\mathrm{sgn}\, Q) |Q| |H|^{-1}) H \\ &= (m + \tau Q)|H|^{-1} H = \tau(m\tau + Q)|H|^{-1} H \\ &= \tau H^2 |H|^{-1} = \tau |H|. \end{aligned} \tag{18}$$

The matrix form of $\tau|H|$ immediately follows from (12), (13) and $|H| = \sqrt{H^2}$. This completes the proof of Theorem 1.

On $\mathrm{Ker}\, Q = \mathrm{Ker}\, D \oplus \mathrm{Ker}\, D^*$ we have $a_+ = 1$, $a_- = 0$, hence $U$ is just the identity on $\mathrm{Ker}\, Q$. We also note that $U^4$ is just the Cayley transform of $iQ\tau/m$ (see Ref. [12]), hence $U$ can also be written in the form

$$U = e^{\frac{i}{2} \arctan(iQ\tau/m)}. \tag{19}$$

For some special examples of supersymmetric Dirac operators $H$ the Foldy-Wouthuysen transformation has been known since the 1950's [5], and has been studied intensively in the context of relativistic quantum mechanics (see, e.g., Ref. [1] for a review).

Since the operator $D$ is densely defined and closed, $D^*D$ and $DD^*$ are both densely defined, self-adjoint, and positive by von Neumann's theorem. By polar decomposition we may write $D = S(D^*D)^{1/2} = (DD^*)^{1/2} S$, where $S$ is a partial isometry from $(\mathrm{Ker}\, D)^\perp$ to $(\mathrm{Ker}\, D^*)^\perp$. Therefore we find easily $DD^* = SD^*DS^*$ which implies that $D^*D$ and $DD^*$ have the same spectrum except possibly at 0. From this we conclude immediately the following result

**Corollary.** *The spectrum of any supersymmetric Dirac operator $H$ is symmetric with respect to 0 (except possibly at $\pm m$), has a gap from $-m$ to $+m$, and is determined by the spectrum of the "nonrelativistic" operator $D^*D$ (except at $-m$). The point $+m$ (resp. $-m$) is an eigenvalue of $H$, iff 0 is an eigenvalue of $D$ (resp. $D^*$).*

We finally note that Dirac operators have supersymmetry in a number of cases including the neutron in an electric field, a particle in a Lorentz-scalar field or, most important, in case of the hydrogen atom in a subspace with fixed angular momentum, see Ref. [12]. Our results can be applied to all these examples.

## 3. Supersymmetric scattering theory

Next we solve the problem of concluding existence of the relativistic wave operators from the existence of the corresponding nonrelativistic operators. The supersymmetric structure can be used to reformulate the existence problem in such a way that the invariance principle of wave operators ([10], p.49ff) can be applied (The calculation in [10] on p.53f does not take into account the negative energy subspace of the Dirac operator. In the standard representation the Dirac operator is not an admissible function of the Pauli operator). The following result is similar to Theorem 4 in Ref. [12], but is proved here under different assumptions, which are more convenient for the applications. By $F$ we denote the projection operator to the subspace belonging to the indicated region of the spectrum of a self-adjoint operator.

**Theorem 2.** Let $H = Q + m\tau$, $H_0 = Q_0 + m\tau$ be two Dirac operators with supersymmetry. Assume that for all $0 < a < b < \infty$ and for $\Psi$ in some dense subset of $F(a < Q_0^2 < b)\mathcal{H}_{a.c.}(Q_0^2)$ the following condition is satisfied with $k = 1,2$

$$\|(Q^k - Q_0^k)e^{-iQ_0^2 t}\Psi\| \leq \text{const.}(1 + |t|)^{1-k-\delta}. \qquad (20)_k$$

Then the wave operators $\Omega_\pm(H, H_0)$ exist, and

$$\Omega_\pm(H, H_0) = \Omega_\pm(Q^2, Q_0^2)F(H_0 > 0) + \Omega_\mp(Q^2, Q_0^2)F(H_0 < 0). \qquad (21)$$

**Proof.** The assumption $(20)_2$ implies existence of the "nonrelativistic" wave operators $\Omega_\pm^{nr} \equiv \Omega_\pm(Q^2, Q_0^2)$ by the following argument due to Cook (see [10], p.20). First note that the set of states $\Psi \in \mathcal{H}_{a.c.}(Q_0^2)$ for which there exist constants $a$ and $b$ such that $\Psi = F(a < Q_0^2 < b)\Psi$ is a dense subset of $\mathcal{H}_{a.c.}(Q_0^2)$. For $\Psi$ in this subset we have

$$\lim_{s \to \infty} \sup_{t \geq 0} \|(e^{iQ^2 t}e^{-iQ_0^2 t} - e^{iQ^2 s}e^{-iQ_0^2 s})\Psi\|$$

$$= \lim_{s \to \infty} \sup_{t \geq 0} \|\int_s^t e^{iQ^2 t'}(Q^2 - Q_0^2)e^{-iQ_0^2 t'}\Psi\| dt'$$

$$\leq \lim_{s \to \infty} \int_s^\infty \|(Q^2 - Q_0^2)e^{-iQ_0^2 t'}\Psi\| dt' = 0. \qquad (22)$$

We have used that by $(20)_2$, the integrand in the last expression decays integrably in time. But (22) is just the Cauchy criterion for the existence of $\Omega_+^{nr}$. A similar argument proves existence of $\Omega_-^{nr}$. The operators $UHU^* = \tau|H|$ and $U_0 H_0 U_0^* = \tau|H_0|$ are admissible functions of $Q^2$ and $Q_0^2$, respectively, if we restrict them to the subspaces $\frac{1}{2}(1 \pm \tau)\mathcal{H}$. Therefore, we can apply the invariance principle to conclude the existence of the wave operators

$$\Omega_\pm^{rel} \equiv \Omega_\pm(\tau|H|, \tau|H_0|) = \Omega_\pm^{nr}\frac{1}{2}(1 + \tau) + \Omega_\mp^{nr}\frac{1}{2}(1 - \tau). \qquad (23)$$

It remains to show that for all $\Psi \in \mathcal{H}_{a.c.}(H_0)$ we can find $\Phi_\pm \in \mathcal{H}$ such that

$$0 = \lim_{t \to \pm\infty} \|e^{iHt}e^{-iH_0 t}\Psi - \Phi_\pm\| \tag{24}$$

$$\leq \lim_{t \to \pm\infty} \|e^{i\tau|H|t}e^{-i\tau|H_0|t}U_0\Psi - U\Phi_\pm\| \tag{25}$$

$$+ \lim_{t \to \pm\infty} \|(UU_0^* - 1)e^{-i\tau|H_0|t}U_0\Psi\|. \tag{26}$$

From the existence of (23) we conclude that (25) vanishes, if we choose

$$\Phi_\pm = U^* \Omega_\pm^{rel} U_0 \Psi. \tag{27}$$

Using the explicit forms of $U$ and $U_0$ given in Eq. (14) we can estimate the term (26)

$$\|(U - U_0)e^{-i\tau|H_0|t}\Psi\| \tag{28}$$

$$\leq \|(a_+ - a_+^0)e^{-i\tau|H_0|t}\Psi\| + \|(a_- - a_-^0)e^{-i\tau|H_0|t}\Psi\| \tag{29}$$

$$+ \|(\operatorname{sgn} Q - \operatorname{sgn} Q_0)e^{-i\tau|H_0|t}a_-^0 \Psi\|. \tag{30}$$

The operators $a_\pm$ are bounded functions of $|H| = |UHU^{-1}|$, and $a_\pm^0$ are defined in the same way with $|H_0|$. Hence we can apply the intertwining relations ([10], p.17)

$$a_\pm \Omega^{rel} = \Omega^{rel} a_\pm^0, \tag{31}$$

where $\Omega^{rel}$ is either $\Omega_+^{rel}$ or $\Omega_-^{rel}$ to conclude that (29) vanishes, as $|t| \to \infty$. Since in (30) the operator $\operatorname{sgn} Q$ is not simply a bounded function of $\tau|H|$, we have to be a little bit more careful. First we note that for $\chi = |Q_0|\Psi$

$$\|(\operatorname{sgn} Q - \operatorname{sgn} Q_0)e^{-iQ_0^2 t}\chi\| \tag{32}$$

$$\leq \|(Q - Q_0)e^{-iQ_0^2 t}\Psi\| + \|(|Q| - |Q_0|)e^{-iQ_0^2 t}\Psi\|. \tag{33}$$

Here the last summand of (33) vanishes for $\Psi \in \mathcal{D}(|Q_0|)$ in the limit $|t| \to \infty$ because of the intertwining relations for $\Omega_\pm^{nr}$. The first summand vanishes by assumption $(20)_1$ for $\Psi \in F(a < Q_0^2 < b)\mathcal{H}_{ac}$. The set of all vectors of the form $\chi = |Q_0|\Psi$, $\Psi \in F(a < Q_0^2 < b)\mathcal{H}_{ac}$ with arbitrary $0 < a < b < \infty$ is dense in $\mathcal{H}_{ac}$, hence we have shown for all $\Psi \in \mathcal{H}_{ac}$

$$0 = \lim_{|t| \to \infty} \|(\operatorname{sgn} Q - \operatorname{sgn} Q_0)e^{-iQ_0^2 t}\Psi\|$$

$$= \|(\operatorname{sgn} Q \,\Omega_\pm^{nr} - \Omega_\pm^{nr} \operatorname{sgn} Q_0)\Psi\| \tag{34}$$

or

$$\operatorname{sgn} Q \,\Omega_\pm^{nr} = \Omega_\pm^{nr} \operatorname{sgn} Q_0. \tag{35}$$

With the help of (23) we can express $\Omega_\pm^{nr}$ in terms of $\Omega_\pm^{rel}$. Taking into account

$$\operatorname{sgn} Q \tfrac{1}{2}(1 \pm \tau) = \tfrac{1}{2}(1 \mp \tau)\operatorname{sgn} Q \tag{36}$$

and the same result with $Q_0$, we obtain from (35)

$$\operatorname{sgn} Q\, \Omega_\pm^{rel} = \Omega_\mp^{rel}\, \operatorname{sgn} Q_0. \tag{37}$$

But this implies almost immediately for all $\chi \in \mathcal{H}_{ac}$

$$\lim_{|t|\to\infty} \|(\operatorname{sgn} Q - \operatorname{sgn} Q_0)\, e^{-i\tau|H_0|t} a_-^0 \chi\| = 0. \tag{38}$$

This completes the proof of existence of $\Omega_\pm(H, H_0)$. In order to show (21) we note that by (31) and (35)

$$U^* \Omega_\pm^{nr} = \Omega_\pm^{nr} U_0^*. \tag{39}$$

Now we calculate, using (27)

$$\begin{aligned}\Omega_\pm(H, H_0) &= U^* \Omega_\pm^{rel} U_0 \\ &= \Omega_\pm^{nr} U_0^* \tfrac{1}{2}(1+\tau) U_0 + \Omega_\mp^{nr} U_0^* \tfrac{1}{2}(1-\tau) U_0.\end{aligned} \tag{40}$$

Finally, the relation

$$U_0^* \tfrac{1}{2}(1 \pm \tau) U_0 = \tfrac{1}{2}(1 \pm H/|H|) = F(H \gtrless 0) \tag{41}$$

completes the proof of Theorem 2.

*Remark:* Instead of assuming $(20)_1$, it would have been sufficient to require vanishing of (32) in the limit of large times. In the applications, however, it is usually easier to verify $(20)_1$.

## 4. Application

Now we apply Theorem 2 to the special case of the Dirac operator in an external magnetic field. The condition $(20)_1$ in Theorem 2 is not very restrictive. In case of the magnetic field scattering problem it just says that we have to choose a gauge in which the vector potential decays at infinity like $|x|^{-\delta}$. Then, since $H(A) - H(0) = -\vec{\alpha} \cdot A$, $(20)_1$ simply becomes

$$\|A(x)\, e^{-ip^2 t}\, \Psi\| \leq \operatorname{const.}(1 + |t|)^{-\delta} \tag{42}$$

for suitable $\Psi$, which can easily be shown by stationary phase arguments (see [10], Appendix 1 to XI.3, and [3]). The condition $(20)_2$ is more restrictive. It can be satisfied by the following assumption [7].

Let the magnetic field strength $B$ decay at infinity, such that for some $\delta > 0$

$$B(x) \leq \text{const.}(1+|x|)^{-3/2-\delta}. \tag{43}$$

Choose the transversal (or Poincaré) gauge

$$A(x) = \int_0^1 s\, B(xs) \wedge x\, ds. \tag{44}$$

Then we have $A(x) \cdot x = 0$, and $A(x)$ decays like $|x|^{-1/2-\delta}$, as $|x| \to \infty$. Hence the expressions $\text{div} A$, $A^2$, $\vec{\Sigma} \cdot B$ occurring in $Q^2 - Q_0^2$ are all of short-range. Hence the crucial long-range term is $A(x) \cdot p$. It can be written as $A(x) \cdot p = G(x) \cdot (x \wedge p)$, where

$$G(x) = \int_0^1 s\, B(xs)\, ds \tag{45}$$

satisfies,

$$|G(x)| \leq \text{const.}(1+|x|)^{-3/2-\delta)} \tag{46}$$

and since the angular momentum $L = x \wedge p$ remains constant under the nonrelativistic free time evolution $\exp(-iQ_0^2 t)$, we easily obtain by a stationary phase argument for $\Psi$ in a suitable dense set

$$\| A(x) \cdot p\, e^{-ip^2 t}\, \Psi \| \leq \text{const.}(1+|t|)^{-3/2-\delta}, \tag{47}$$

so that all assumptions of Theorem 2 are satisfied. Hence we have proven the following theorem

**Theorem 3.** *Let $H(A)$ and $H(0)$ be given as in (2) and assume that the magnetic field strength $B$ satisfies (43). Then both the relativistic and the nonrelativistic wave operators exist in the transversal gauge, and are related by (21).*

*Remark:* In order to appreciate the use of Theorem 2 let us briefly discuss the direct proof of the existence of relativistic wave operators in the magnetic field example, which has been given in Ref [8]. There is the following difficulty. In order to apply the Cook argument, which worked well in the nonrelativistic case (see Eq. (22), together with Eq. (47)), one would have to verify, that

$$\| \alpha \cdot A(x)\, e^{-iH(0)t}\, \Psi \| \tag{48}$$

decays integrably in time for $\Psi$ in a suitable dense subset of $\mathcal{H}_{ac}(H(O))$. But this is wrong, because $A(x)$ decays only like $|x|^{-1/2-\delta}$, at least if we choose the transversal gauge. In general, $A(x)$ cannot be made short-range by some other clever choice of gauge. This can be seen already in very simple cases. Consider, for example, in two dimensions a magnetic field which has compact support and nonvanishing flux. Now, apply Stoke's Theorem to a large circle surrounding the support of $B$ to find that for this perfectly short-range situation $A(x)$ cannot decay faster than $|x|^{-1}$ (which is of long-range).

For a long range potential matrix of the form $\alpha \cdot A$ the Cook argument does not work, because it cannot take into account an effect which is known as "Zitterbewegung". This means that the operator $\alpha(t) = e^{iH_0 t} \alpha e^{-iH_0 t}$ oscillates whithout damping around a mean value $pH_0^{-1}$, which corresponds to the classical relativistic velocity $v = p/E$.

## 5. Asymptotic completeness

Asymptotic completeness is the statement that the range of the wave operators Ran $\Omega_\pm$ equals $\mathcal{H}_{ac}(H)$, the absolutely continuous spectral subspace of $H$. This means that every state $\Psi$ in the continuous spectral subspace of $H$ is an asymptotically free scattering state. A consequence is that the continuous spectrum of $H$ is purely absolutely continuous. By [10], p.19, asymptotic completeness is equivalent to the existence of the adjoint wave operators $\Omega_\pm(H_0, H)$. Hence we simply have to exchange the roles of $H$ and $H_0$ in Theorem 2 in order to get conditions for the asymptotic completeness of the relativistic supersymmetric system, which are formulated entirely in terms of the nonrelativistic time evolution.

Unfortunately, these conditions are of little use for long range magnetic fields. In particular, in order to verify condition $(20)_2$ we would have to show decay in time of the expression

$$\|A(x) \cdot p \, e^{-i(H(A)^2 - m^2)t} \Psi\| = \|G(x) \cdot L \, e^{-i(H(A)^2 - m^2)t} \Psi\|. \tag{49}$$

This cannot be done as before, because magnetic fields are usually not spherically symmetric (in three dimensions, a singularity free magnetic field is never spherically symmetric). Hence the angular momentum $L$ does not commute with the time evolution $\exp(-iH(A)t)$ in a magnetic field. In fact, the time decay of Eq. (49) is not at all obvious (see Ref. [7]). Nevertheless, the desired result can be obtained using a different method.

**Theorem 4.** *Let $H(A)$ and $H(0)$ be given as in (2) and assume that the magnetic field strength $B$ satisfies*

$$D^\gamma B(x) \leq \text{const.}(1 + |x|)^{-3/2 - \delta - \gamma}, \tag{50}$$

*for some $\delta > 0$ and multiindices $\gamma$ with $|\gamma| = 0, 1, 2$. Then the relativistic wave operators in the transversal gauge are asymptotically complete.*

**Proof.** Since a detailed proof appeared in Ref. [8] we only give a sketch here and show how the Zitterbewegung can be controlled. To prove asymptotic completeness for $\Omega_+$ it suffices to show for a sequence of times $\tau_n \to \infty$

$$\lim_{\tau_n \to \infty} \sup_{t \geq 0} \| \{ e^{-iH(A)t} - e^{-iH(0)t} \} e^{-iH(A)\tau_n} \Psi \| = 0, \tag{51}$$

For all $\Psi$ in the continuous spectral subspace $\mathcal{H}_{cont}(H)$. We write $\mathcal{H}_{cont}(H)$ as an orthogonal direct sum of $\mathcal{H}^+_{cont}$ and $\mathcal{H}^-_{cont}$, the subspaces of positive, resp. negative energy scattering states. It is sufficient to prove (51) on each subspace separately. Assume that $\Psi \in \mathcal{H}^+_{cont}$, the proof for $\mathcal{H}^-_{cont}$ is completely analogous. In Ref. [8] it is shown that one can approximate $\exp{-iH(A)\tau_n}\Psi$ by a sequence of states $\Phi_k(\tau_n)$ with the following properties. $\Phi_k(\tau_n)$ consist of a finite sum of well localized states which have all positive kinetic energy, and angular momentum slowly increasing with $\tau_n$. More precisely, let $P_0^+$ denote the projection onto the positive energy subspace of the free Dirac operator $H(0)$, and let $u(p) = p|H(0)|^{-1} = P_0^+ \alpha P_0^+$ be the velocity operator for a free particle with positive energy. Then we have the following result.

We can choose a sequence of times $\tau_k$, $k = 1, 2, \ldots$ such that for each $\Psi \in \mathcal{H}^+_{cont}$ there are $N$ functions $f_i \in C_0^\infty$, each with support in a ball away from the origin, and $N$ states $\phi_k^i(\tau_n)$ with the following properties.

1. For any $\epsilon > 0$ there exists $k = k(\epsilon)$ such that for all $\tau_n \geq \tau_k$

$$\| e^{-iH(A)\tau_n} \Psi - \sum_{i=1}^N \Phi_k^i(\tau_n) \| \leq \epsilon \tag{52}$$

where

$$\Phi_m^i(\tau_n) \equiv P_0^+ f_i(u(p)) f_i(x/\tau_n) \phi_k^i(\tau_n). \tag{53}$$

2. Each component of the angular momentum $L = x \wedge p$ satisfies

$$\| L \Phi_k^i(\tau_n) \| \leq C_k \tau_n^{1/2-\delta}, \quad i = 1, \ldots, N, \tag{54}$$

where $C_k$ is a constant which may depend on $k$. The functions $f$ express a localization in phase space. Due to $f(u(p))$ each of the approximating states has velocities in a small region (away from the origin) around some mean velocity $u_i$. Furthermore, the states are approximately localized in the support of $f(x/\tau_n)$, which is far away, if $\tau_n$ is large. The conditions on the derivatives of $B$ are needed in the proof of 1 and 2. This proof uses the asymptotic observable theory, which has been developed for the Dirac equation in Ref. [4], and the idea of an approximating time evolution which has been applied to the case of long-range magnetic fields in Ref. [7]. By the asymptotic observable theory, one can replace a scattering state by a finite sum of well localized states. On each of these localized states one replaces the true time evolution by an approximating time evolution, which is similar to a Trotter product, but with increasing time intervals. The approximating time evolution leaves the states localized in phase space and allows to estimate the increase in time of the angular momentum operator with the help of the Gronwall lemma. See Refs. [7] and [8] for details.

Using (52) we can prove (51) by showing vanishing as $n \to \infty$ of

$$\| \{e^{-iH(A)t} - e^{-iH(0)t}\} e^{-iH(A)\tau_n} \Phi_k^i(\tau_n) \| \tag{55}$$

for all $k$, $i$, and $t \geq 0$. Writing $\{\dots\}$ in (55) as the integral of its derivative, we see that we have to estimate

$$\left\| \int_0^t ds\, e^{iH(A)s}\, A \cdot \alpha\, e^{-iH(0)s}\, \Phi_k^i(\tau_n) \right\|. \tag{56}$$

Next we choose a cut-off function $g \in C_0^\infty$ with $g(x) = 1$ for $|x| \geq 1$, and $g(x) = 0$, for $|x| \leq \frac{1}{2}$. Let $u_0 < \mathrm{dist}(\mathrm{supp} f_i, O)$, all $i$, and define

$$A^t(x) = A(x)\, g(x/u_0 t). \tag{57}$$

By the stationary phase method we can find a constant $D_K$ for all positive integers K, such that

$$\left\| \{1 - g(x/u_0(\tau_n + s))\}\, e^{-iH_0 s}\, \Phi_k^i(\tau_n) \right\| \leq D_K (1 + s + \tau_n)^K. \tag{58}$$

Hence we can replace $A$ in (56) by $A^{\tau_n+s}$. Next we write

$$\alpha\, e^{-iH_0 s} = e^{-iH_0 s} F(s) + p H_0^{-1} e^{-iH_0 s}. \tag{59}$$

The operator $F(s) = \exp(iH_0 s)\, F \exp(-iH_0 s)$, where

$$F = \alpha - p H_0^{-1} \tag{60}$$

describes the difference between the velocity operator $\alpha$ of free Dirac particles and the classical velocity operator $p H_0^{-1}$, which corresponds to the classical expression for the velocity of a relativistic particle. From $FH_0 = -H_0 F$ we obtain

$$F(s) = e^{2iH_0 s} F, \quad \int_0^s F(s')\, ds' = \frac{1}{2iH_0} \{e^{2iH_0 s} - 1\} F. \tag{61}$$

These operators are all bounded uniformly in s. According to (59), Eq. (56) splits into two parts. The second summand is

$$\left\| \int_0^t ds\, e^{iHs}\, A^{\tau_n+s} \cdot p H_0^{-1}\, e^{-iH_0 s}\, \Phi_k^i(\tau_n) \right\|$$

$$\leq \mathrm{const.} \int_0^\infty ds\, \|G^{\tau_n+s}\|\, \|L\, \Phi_k^i(\tau_n)\|. \tag{62}$$

Here we have used $A \cdot p = G \cdot L$, $\|H_0^{-1}\| = \frac{1}{m}$, and $L H_0 P_0^+ = |H_0| L P_0^+$. Hence by (54) and (46) we can estimate (62) by $C'_m \tau_n^{-2\delta}$ with some suitable constant $C'_m$. It remains to control

$$\left\| \int_0^t ds\, e^{iHs}\, A^{\tau_n+s}\, e^{-iH_0 s}\, F(s) \right\|$$

$$\leq \left\| e^{iHt}\, A^{\tau_n+t}\, e^{-iH_0 t} \int_0^t F(s)\, ds \right\| \tag{63}$$

$$+ \left\| \int_0^t ds\, \frac{d}{ds}\left( e^{iHs}\, A^{\tau_n+s}\, e^{-iH_0 s} \right) \int_0^s F(s')\, ds' \right\|. \tag{64}$$

Since (61) is bounded, we can estimate (63) by

$$\text{const.} \|A^{\tau_n + t}\| \leq \text{const.} \tau_n^{-1/2 - \delta} \tag{65}$$

for all $t \geq 0$. The term (64) is easily seen to be bounded by

$$\text{const.} \int_0^t ds \left\{ \|[H_0, A^{\tau_n + s}]\| + \|\alpha \cdot A A^{\tau_n + s}\| + \|\tfrac{d}{ds} A^{\tau_n + s}\| \right\}$$
$$\leq \text{const.} \tau_n^{-2\delta}. \tag{66}$$

Now, since $\delta > 0$, all the expressions (62), (65), and (66) vanish, as $n$ tends to $\infty$. This proves (51).

## 6. Discussion

The usual formalism of scattering theory is expected to hold for "short-range potentials", where (each component of) the potential matrix $V$ satisfies

$$|V(x)| \leq \text{const}(1 + |t|)^{-1-\delta}. \tag{67}$$

A famous counter example is the electrostatic Coulomb potential, where $|V(x)|$ decays like $|x|^{-1}$. In this case the wave operators do not exist [2] and one has to introduce modifications of the asymptotic time evolution. For the magnetic fields in Theorem 4 the potential matrix $-\vec{\alpha} \cdot A$ has a much slower decay. Indeed, previous results in the literature have been obtained only by introducing modifications of the wave operators (see, e.g., Ref.[11] and the references therein). Asymptotic completeness in the sense of Theorem 4 is obviously due to the transversality of $A$. In another gauge (e.g., if $\nabla g$ is long-range), the unmodified wave operators (1) possibly do not exist. Instead, our considerations show that the correspondingly modified wave operators $\Omega_\pm(H(A), H(\nabla g))$ exist and are asymptotically complete. These remarks might be of importance, because in physics for time independent problems the Coulomb gauge div $A = 0$ is used almost exclusively instead of the transversal gauge, which is best adapted to scattering theory. Note that although the wave and scattering operators depend on the choosen gauge, the physically observable quantities like scattering cross sections are gauge independent.

In situations like the Aharonov Bohm effect one has used the free asymptotics (e.g., plane waves for the asymptotic description of stationary scattering states), together with the Coulomb gauge, although the vector potential is long-range. But in this case the calculations are justified, because in two dimensions and for rotationally symmetric fields the Coulomb gauge happens to coincide with the transversal gauge (see also Ref. [9], for a discussion).

Under weaker decay conditions on the magnetic field strength the wave operators would not exist in that form, because then the term $A^2$ occurring in $Q^2 - Q_0^2$ would become long-range. In this case one really needs modified wave operators, similar to the Coulomb case.

Asymptotic completeness is also true in relativistic and nonrelativistic quantum mechanics, if one adds a short-range electric potential to the magnetic field [4]. Of course, the resulting Dirac operator has no supersymmetry, but the method of the proof in Theorem 4 still works.

The scattering problem is nontrivial even in classical mechanics. From special examples we know that classical paths of particles in magnetic fields satisfying our requirements even do not have asymptotes. It is easy to see that the velocity of the particles is asymptotically constant. But if we compare the asymptotic motion of a particle in a magnetic field with a free motion, one would have to add a correction which is transversal to the asymptotic velocity and which increases for $\delta < 1/2$ like $|t|^{1/2-\delta}$. Thus the situation seems to be worse than in the Coulomb problem. There the interacting particles also cannot be asymptotically approximated by free particles, but at least the classical paths do have asymptotes. (The correction in the Coulomb problem increases like $\ln |t|$ and is parallel to the asymptotic velocity). A discussion of classical scattering theory with magnetic fields is given in Ref. [7].

## References

[1] deVries, E. *Foldy-Wouthuysen transformations and related problems*, Fortschr. d. Physik **18** (1970), 149–182.

[2] Dollard, J. and Velo, G., *Asymptotic behaviour of a Dirac particle in a Coulomb field*, Il Nuovo Cimento **45** (1966), 801–812.

[3] Enss, V., *Propagation properties of quantum scattering states*, J. Func. Anal. **52** (1983), 219–251.

[4] Enss, V. and Thaller, B., *Asymptotic observables and Coulomb scattering for the Dirac equation*, Ann. Inst. H. Poincaré **45** (1986), 147–171.

[5] Foldy, L. L. and Wouthuysen, S. A., *On the Dirac theory of spin-1/2 particles and its nonrelativistic limit*, Phys. Rev. **78** (1950), 29–36.

[6] Helffer, B., Nourrigat, J. and Wang, X. P., *Sur le spectre de l'equation de Dirac avec champ magnetique*, Preprint (1989).

[7] Loss, M. and Thaller, B., *Scattering from magnetic fields*, Proceedings of the XXVI. Internationale Universitätswochen für Kernphysik, H. Mitter and L. Pittner eds., Springer Verlag, Berlin, Heidelberg, New York., pp. 317–321, Schladming, Austria 1987.

[8] Loss, M. and Thaller, B., *Short-range scattering in long-range magnetic fields: The relativistic case*, J. Diff. Eq. **73** (1988), 225–236.

[9] Perry, P., *Scattering Theory by the Enss Method*, Math. Rep. **1**, Harwood academic publishers, New York, 1983.

[10] Reed, M. and Simon, B., *Methods of modern mathematical physics III, Scattering theory*, Academic Press, New York, San Francisco, London, 1979.

[11] Thaller, B., *Relativistic scattering theory for long-range potentials of the nonelectrostatic type*, Lett. Math. Phys. **12** (1986), 15–19.

[12] Thaller, B., *Normal forms of an abstract Dirac operator and applications to scattering theory*, J. Math. Phys. **29** (1988), 249–257.


Bernd Thaller
Institut für Mathematik
Karl–Franzens–Universität Graz
Heinrichstraße 36
A-8010 Graz, Austria

# Index

## A

absolutely continuous spectrum  161, 166, 170
age models  280
AIDS  47
analytic semigroup  25, 27, 28, 130, 245, 263
approximate system  23
asymptotic behavior  116, 252
asymptotic completeness  321

## B

beam  227
Bessel function  118
bilinear form  3, 7

## C

Cauchy problem  79, 129
chain trickery  271, 277, 278
characteristic  96
characteristic equation  58
clamped boundary conditions  16
classical solution  186, 192, 232
coercive  16, 18, 24

cohort 96
commutation method 139
commutation relation 141
consistent approximation 108
convergence 108
critical Schrödinger operator 144, 146, 151, 157

## D

damped plate equations 242
damping operator 26
*Daphnia magna* 269, 271
density function 93
differentiable semigroup 214
diffusion-reaction delay system 259
Dirac equation 291, 297, 295
Dirac operator 141, 152, 169, 179, 313, 315, 317
Dirichlet boundary condition 39, 197
Dirichlet eigenfunction 175
Dirichlet operator 171
discrete spectrum 161
dissipative 215, 305, 310
dissipative boundary condition 230
distributional solution 141, 142, 143, 146, 165
dual semigroup 275

## E

eigenvalues 152, 161, 282
elastic plates 1
elliptic operator 72
elliptic problem 307
energy 229
energy space 116
environmental state variables 93, 94
Escalator boxcar train 91, 94, 95
essential spectrum 161
Euler Bernouilli beam 30, 31
evolution equations 67
evolution operator 242, 245
exponential decay 254
exponentially stable 73

## F

feedback control 230
Floquet determinant 166

Floquet function  175
Floquet multiplier  167
Floquet solution  157, 168
Floquet theory  154, 157
Foldy-Wouthuysen transformation  315
fractional power  244

# G

Galerkin approximation  25
gap of the spectrum  166, 316
generator  131, 189, 190, 191, 210, 218
Green's formula  4, 14
Green's function  155
ground state solution  219, 221
growth bound  116, 118, 119
growth laws  285

# H

Hille-Yosida estimate  186, 198

# I

i-state  91, 93, 95, 100, 272, 279, 280
incidence matrix  37
infinitesimal generator  27, 28, 68, 70, 81, 130
initial boundary value problem  301, 302
integral solution  186, 193
integrodifferential equation  209
interpolation space  130
inverse problem  21, 23
isospectral manifold  173

# J

Jost functions  150, 161, 163

# K

Kirchhoff law  38
Kirchhoff shear force  2
Klein-Gordon equation  291, 295
Kolmogorov backward equation  273
Korteweg-de Vries equation  139, 140, 143

## L

Laplace operator  241, 259
Laplace-Beltrami operator  204
Lax pair  141, 143
limit solution  302, 304, 311
local nodal domain  201

## M

M-dissipative  306
M-matrix  260
M-soliton solution  152
magnetic field  313, 319, 321
maximum-minimum principle  40
measurements  22
method stability  23
mild solution  263, 293
Miura's transformation  139, 140, 142, 144, 153
mixing function  48, 49, 51, 56
modified Korteweg-de Vries equation  139, 140, 143
monodromy matrix  167

## N

n-times integrated semigroup  186, 189
network  37
Neumann boundary conditions  241
Neumann mapping  248
Neumann operator  171, 243
nondegenerate integrated semigroup  187, 189
nonlinear beam model  227

## O

observation operator  21
obstacle problem  16
operator matrix  166
operator pencil  130, 136
Osgood condition  43

## P

P-coercive  7
parabolic  22, 130
parabolicity  38
parameter convergence  23

parameter dependent system  22
Pauli matrices  314
periodic boundary conditions  171
Perron Frobenius  69
persistent solution  261
plate  2, 33, 242
point spectrum  161
polynomial operator matrices  115
population model  91
positivity  68, 69, 75
potential  152
preferred mixing  50
proportionate mixing  49, 50, 51

# R

reaction-diffusion equations  37
regularization  22
regular solution  129
resolvent  68, 73, 74, 210
resolvent set  79
Riccati type equation  178

# S

scattering matrix  162, 164, 165
Schrödinger equation  201
Schrödinger operator  140, 144, 160, 314
Schur complements  73, 74
second order Cauchy problem  25, 115
semilinear equation  38
separable mixing  50, 55
sesquilinear form  24, 26, 28
shear  228
singular continuous  161
SIR model  48
size structured model  269
soliton-like solution  139, 151
soliton solution  145, 147, 150
spectral bound  68, 75, 116, 118
spectral multiplicity  161, 166
spectrum  68, 116, 117, 142, 166, 244, 246
stability  69
stabilization  231, 241
stiffness operator  26
strains  228

strict solution  80, 86, 88, 134
string  115
strong solution  80, 83, 84, 87
strongly continuous group  117
strongly continuous semigroup  26, 70, 73, 81, 118, 119, 185, 210, 218
subcritical Schrödinger operator  144, 146, 157

## T

T-periodic  241
transversal gauge  320
two sex mixing function  55

## U

uniform exponential stability  119
unilateral boundary conditions  10
unitary operator  140, 141

## V

variational inequalities  1, 5, 13, 16, 18
vibration  115
Volterra equation  142
Volterra integrodifferential equation  72, 185

## W

wave equation  197, 198, 241
wave operators  317, 320
Weierstrass P-function  159

## Y

Yosida approximation  81